SCIENCE AWAKENING

I

B. L. VAN DER WAERDEN

SCIENCE
AWAKENING

I

English translation

by

Arnold Dresden

with additions of the author

Fourth edition

KLUWER ACADEMIC PUBLISHERS, DORDRECHT, THE NETHERLANDS
SCHOLAR'S BOOKSHELF, PRINCETON JUNCTION, NEW JERSEY, U.S.A.

Hardcover edition published throughout the world, exclusive of North America, by Kluwer Academic Publishers, Spuilboulevard 50, P.O. Box 17, 3300 AA Dordrecht, The Netherlands

Paperback edition published throughout the world by The Scholar's Bookshelf, 51 Everett Drive, P.O. Box 179, Princeton Junction, New Jersey 08550, United States of America

Hardcover edition published in North America by The Scholar's Bookshelf

Publishing, a division of Kluwer Academic Publishers, Dordrecht, The Netherlands

First edition 1954
Second edition 1961
Third edition 1969
Fourth edition 1975
Fifth edition 1988

Kluwer Academic Publishers

ISBN-13: 978-94-010-7115-4   e-ISBN-13: 978-94-009-1379-0
DOI: 10.1007/978-94-009-1379-0

First Scholar's Bookshelf hardcover printing, 1988
First Scholar's Bookshelf paperback printing, 1988

Soon after the publication of my "Ontwakende Wetenschap" the need for an English translation was felt. We were very glad to find a translator fully familiar with the English and Dutch languages and with mathematical terminology. The publisher, Noordhoff, had the splendid idea to ask H. G. Beyen, professor of archeology, for his help in choosing a nice set of illustrations. It was a difficult task. The illustrations had to be both instructive and attractive, and they had to illustrate the history of science as well as the general background of ancient civilization. The publisher encouraged us to find better and still better illustrations, and he ordered photographs from all over the world, with never failing energy and enthusiasm. Mr. Beyen's highly instructive subscripts will help the reader to see the inter-relation between way of living, art, and science of the ancient world.

Thanks are due to many correspondents, who have suggested additions and pointed out errors. Sections on Astrolabes and Stereographic Projection and on Archimedes' construction of the heptagon have been added. The sections on Perspective and on the Anaphorai of Hypsicles have been enlarged.

In the second English edition I have incorporated an important discovery of P. Huber, which sheds new light upon the role of geometry in Babylonian algebra (see p. 73). The section on Heron's Metrics (see p. 277) was written anew, following a suggestion of E. M. Bruins.

Zürich, 1961   B. L. VAN DER WAERDEN

Why History of Mathematics?

Every one knows that we are living in a technological era. But it is not often realized that our technology is based entirely on mathematics and physics.

When we ride home on the streetcar in the evening, when we turn on the electric light and the radio, everything depends on cleverly constructed physical mechanisms based on mathematical calculations. But more than that! We owe to physics not only these pleasant articles of luxury, but, to a large extent, even our daily bread. Apart from the fact that our grains come to us, chiefly by steamer from overseas, our own agriculture would be far less productive without artificial fertilizers. Such fertilizers are chemical products, and chemistry depends on physics. [1]

But science has not brought us blessings only. The destructive armaments which mankind uses at the present time to knock its own civilization to pieces are also products to which the development of mathematics and physics have inevitably led.

Our spiritual life is also influenced by science and technology, in a measure but rarely fully understood. The unprecedented growth of natural science in the 17th century was followed ineluctably by the rationalism of the 18th, by the deification of reason and the decline of religion; an analogous development had taken place earlier, in Greek times. In a similar manner, the triumphs of technology in the 19th century were followed in the 20th by the deification of technology. Unfortunately, man seems to be overly inclined to deify whatever is powerful and successful.

These considerations indicate that science has put its stamp on the whole of our life, material and spiritual, in its beneficent and in its evil aspects. Science is the most significant phenomenon of modern times, the principal ingredient of our civilization — alas!

But if this be true, then the most important question for the history of culture is: How did our modern natural science come about?

It will be conceded that most historical writings either do not consider this question at all, or else deal with it in a very unsatisfactory manner. For example, which are the histories of Greek culture that mention the names of Theaetetus and of Eudoxus, two of the greatest mathematicians of all times? Who realizes that, from the historical point of view, Newton is the most important figure of the 17th century?

Every physicist will admit that the mechanics of Newton are the foundation of modern physics. Every astronomer knows that modern astronomy begins with Kepler and Newton. And every mathematician knows that the largest domain of

---

[1] The reader should bear in mind that this book was addressed originally to the public of the Netherlands.

modern mathematics, the part most important for physics, is Analysis, which has its roots in the Differential and Integral Calculus of Newton. Thus the work of Newton constitutes the foundation for by far the greater part of modern exact science.

It was Newton who discovered the fundamental laws of motion, to which terrestrial as well as celestial objects are subject. He placed the crown on the task of renovating antique astronomy, begun by Copernicus and Kepler. He discovered a general method for solving all problems of differentiation and integration, whereas Archimedes, the greatest genius of antiquity, had not progressed beyond special methods for particular problems.

The work of Newton can not be understood without a knowledge of antique science. Newton did not create in a void. Without the stupendous work of Ptolemy, which completed and closed antique astronomy, Kepler's Astronomia Nova, and hence the mechanics of Newton, would have been impossible. Without the conic sections of Apollonius, which Newton knew thoroughly, his development of the law of gravitation is equally unthinkable. And Newton's integral calculus can be understood only as a continuation of Archimedes' determination of areas and volumes. The history of mechanics as an exact science begins with the laws of the lever, the laws of hydrostatics and the determination of mass centers by Archimedes.

In short, all the developments which converge in the work of Newton, those of mathematics, of mechanics and of astronomy, begin in Greece.

The History of Greek Mathematics,

from Thales to Apollonius, covers the four centuries from 600 B.C. to 200 B.C. Until recently, the first three of these four centuries were enveloped in twilight, because we possess only two original texts from this period: the fragment concerning the lunules of Hippocrates and that of Archytas on the duplication of the cube. To this can be added two brief fragments of Archytas, a number of scattered communications of Plato, Aristotle, Pappus, Proclus and Eutocius, and a self-contradictory set of Pythagorean legends. For this reason, the older works, such as Cantor's Geschichte der Mathematik, contain little more about this period than speculations concerning things of which we really do not know anything, such as, for example, the "Theorem of Pythagoras"

In recent times however more light has penetrated into the darkness. In the first place, as a result of the indefatigable industry of Otto Neugebauer and his collaborators, we know now the mathematical cuneiform texts, which have thrown an entirely new light not only on the Theorem of Pythagoras, but especially on the earliest history of arithmetic and of algebra. Neugebauer, following in the tracks of Zeuthen, succeeded in discovering the hidden algebraic element in Greek mathematics and in demonstrating its connection with Babylonian algebra. No longer does the history of algebra begin with Diophantus; it starts 2000 years earlier in Mesopotamia. And, as to arithmetic, in 1937 Neugebauer wrote: "What is called Pythagorean in the Greek tradition, had probably better be called Babylonian"; and, a cuneiform text, concerning "Pythagorean numbers", discovered in 1943, showed that he was entirely right.

A second new impulse came from philosophically oriented philology. In 1927, Stenzel and Toeplitz, with Neugebauer, established the periodical "Quellen und Studien zur Geschichte der Mathematik, Astronomie und Physik". It was their purpose to get to know more about the philosophy of Plato by an analysis of the fundamental concepts of Greek mathematics, and, reciprocally, to learn more about Greek mathematics by means of an analysis of Plato. This method has enabled Becker, Reidemeister and others to obtain highly important results. At an earlier date, Eva Sachs had rescued the excellent mathematician Theaetetus from oblivion.

Another very fertile method was the analysis of the Elements of Euclid. This work, written about 300 B.C., proves to be largely a compilation of mathematical fragments, quite diverse in calibre and quite varied in age. By carefully taking these fragments apart, by dusting them off and then replacing them in the mathematical historical environment from which they had originally come, it has become possible to obtain a considerably clearer picture of Greek mathematics of the years 500—300 B.C.

These different things have not as yet been brought together in a book. We have, it is true, an excellent book by O. Neugebauer on "Vorgriechische Mathematik", and there are excellent works on Euclid, Aristarchus, Archimedes and Apollonius. As to these latter writers, we shall therefore be able to confine ourselves to the most interesting and the most important things; we shall pick tidbits here and there from their works and we shall try to serve them in as tasteful a manner as possible. But the principal purpose of the present book is to explain clearly

how Thales and Pythagoras took their start from Babylonian mathematics but gave it a very different, a specifically Greek character;

how, in the Pythagorean school and outside, mathematics was brought to higher and ever higher development and began gradually to satisfy the demands of stricter logic;

how, through the work of Plato's friends Theaetetus and Eudoxus, mathematics was brought to the state of perfection, beauty and exactness, which we admire in the elements of Euclid.

We shall see moreover that the mathematical method of proof served as a prototype for Plato's dialectics and for Aristotle's logic.

The history of mathematics should not be detached from the general history of culture. Mathematics is a domain of intellectual activity, intimately related not only to astronomy and mechanics, but also to architecture and technology, to philosophy, and even to religion (Pythagoras!).

Political and social conditions are of very great importance for the flowering of science and for its character. This will become very clear in Chapter VII when the mathematics of Alexandria is compared with that of the classical period, and still more in Chapter VIII in the discussion of the causes for the decay of Greek mathematics.

The plan of this book.

It is the intention to make this book scientific, but at the same time accessible to any one who has learned some mathematics in school and in college, and who is interested in the history of mathematics. It is to be scientific in the sense that it is to be based on a study of the sources and that its conclusions are to be supported by arguments, so as to enable the reader to judge the conclusions for himself.

The naive reader may take the use of such a method for granted. But — how often has it been sinned against! How frequently it happens that books on the history of mathematics copy their assertions uncritically from other books, without consulting the sources! How many fairy tales circulate as "universally known truths"!

Let us quote an example. In 90 % of all the books, one finds the statement that the Egyptians knew the right triangle of sides 3, 4 and 5, and that they used it for laying out right angles. How much value has this statement? None! What is it based on? On two facts and an argument of Cantor. The facts are the following: "rope-stretchers" took part in laying out an Egyptian temple, and the angles at the base of temples and pyramids are nearly always, very accurately, right angles. Now Cantor reasons as follows: these right angles must have been constructed by the rope-stretchers, and I (Cantor) can not think of any other way of constructing a right angle by means of stretched ropes than by using three ropes of lengths 3, 4 and 5, forming a right triangle. Therefore the Egyptians must have known this triangle.

Is this not incredible? Not that Cantor at one time formulated this hypothesis, but that repeated copying made it a "universally known fact". This is nevertheless the fact.

To avoid such errors, I have checked all the conclusions which I found in modern writers. This is not as difficult as might appear, even if, as is my case, one cannot read either the Egyptian characters or the cuneiform symbols, and one is not a classical philologist. For reliable translations are obtainable of nearly all texts. For example, Neugebauer has translated and published all mathematical cuneiform texts. The Egyptian mathematical texts have all been translated into English or German. Plato, Euclid, Archimedes, . . . of all these, good translations exist in French, German and English. Only in a few doubtful cases it became necessary to consult the Greek text.

Not only is it more instructive to read the classical authors themselves (in translation if necessary), rather than modern digests, it also gives much greater enjoyment. If my book should lead the reader to do this, it will fully have accomplished its purpose. For this reason, I advise the reader emphatically not to accept anything on my say-so, but to verify everything.

I have tried to consider the great mathematicians as human beings living in their own environment and to reproduce the impression which they made on their contemporaries. In some cases, the scarcity of source material made this impossible, but striking personalities such as Pythagoras, Archytas, Theaetetus and Archimedes can be made to stand out clearly. It is also possible to get an impression of the character of Thales, Eudoxus and Eratosthenes. Of the Egyptian and Babylonian mathematicians not even the names are known.

What is new in this book,

In Chapter II.
A hypothesis of Freudenthal on Indian number symbols.

In Chapter III.
Freudenthal's interpretation of a Babylonian textbook.

In Chapter IV.
A new way of looking at the mathematics of Thales.

In Chapter V.
Reconstruction of the Pythagorean theory of numbers from the arithmetical books of the Elements.
Connections between the Babylonian and the Greek mathematics, particularly the Pythagorean mathematics.
The irrationality proofs of Theodorus of Cyrene.

In Chapter VI.
The feeble logic of Archytas of Taras.
Mathematics and the theory of harmony in the Epinomis.
An analysis of Book X of the Elements and a reconstruction of the mathematical work of Theaetetus.

In Chapter VII.
The history of the Delian problem, actually and according to the dialogue Platonicus.

In Chapter VIII.
The cause of the decay of Greek mathematics.

Acknowledgements.

I thank Dr. Brinkman, Professor Freudenthal and especially Dr. Dijksterhuis, who have read the manuscript critically and have made numerous useful remarks.

I thank the many others who have helped with brief observations or with technical advice. I am very much obliged to Mr. Wijdenes who has taken care of the diagrams in his well-known careful manner, assisted by his excellent draughtsman, Mr. Bousché. In conclusion, I thank the publishers for the generous way in which they have taken all my wishes into account.

TABLE OF CONTENTS

CHAPTER I

THE EGYPTIANS

Chronological Summary

General History	History of Civilization	History of Science
3000 Menes The Old Kingdom	Hieroglyphics Pyramids	Number symbols to 100,000
2000—1800 The Middle Kingdom	Literature The goldsmith's art	Rhind papyrus and Moscow papyrus Star calendars on sarcophagi
1700 The Hyksos domination		Ahmes copies the Rhind papyrus
1600—1100 The New Kingdom	New theology (Echnaton) Architecture Sculpture	Very primitive astronomy (Senmuth's tomb)
300 B.C. — 300 A.D. Hellenism	Alexandria as center of Greek art and science Rise of astrology	Highest development of Greek science Egyptian arithmetic and astronomy remain very primitive

The Egyptians as "inventors" of geometry

The Greeks generally assumed that mathematics had its origin in Egypt. For instance, Aristotle writes (Metaphysics A 1)[1]: "Thus the mathematical sciences originated in the neighborhood of Egypt, because there the priestly class was allowed leisure." Herodotus, who knew Egypt better, looked at the more practical side of the matter. When the Nile had flooded an agricultural tract, it became necessary for the purpose of taxation, to determine how much land had been lost; "from this, to my thinking, the Greeks learned the art of measuring land" (Herodotus II 109). And Democritus writes: "No one surpasses me in the construction of lines with proofs, not even the so-called rope-stretchers among the Egyptians." The rope-stretchers ("harpedonaptai"), to whom Democritus here refers, are probably surveyors, whose principal measuring instrument is everywhere the stretched cord. [2]

In view of the fact that the mathematical abilities of the Egyptians are so highly praised by the Greeks, it is certainly worth while to have a look at the Egyptian mathematical texts. The largest and most famous of these is

---

[1] All references to classical writers, mentioned here and in the sequel, follow the translation of the corresponding author in Loeb's Classical Library, unless otherwise noted.

[2] See S. Gandz, *Die Harpenodapten, Quellen und Studien zur Geschichte der Mathematik*, B I, 255.

The Rhind papyrus,

named after Mr. A. H. Rhind, who bought this text in Luxor and then willed it
to the British Museum. [1]

This papyrus was written during the period of the domination of Egypt by
the Hyksos (after 1800 B.C.) but, as its writer assures us, it derives from a
prototype which dates from the Middle Kingdom (2000—1800). All other texts
that have mathematical contents and that are known to us, also belong to the
Middle Kingdom. There is therefore reason to expect that we may become acquainted through this text, with the principles of the mathematical knowledge of this
period.

The papyrus starts in a very promising way: "Complete and thorough study
of all things, insight into all that exists, knowledge of all secrets . . ." he undertakes
to teach. But it soon becomes evident that we shall not witness the revelation of
the origin of things, but that we shall merely be initiated into the secrets of
numbers and into the art of calculating with fractions, in order to apply these to
various practical problems with which the officials of the great state had to deal,
such as the distribution of wages among a number of laborers, the calculation of
the amount of grain needed for the production of a given quantity of bread or of
beer, the calculation of areas and volumes, the conversion of different measures
for grains. Among these are also found however some purely theoretical questions,
which provide exercises in the difficult art of calculation with fractions.

In view of this, one is led to ask

For whom was the Rhind papyrus written?

Is Aristotle right in asserting that the priests were the actual carriers of mathematical development, that theoretical interests led them to devote their leisure
to mathematics, as Greek scholars did in his day? Or was it, as Herodotus and
Democritus thought, primarily the people concerned with the applications, who
cultivated geometry?

During the Old Kingdom of the Egyptians (3200—2000) and even during
the Middle Kingdom (2000—1800), the period of our texts, there was no well-
organised estate of priests; the task of the priests was usually carried on by lay-
men, along with their ordinary occupations. Accordingly, Aristotle starts from
an entirely wrong presupposition. As we shall see, the geometrical problems in
our texts are entirely designed for the applications. We are not asked to prove
or to construct something, but we are concerned with calculating the area of a
piece of land, or the volume of a barn for storing grain. Problems of this kind
are the concern of the surveyor, or of the official who has to erect a granary, but
not of the priest. Calculation with fractions, taught systematically in the Rhind

---

[1] See the excellent edition of T. E. Peet, *The Rhind Mathematical Papyrus*, London, 1923.

papyrus, is applied in texts on economics, such as the Kahun papyrus.[1] Who needs such calculations? It is

The class of the royal scribes

We get a good idea of the problems of these officials from the papyrus Anastasi I, in which a scribe ridicules another one for his lack of skill. One has obviously to think of the person addressed as some one who occupies an important position as a "scribe of the army", who has to calculate, e.g., how many soldiers are needed for digging a ditch, how many bricks are needed for a certain structure, etc. The writer of the letter reproaches him with his inability to solve these problems without his help.

"I will cause you to know how matters stand with you, when you say 'I am the scribe who issues commands to the army'. You are given a lake to dig. You come to me to inquire concerning the rations for the soldiers, and you say 'reckon it out'. You are deserting your office, and the task of teaching you to perform it falls on my shoulders" ... "I cause you to be abashed(?) when I disclose to you a command of your lord, you, who are his Royal Scribe" ... "the clever scribe who is at the head of the troops. A (building-)ramp is to be constructed, 730 cubits long, 55 cubits wide, containing 120 compartments, and filled with reeds and beams; 60 cubits high at its summit, 30 cubits in the middle, with a batter of twice 15 cubits and its pavement 5 cubits. The quantity of bricks needed for it is asked of the generals, and the scribes are all asked together, without one of them knowing anything. They all put their trust in you and say 'You are the clever scribe, my friend! Decide for us quickly!' Behold your name is famous ... Answer us how many bricks are needed for it?"[2]

The problems in the Rhind papyrus are of exactly the same general character; often they gradually increase in difficulty. It seems clear that this papyrus was intended for use in a school for scribes.

Let us now have a closer look at the papyrus.

To start with, we have to familiarize ourselves with

The technique of calculation,

which is indeed the first topic dealt with in the Rhind papyrus.

The number system of the Egyptians is as simple and as primitive as that of the Romans; it is a strictly decimal system. In hieroglyphics:

$\mid$	= one	$\mid\mid\mid$	therefore	= 3
$\cap$	= ten	$\cap\cap$ $\cap\cap$	therefore	= 40
$\wp$	= one hundred			
$\mathring{\mathsf{J}}$	= one thousand, etc.			

[1] Griffith, *Hieratic Papyri from Kahun and Gurob*, London 1898, p. 15, Plate VIII.
[2] O. Neugebauer, *The exact sciences in antiquity*. Princeton. New Jersey, 1952, p. 79.

All numbers that arose could be represented by placing these symbols in a row.

There is no difficulty in adding these numbers; it is only necessary to count the numbers of units, of tens, of hundreds, etc. Duplication is a special case of addition, so that this presents no difficulties. But very peculiar is

Multiplication

This is accomplished by doubling and adding the results. As an example we quote from the Rhind papyrus (Peet edition), No. 32, the multiplication of 12 by 12, first in hieroglyphics (to be read from right to left) [1] and then in modern notation

‖∩	∣		1	12
2 1	∣			
‖‖∩∩	‖		2	24
4 2	2			
‖‖‖∩∩∩∩	‖‖ /		/4	48
8 4	4			
‖‖∩∩∩∩∩∩∩	‖‖ /		/8	96 Sum 144
4 4 1 dmd 6 9	8			

Four times 12 and eight times 12 are added to produce 12 times 12. The numbers which are to be added, are indicated by an inclined line on the right (placed on the left in the "translation"). The result 144 is accompanied by the hieroglyph dmd, which represents a scroll with a seal.

To proceed more rapidly, one frequently multiplies by 10; sometimes the 10-fold multiple is halved, as, e.g., in the Kahun papyrus, No. 6, where 16 × 16 is calculated as follows:

```
/ 1    16
/10   160
/ 5    80
sum   256
```

This Egyptian method of multiplication is the foundation of the entire art of calculation. It must be very ancient; but it has been able to maintain itself, without change, into the Hellenistic period. And, in the Greek schools, it was taught as "Egyptian calculation". Even during the Middle Ages, "duplatio" was looked upon as an independent operation. Along with it, the process of halving, "mediatio", was taught.

---

[1] The papyrus itself is not written in hieroglyphics, but in hieratic script, which can readily be changed into hieroglyphics.

Division

is considered by the Egyptians as a kind of multiplication, but formulated inversely. "Multiply 80 (or literally: add beginning with 80) until you get 1120", says No. 69 in the Rhind papyrus, and the solution looks exactly like a multiplication:

1	80
/10	800
2	160
/ 4	320
sum	1120

We would write the result in the form 1120 : 80 = 14. But for the Egyptians, 1120 is the result; it is indicated as such by the symbol of the scroll.

For a fuller understanding of what follows, a mastery of the Egyptian process for division is absolutely indispensable. The interested reader is therefore requested to take a pencil and to carry out a few divisions in accordance with the Egyptian method. It should not take him long to familiarize himself with it and thus to acquire the point of view of the Egyptians.

But what did the Egyptian do when the division did not go "evenly"? He did just what we do; he took recourse to fractions.

Natural fractions and unit fractions

Fractions with numerator and denominator, as we know them, do not exist in Egyptian mathematics. In the first place, there is a limited number of *natural fractions*, which occur in daily life and which are designated by specific names.

In modern life, one rarely uses fractions except ½, ¼ and ¾, and the expression "percent". The stock exchange calculates also with an eighth or a sixteenth of a percent. The French language still has a special word for a third (tiers). Ancient Egypt had special words for ½, ⅓, ⅔, ¼ and ¾, which were the natural fractions. In addition, ⅙ and ⅛ may be counted among the natural fractions.

As the technique of calculation developed, the set of fractions was extended to include the *unit fractions* (fractions which have 1 as a numerator). The Egyptian was not able to write any other fraction, except those which have been mentioned.

Worthy of notice is the verbal expression for ⅔ which means literally "the two parts". The complement, necessary to make a whole out of the two parts is "the third part".

In Greek one also speaks of

"the two parts" ⅔ — "the third part" ⅓
"the three parts" ¾ — "the fourth part" ¼.

It presents quite naturally a concrete image: three parts and then a fourth part combine to make the whole. [1] Analogously we can explain our use of the words third, fourth, fifth, etc. In this representation the fifth part is the last part, which combines with the four other parts to complete the unit. Philologically it does not make sense to speak of two fifths, because there is only *one* fifth part, viz. the last.

The Egyptians did not succumb to the temptation of this philological contradiction. Consequently they did not obtain a convenient notation for fractions which are not unit-fractions. They represent the unit-fractions by writing the denominator, supplied with a special symbol *r* = part:

$$\underset{\text{iin}}{\frown} \; = \; {}^1\!/_{12},$$

and they reduce other fractions to unit fractions, e.g.:

$$\tfrac{3}{8} = \tfrac{1}{4} + \tfrac{1}{8}.$$

The fraction $\tfrac{2}{3}$ is the only one which did not need to be reduced, because it had a name of its own, "the two parts", and a hieroglyphic of its own ꝑ

The ancient symbol for $\tfrac{3}{4}$ was replaced later on by the partition

$$\Longleftarrow\!\mathsf{X} \; = \; \tfrac{1}{2} + \tfrac{1}{4}.$$

In order to follow Neugebauer [2]) in his interpretation of Egyptian calculation with fractions from their own point of view, it is necessary to introduce a notation for fractions based on the Egyptian notation, and not reminiscent of our numerators and denominators. Thus we shall write, as Neugebauer does, $\bar{n}$ for $1/n$ and $\bar{3}$ for $\tfrac{2}{3}$.

Calculation with natural fractions

is very simple for any one who has mastered a few simple formulas, which follow immediately from the meaning of the fractions. Simple examples are the following three, to be found in the "London leather-scroll" [3]: (Plate 3)

$$\bar{6} + \bar{6} = \bar{3},$$
$$\bar{6} + \bar{6} + \bar{6} = \bar{2},$$
$$\bar{3} + \bar{3} = \bar{3}.$$

Further relations, deduced from these, are constantly applied in the Rhind papyrus, e.g.

[1] See K. Sethe, *Von Zahlen und Zahlwörtern bei den alten Ägyptern und was für andere Völker und Sprachen daraus zu lernen ist*, Schriften der wissenschaftlichen Gesellschaft, Strassburg, 1916.
[2] O. Neugebauer, *Arithmetik und Rechentechnik der Ägypter*, Quellen und Studien, B I, p. 301.
[3] See Plate 3. Also A. B. Chace, *The Rhind Mathematical Papyrus*, Ohio 1929.

PLATE 1

PL. 1. Painting of the tomb of Djeserkere-sonb (No. 38), Thebes, Egypt. Upper part: surveying. The New Kingdom, 18th dynasty (1567–1310 B.C.). The mathematical knowledge of the Egyptians, concerned with practical matters, goes along with a form of reproduction in art, attempting to give which aims at giving as complete, or rather as perfect a picture of things as possible. This art is called "ideoplastic", because the Egyptian does not picture things as he sees them ("physioplastic") but as he knows them to be. In relief art and in painting, the artist spreads everything as much as possible in the plane, because this enables him to give the largest number of exact data concerning reality. Up to a certain point, this art must be "read". Yet not entirely so, for this "ideoplastic" conception is joined with a sharp observation of nature (see next plate).

(Photo Lehnert & Landrock, Cairo)

PLATE 2

PL. 2. A writer. Limestone statue from Egypt, Louvre, Paris. The Old Kingdom, fifth dynasty, about 2500 B.C. The statue is full of spirit and "magic" life. This art is not abstract in the sense that it considers the external appearance of life as unimportant. On the contrary, it seeks to perpetuate it. This is shown not only by the plasticity of the body, but also and especially by the almost frightening "realism" of the look. To approximate nature as closely as possible, the eyes have been formed from various materials: white quartz, ebony and rock-cystal.

(Photo Alinari)

PLATE 3

Pl. 3. Leather scroll in the British Museum (BM 10250), containing simple relations between fractions. The scroll is said to have been found together with the Rhind papyrus near the Rammesseum at Thebes. The date is about 1700 B.C. The unrolling of the hardened scroll was a clever piece of work of modern chemistry. At first, the contents were disappointing; every line gave a simple relation between fractions, such as $\bar{9} + \overline{18} = \bar{6}$, $\bar{5} + \overline{20} = \bar{4}$, etc. Nevertheless, the leather scroll has proved valuable inasmuch as it supplies a key for the understanding of the first stages in calculation with fractions. See S. R. K. Glanville, *The mathematical leather roll in the British Museum*, Journ. Egypt. Archeol. 13 (1927) p. 232, and B. L. v. d. Waerden, *Die Entstehungsgeschichte der ägyptischen Bruchrechnung*, Quellen u. Studien Gesch. Math. B 4, p. 359.

PLATE 4

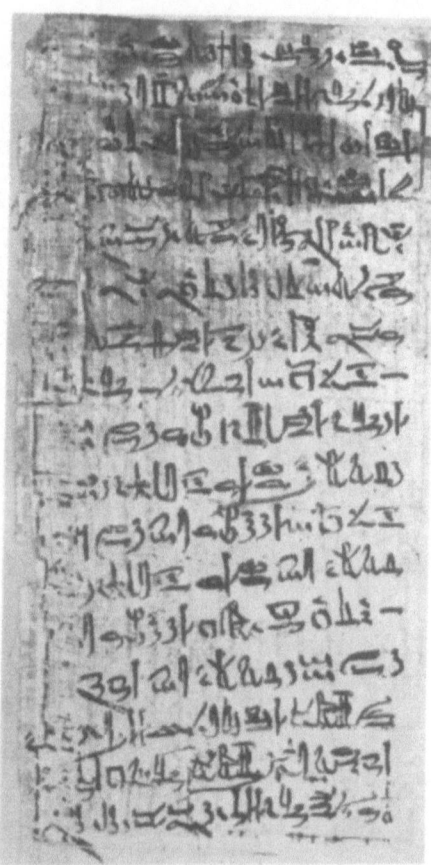

Pl. 4a. Egyptian stele with hieroglyphic writing from the Mastaba of Meri. Louvre, Paris. Old Kingdom, 4th dynasty (about 2600 B.C.).

Pl. 4b. Hieratic writing, papyrus in London, British Museum (about 1480 B.C., early 18th dynasty). Hieratic writing is a simplification of hieroglyphic writing; it already came into use during the Old Kingdom.

(1) $$\bar{3} + \bar{6} = \bar{2},$$
(2) $$\bar{2} + \bar{3} + \bar{6} = 1.$$

If $\bar{6}$ is added to both members of (1), there results the very important formula

(3) $$\bar{3} = \bar{2} + \bar{6}.$$

Again adding $\bar{6}$ to both sides, and writing the result from right to left, we obtain two equivalent representations of $^5/_6$:

(4) $$\bar{2} + \bar{3} = \bar{3} + \bar{6}.$$

Finally, addition of $\bar{6}$ to both sides of (2) gives

(5) $$\bar{3} + \bar{2} = 1 + \bar{6}.$$

These are the rules for the calculation with halves, thirds and sixths which every Egyptian computer had to know by heart; the Rhind papyrus takes them for granted. For example, when in the first step of the division of 2 by 41, we have to determine one half of $13 + \bar{3}$, this is at once written in the form $6 + \bar{3} + \bar{6}$, which follows from $6 + \bar{2} + \bar{3}$ by means of (4). Again, to double $4 + \bar{3} + \bar{4}$ in Rhind No. 31, we do not first find $8 + \bar{3} + \bar{2}$, but immediately $9 + \bar{6}$, by use of (5).

Formula (3) is applied especially when it is required to determine two thirds of a unit fraction. For example, in Rhind, No. 61 we find

$$\bar{3} \text{ of } \bar{11} \text{ is } \bar{22} + \bar{66}.$$

Further relations between fractions

are obtained from (1)-(5), by dividing by 2, 3, etc. By halving, formula (3) of the leather scroll leads to

$$\bar{3} = \bar{4} + \bar{12};$$

similarly (1) gives

$$\bar{6} + \bar{12} = \bar{4}.$$

That the Egyptians used this method systematically becomes clear from the following sequence found in the leather scroll and obtained from (1) upon division by 3, 4, etc.:

$$\bar{9} + \bar{18} = \bar{6},$$
$$\bar{12} + \bar{24} = \bar{8},$$
$$\bar{15} + \bar{30} = \bar{10},$$
$$\bar{18} + \bar{36} = \bar{12},$$
$$\bar{21} + \bar{42} = \bar{14},$$
$$\bar{24} + \bar{48} = \bar{16},$$
$$\bar{30} + \bar{60} = \bar{20},$$
$$\bar{45} + \bar{90} = \bar{30},$$
$$\bar{48} + \bar{96} = \bar{32},$$
$$\bar{96} + \bar{192} = \bar{64}.$$

Moreover, the same method makes it possible to double a unit-fraction whose denominator is divisible by 3. For, if we write (3) in the form

$$\bar{3} + \bar{3} = \bar{2} + \bar{6},$$

and then divide by 3, 5, etc., we obtain

$$\bar{9} + \bar{9} = \bar{6} + \overline{18},$$
$$\overline{15} + \overline{15} = \overline{10} + \overline{30},$$
$$\overline{21} + \overline{21} = \overline{14} + \overline{42}, \text{ etc.}$$

This

Duplication of unit fractions

is of great importance for the Egyptian computer; indeed, for him, multiplication depends upon duplication.

Doubling a fraction with even denominator is very easy; it is done by simply halving the denominator. And we have already seen how to double a fraction whose denominator is divisible by 3, namely, as shown above, on the model of

$$\bar{9} + \bar{9} = \bar{6} + \overline{18}.$$

At this point the reader may well ask: why not leave well enough alone and be satisfied with $\bar{9} + \bar{9}$?

The answer to this question is found by considering what would happen in the further development of the calculation. In doubling $\bar{9} + \bar{9}$, one would get $\bar{9} + \bar{9} + \bar{9} + \bar{9}$, a third duplication would lead to $\bar{9} + \bar{9} + \bar{9} + \bar{9} + \bar{9} + \bar{9} + \bar{9} + \bar{9}$. Obviously, there is little of value in such a senseless accumulation of unit fractions. The Egyptian proceeds however as follows:

$$
\begin{array}{ll}
1 & \bar{9}, \\
2 & \bar{6} + \overline{18}, \\
4 & \bar{3} + \bar{9}, \\
8 & \bar{3} + \bar{6} + \overline{18}.
\end{array}
$$

He obtains expressions which are easily managed and which contain but few unit-fractions with larger denominators, in addition to natural fractions, like $\bar{3}$, $\bar{3}$, $\bar{6}$, whose order of magnitude is readily kept in mind and which are dealt with easily.

To this one might make the rejoinder that the Egyptians might have thought of a shorter way to write $\bar{9} + \bar{9}$ and $\bar{9} + \bar{9} + \bar{9} + \bar{9}$, by using something like our numerators and denominators.

Our reply would be that we have to accept the fact that the Egyptians were not like our modern mathematicians, who immediately introduce an abbreviation for every expression that occurs frequently. They were extremely conservative, strictly observing ancient traditions and adhering to current notations. An example is found in the fact that for thousands of years Egyptian documents continue to speak of "the two kingdoms", although Upper- and Lower-Egypt had long ago been united into one empire.

What was done when the denominator was not divisible by 2 or by 3? To understand this, we have to consider

Division once more.

The following three are divisions from the Rhind papyrus, which do leave a remainder:

(No. 24)	19 : 8	(No. 25)	16 : 3	(No. 21)	4 : 15
	1 8	/	1 3		1 15
/	2 16		2 6		$\bar{10}$ $1 + \bar{2}$
	$\bar{2}$ 4	/	4 12	/	$\bar{5}$ 3
/	$\bar{4}$ 2		$\bar{3}$ 2	/	$\bar{15}$ 1
/	$\bar{8}$ 1	/	$\bar{3}$ 1		
Quotient: $2 + \bar{4} + \bar{8}$		Quotient: $5 + \bar{3}$		Quotient: $\bar{5} + \bar{15}$.	

In No. 24, one is asked "to operate with 8 so as to obtain 19". Doubling 8 gives 16; we lack 3. Now we take one half of 8, and again, and again. By adding together the fourth and the eighth, we obtain exactly the 3 we need.

It is customary to use, instead of the sequence of fractions $\bar{2}$, $\bar{4}$, $\bar{8}$, . . . employed here, the other sequence $\bar{3}$, $\bar{3}$, $\bar{6}$, . . . This is seen in No. 25 where we take first the double and the quadruple, and then $\frac{2}{3}$ and $\frac{1}{3}$ of 3. It is a peculiarity of the Egyptian calculators, not to write down $\bar{3}$ first, but to obtain it from $\bar{3}$. The sequence of fractions $\bar{3}$, $\bar{3}$, $\bar{6}$ . . . had evidently become entrenched in an established tradition.

In No. 21, perhaps in order to reach smaller numbers more rapidly, we take first $1/_{10}$, this is doubled and then increased by $1/_{15}$. Thus the required sum $3 + 1 = 4$ is obtained in an elegant manner.

Perhaps the reader will find it of interest to use this method to answer questions 3 to 6 of the Rhind papyrus! The problems are: to divide 6, 7, 8, or 9 loaves among 10 men. The Egyptian answers are as follows: $\bar{2} + \bar{10}$ each, $\bar{3} + \bar{30}$ each, $\bar{3} + \bar{10} + \bar{30}$ each, $\bar{3} + \bar{5} + \bar{30}$ each.

The (2 : n)-table.

At the beginning of the Rhind papyrus, the (2 : n)-table answers the question as to the duplication of fractions, like $\bar{5}$ or $\bar{7}$, which was raised above. The solution is based on a principle, which is as simple as it is fertile: To calculate 2 . $\bar{n}$, divide 2 by n.

Perhaps it appears to us as a matter of course that $^2/_5$ and 2 : 5 are the same thing. For the Egyptians this was not the case, since the problem of duplicating $\bar{5}$ is totally different from the division problem which amounts to calculating with 5 until you obtain 2. It was therefore indeed a new idea to duplicate $\bar{n}$ by dividing 2 by n.

We do not know in whose brain this thought arose for the first time, nor when this happened. It certainly occurred long before the era of our texts, for the $(2 : n)$-table of the Rhind papyrus, which includes all the odd numbers from $n = 3$ to $n = 101$, was not constructed all at one time; its separate parts were computed by different methods. The oldest section contains the denominators which are divisible by 3; without exception, they all proceed according to the same rule:

$$2 : 9 = \bar{6} + \overline{18}$$
$$2 : 15 = \overline{10} + \overline{30}$$
$$2 : 21 = \overline{14} + \overline{42}.$$

In these cases the division $2 : 3k$ is simply a confirmation of a known result. In the other cases (certainly from $n = 11$ on), the duplication appears to have been obtained by actually carrying out the division of 2 by n. The text exhibits the divisions more or less explicitly, as in the following examples $(2 : 5$ and $2 : 7)$.

What part is 2 of 5? **$\bar{3}$** is $1 + \bar{3}$, **$\overline{15}$** is $\bar{3}$,

i.e. a third of 5 is $1 + \bar{3}$, a fifteenth is $\bar{3}$; these add up to 2. The result of the division is therefore $\bar{3} + \overline{15}$; the terms $\bar{3}$ and $\overline{15}$ are clearly visible because they are written in red. In our "translation" the red symbols have been printed in bold-face type.

Computation: 1   5
                      $\bar{3}$ $3 + \bar{3}$
                    / $\bar{3}$ $1 + \bar{3}$
                    / $\overline{15}$ $\bar{3}$

What part is 2 of 7? **$\bar{4}$** is $1 + \bar{2} + \bar{4}$, **$\overline{28}$** is $\bar{4}$.

Computation: 1   7
                      $\bar{2}$ $3 + \bar{2}$ 1   7
                    / $\bar{4}$ $1 + \bar{2} + \bar{4}$ 2   14
                  / 4 $\overline{28}$ $\bar{4}$ / 4   28.

In this manner the work proceeds. In dividing 2 by 5, 9, 11, 17, 23, 29 and a few of the larger integers, the $\bar{3}$-sequence is used, i.e. the sequence of fractions $\bar{3}, \bar{3}, \bar{6}, \overline{12}, \ldots$; but the division by 7 and 13 employs only the $\bar{2}$-sequence $(\bar{2}, \bar{4}, \bar{8}, \ldots)$. It turns out that only in these two cases the $\bar{2}$-sequence produces a simpler result than the $\bar{3}$-sequence. For instance, the use of the $\bar{2}$-sequence would, in calculating $2 : 11$, lead to the result $2 : 11 = \bar{8} + \overline{22} + \overline{88}$, while the $\bar{3}$-sequence gives $2 : 11 = \bar{6} + \overline{66}$, which, having fewer terms and smaller denominators, is obviously to be preferred.

The calculations which have been reproduced here certainly tell their own story. In the case $2 : 7$, the number 4, placed in front of $\overline{28}$, indicates where 28 comes from, viz. from 4×7, the further details being shown in an auxiliary column.

The results of the divisions $2 : n$ are summarized in the following table, which does not include divisors that are divisible by 3, all of which follow the rule $2 : 3k = \overline{2k} + \overline{6k}$.

$2 : \ 5 = \ \overline{3} + \ \overline{15}$	$2 : \ 53 = \ \overline{30} + \overline{318} + \overline{795}$	
$2 : \ 7 = \ \overline{4} + \ \overline{28}$	$2 : \ 55 = \ \overline{30} + \overline{330}$	
$2 : 11 = \ \overline{6} + \ \overline{66}$	$2 : \ 59 = \ \overline{36} + \overline{236} + \overline{531}$	
$2 : 13 = \ \overline{8} + \ \overline{52} + \overline{104}$	$2 : \ 61 = \ \overline{40} + \overline{244} + \overline{488} + \overline{610}$	
$2 : 17 = \ \overline{12} + \ \overline{51} + \ \overline{68}$	$2 : \ 65 = \ \overline{39} + \overline{195}$	
$2 : 19 = \ \overline{12} + \ \overline{76} + \overline{114}$	$2 : \ 67 = \ \overline{40} + \overline{335} + \overline{536}$	
$2 : 23 = \ \overline{12} + \ \overline{276}$	$2 : \ 71 = \ \overline{40} + \overline{568} + \overline{710}$	
$2 : 25 = \ \overline{15} + \ \overline{75}$	$2 : \ 73 = \ \overline{60} + \overline{219} + \overline{292} + \overline{365}$	
$2 : 29 = \ \overline{24} + \ \overline{58} + \overline{174} + \overline{232}$	$2 : \ 77 = \ \overline{44} + \overline{308}$	
	$2 : \ 79 = \ \overline{60} + \overline{237} + \overline{316} + \overline{790}$	
$2 : 31 = \ \overline{20} + \overline{124} + \overline{155}$	$2 : \ 83 = \ \overline{60} + \overline{332} + \overline{415} + \overline{498}$	
$2 : 35 = \ \overline{30} + \overline{42}$	$2 : \ 85 = \ \overline{51} + \overline{255}$	
$2 : 37 = \ \overline{24} + \overline{111} + \overline{296}$	$2 : \ 89 = \ \overline{60} + \overline{356} + \overline{534} + \overline{890}$	
$2 : 41 = \ \overline{24} + \overline{246} + \overline{328}$	$2 : \ 91 = \ \overline{70} + \overline{130}$	
$2 : 43 = \ \overline{42} + \ \overline{86} + \overline{129} + \overline{301}$	$2 : \ 95 = \ \overline{60} + \overline{380} + \overline{570}$	
$2 : 47 = \ \overline{30} + \overline{141} + \overline{470}$	$2 : \ 97 = \ \overline{56} + \overline{679} + \overline{776}$	
$2 : 49 = \ \overline{28} + \overline{196}$		
$2 : 51 = \ \overline{34} + \overline{102}$	$2 : 101 = \overline{101} + \overline{202} + \overline{303} + \ \overline{606}$	

Beginning with $2 : 31$, the form of presentation changes; the calculations are given in abbreviated form. But, what is more important, the method of calculation changes; another idea is introduced. While up to this point, all divisions were carried out by means of the $\overline{2}$-sequence and the $\overline{3}$-sequence, the divisions $2 : 31$ and $2 : 35$ proceeded quite differently, as is seen from the following examples:

What part is 2 of 31? $\overline{20}$ is $1 + \overline{2} + \overline{20}$, $\overline{124}$ is $\overline{4}$, $\overline{155}$ is $\overline{5}$.

Computation:

$$
\begin{array}{lcl}
 & 1 & 31 \\
 & \overline{20} & 1 + \overline{2} + \overline{20} \\
/4 & \overline{124} & \overline{4} \\
/5 & \overline{155} & \overline{5}
\end{array}
$$

What part is 2 of 35? $\overline{30}$ is $1 + \overline{6}$, $\overline{42}$ is $\overline{3} + \overline{6}$

Computation:

$$
\begin{array}{lcl}
 & 1 & 35 \\
 & \overline{30} & 1 + \overline{6} \\
 & \overline{42} & \overline{3} + \overline{6}
\end{array}
$$

The start of the computation of $2 : 31$ is easy to account for, since division of 31 by 10, and halving of the result shows that $1/_{20}$ of 31 is $1 + \overline{2} + \overline{20}$. This fraction is to be increased so as to produce 2. How did the calculator hit upon the idea that this requires $\overline{4} + \overline{5}$? It checks; for the leather scroll has the relation

$\overline{20} + \overline{5} = \overline{4}$, and $1 + \overline{2} + \overline{4} + \overline{4}$ is of course equal to 2. How does one obtain the fractions needed to increase $1 + \overline{2} + \overline{20}$ to 2? This requires a new procedure viz.

The red auxiliaries.

In the calculation of 2 : 35, reproduced above, auxiliary numbers occur whose significance must not be underestimated, viz. the red (bold-faced) 6 and the numbers 7 and 5 which follow it. They are inversely proportional to the numbers 35, 30 and 42 under which they are placed, and therefore proportional to $\overline{35}$, $\overline{30}$ and $\overline{42}$. In our modern arithmetic, these numbers 6, 7 and 5 would appear as numerators when the three unit-fractions are reduced to the common denominator 210. In modern notation, these considerations lead to

$$\frac{2}{35} = \frac{12}{210} = \frac{7}{210} + \frac{5}{210} = \frac{1}{30} + \frac{1}{42}$$

which corresponds to the Egyptian division

$$2 : 35 = \overline{30} + \overline{42}.$$

It appears therefore that these auxiliary numbers are to be looked upon as the numerators of fractions which have been reduced to a common denominator.

But these modern ideas should not simply be read into the old texts. Something further about these auxiliary numbers and about the role they play, is learned from the complementary calculations (śekem-calculations) of the Rhind papyrus; we proceed to consider these more fully.

Complementation of a fraction to 1

is a problem which occurs over and over again in the Egyptian divisions. In the division of 2 by 31, given above, the sum $\overline{2} + \overline{20}$ has to be complemented to 1; the solution $\overline{4} + \overline{5}$ is not at all obvious. It is not surprising therefore that such complementations receive special attention in the Rhind papyrus.

In No. 21 we find:

How is $\overline{3} + \overline{15}$ complemented to 1?

     10   **1**   sum 11, remainder 4.

Calculate with 15 till you find 4 (i.e. divide 4 by 15)

1	15
$\overline{10}$	$1 + \overline{2}$
/ $\overline{5}$	3
/ $\overline{15}$	1

sum 4. Hence $\overline{5} + \overline{15}$ is the desired complement.

This is followed by a cheek.

The red auxiliaries 10 and 1 are the numerators of the fractions $\bar{3}$ and $\overline{15}$, when they are reduced to the denominator 15. Since their sum is 11, there is lacking $^4/_{15}$ to obtain the total of $^{15}/_{15}$. The calculation of $^4/_{15}$ then follows.

It appears however from No. 23 that our interpretation of the auxiliaries as "numerators" is not entirely valid; for fractions appear there as auxiliaries:

$\bar{4}$ $+$ $\bar{8}$ $+$ $\overline{10} + \overline{30} + \overline{45}$ has to be complemented to $\bar{3}$.

$11 + \bar{4}$ $5 + \bar{2} + \bar{8}$ $4 + \bar{2}$ $1 + \bar{2}$ 1

The required complement is $\bar{9} + \overline{40}$.

Check:

$\bar{4}$ $+$ $\bar{8}$ $+$ $\bar{9} +$ $\overline{10} +$ $\overline{30} +$ $\overline{40}$ $+ \overline{45} +$ $\bar{3}$

$11 + \bar{4}$ $5 + \bar{2} + \bar{8}$ 5 $4 + \bar{2}$ $1 + \bar{2}$ $1 + \bar{8}$ 1 $15.$ Sum 1.

The computation is not given, but it is readily supplied. When the given fractions are reduced to the "denominator" 45, the sum of the numerators is $(11 + \bar{4}) + (5 + \bar{2} + \bar{8}) + (4 + \bar{2}) + (1 + \bar{2}) + 1 = 23 + \bar{2} + \bar{4} + \bar{8}$. To obtain 30, we lack $6 + \bar{8}$; division of $6 + \bar{8}$ by 45 gives the quotient $\bar{9} + \overline{40}$.

As a rule, the denominator of the smallest fraction, in our case 45, is taken as the new denominator. A "numerator", in the sense of an integer, that "counts" the number of forty-fifths, does not occur. It is more accurate to describe the state of affairs as follows: *When a somewhat complicated sum of fractions has to be compared with another such sum, or has to be complemented to 1, the smallest of the fractions is taken as a new unit and the other fractions are then expressed in terms of it.* Or, even simpler and very readily applicable: *The transformation of the given quantities to the auxiliaries is accomplished by multiplying by the largest of the denominators (in our case 45), and the inverse process is carried out by division by this number.* An understanding of this simple rule gives complete control of the calculations by means of auxiliaries.

This calculation with auxiliaries completes and puts the crown on the Egyptian computation technique. By means of it, every division, no matter how complicated, can be carried out.

"Aha-calculations".

The Egyptian word 'h', that used to be pronounced, incorrectly, "hau", and at the present time somewhat less incorrectly as "aha", means a quantity, a collection. These aha-calculations are quite like our linear equations in one unknown. A simple example is found in Rhind, No. 26:

"A quantity and a fourth part of it give together 15".

The Egyptian solution begins as follows:

"Calculate with 4, of this you must take the fourth part, namely 1; together 5".

Then the division $15 : 5 = 3$ is carried out, finally a multiplication, $4 \cdot 3 = 12$. The required "quantity" is therefore 12, the fourth part is 3, together 15.

It is clear that the method followed here is that of the "false assumption":

one starts with an arbitrarily chosen number as the required quantity, in our case 4, because this makes the computation of the fourth part easy. Four and a fourth part of four give 5. But the required result is 15; hence the quantity has to be multiplied by $15 : 5 = 3$.[1]

Frequently the quantity is at first taken to be 1, e.g. in Rhind, No. 37.

I go three times in a bushel; my third part and a third of my third part, and my ninth part are added to me and I come out entirely (i.e. the bushel is entirely filled). Who says this?

Answer:

$$
\begin{array}{rcl}
 & 1 & 1 \\
 & 2 & 2 \\
 & \bar{3} & \bar{3} \\
\text{3 of his } & \bar{3} & \bar{9} \\
\text{his } & \bar{9} & \bar{9} \\
\text{sum} & & 3 + \bar{2} + \overline{18} \ (\text{because. } \bar{9} + \bar{9} = \bar{6} + \overline{18}).
\end{array}
$$

Divide 1 by $3 + \bar{2} + \overline{18}$:

$$
\begin{array}{rl}
1 & 3 + \ \bar{2} + \overline{18} \\
\bar{2} & 1 + \ \bar{2} + \ \bar{4} + \ \overline{36} \\
/\ \bar{4} & \bar{2} + \ \bar{4} + \ \bar{8} + \ \overline{72} \\
\bar{8} & \bar{4} + \ \bar{8} + \overline{16} + \overline{144} \\
\overline{16} & \bar{8} + \overline{16} + \overline{32} + \overline{288} \\
/\overline{32} & \overline{16} + \overline{32} + \overline{64} + \overline{576}
\end{array}
$$

Now the auxiliaries enter the field: the sum $\bar{2} + \bar{4} + \bar{8} + \overline{72} + \overline{16} + \overline{32} + \overline{64} + \overline{576}$ must be complemented to 1. The auxiliaries show that the last five fractions add up to $\bar{8}$ exactly:

$$
\begin{array}{cccccl}
\overline{72} & \overline{16} & \overline{32} & \overline{64} & \overline{576} & \text{sum } \bar{8} \\
8 & 36 & 18 & 9 & 1 & \quad 72.
\end{array}
$$

Hence the sum of all the fractions is already 1; so that the result of the division is $\bar{4} + \overline{32}$. To check this result, three times $\bar{4} + \overline{32}$, a third, a third of a third, and a ninth of $\bar{4} + \overline{32}$ are determined; then it is shown by means of auxiliaries that their sum is indeed 1.

The group of aha-calculations also includes the first problem of the Berlin

---

[1] The above explanation seems to me to be the simplest one. But I must not fail to mention that Neugebauer considers the "false assumption" as a legend, and that he interprets the beginning of the computation ("calculate with 4, take a fourth part, equal 1, together 5") differently, viz. as a calculation with auxiliaries:

$$
\begin{array}{c}
1 + \bar{4} \\
4 + 1 = 5
\end{array}
$$

(O. Neugebauer, *Quellen und Studien*, B 1, 333). On the other hand, there seems to be little justification for the auxiliaries, since the division 15: (1 + 4) can very easily be carried out without using them.

papyrus 6619[1], whose solution requires even the extraction of square roots. The text is as follows:

"A square and a second square, whose side is ¾ (in the text, $\bar{2} + \bar{4}$) of that of the first square, have together an area of 100. Show me how to calculate this".

It is perfectly clear that the solution starts with a false assumption: "Take a square of side 1, and take ¾ of 1, that is to say $\bar{2} + \bar{4}$, as the side of the other area. Multiply $\bar{2} + \bar{4}$ by itself, this gives $\bar{2} + \overline{16}$. Hence, if the side of one of the areas is taken to be 1, that of the other as $\bar{2} + \bar{4}$, then the addition of the areas gives $1 + \bar{2} + \overline{16}$. Take the square root of this; it is $1 + \bar{4}$. Take the square root of the given number 100; it is 10. How many times is $1 + \bar{4}$ contained in 10? 8 times". From here on the text becomes undecipherable, but we can easily make a guess at the rest: $8 \cdot 1 = 8$ and $8 \cdot (\bar{2} + \bar{4}) = 6$ are the sides of the required squares.

The aha-calculations constitute the climax of Egyptian arithmetic. The Egyptians could not possible get beyond linear equations and pure quadratics with one unknown, with their primitive and laborious computing technique.

The aha-calculations are not based on practical problems; they bear witness to the purely theoretical interests of the Egyptian computers. They have obviously been set up by people who enjoyed pure calculations and who wanted to drill their pupils on really hard problems. Like every art, arithmetic strives for its highest development.

Much space is taken up in the Rhind papyrus by

Applied calculations,

such as, e.g., the "pesu-calculations" which are concerned with determining the amounts of grain needed for making beer or bread. The technical term "pesu" = "baking value" designates the number of loaves of bread or the number of jugs of beer that can be made from a bushel of grain. Thus we have the following equation

quantity of grain × pesu = number of loaves (jugs of beer)
pesu = number / amount of grain.

The reciprocal quotient is the grain-density or the "strength" of the loaf of bread or the jug of beer.

These simple relations would suffice to solve all pesu-problems, if it were not for complications arising from the non-equivalence of different grains; this requires the determination of reduction-coefficients which the computer has to know. The problems can then be complicated still further by trading loaves of one "baking value" against those of another value, etc.

Other problems are concerned with the reduction of bushels to other units of

[1] Schack-Schackenburg, Äg. Z. *38* (1900), p. 138 and *40* (1900), p. 65.

measure, with the calculations of quantities of food, the distribution of wages, and so forth.

Let us now try to survey

The development of the computing technique

and to assign it a place in the history of civilization.

Initially, the Egyptians, like other nations, certainly disposed only of a restricted number of integers, sufficient for daily life, and of a similarly restricted number of "natural fractions": $\frac{1}{2}$, $\frac{1}{3}$, $\frac{2}{3}$, $\frac{1}{4}$, $\frac{3}{4}$, $\frac{1}{6}$, $\frac{1}{8}$. This primitive stage belongs however to pre-history; the history of the technique of computing begins with the extension of this primitive arsenal of numbers in both directions. Such an extension became a necessity at a time when an empire had to be organised, an army administered, and taxes collected. Indeed, on a monument of the first dynasty, we already find the symbol for 100,000. During the Old Kingdom, a symbol for a million existed; but during the New Kingdom it disappeared.

The dominant element in the thought of the Egyptian calculator is addition. He writes fractions as sums of unit-fractions. Multiplication is for him a kind of addition; the technical term for multiplying is "add, beginning with . . ."

Egyptian multiplication is markedly a written operation. It can not antedate the notation for numbers therefore. Neither can it be much younger than the art of writing; for obviously to calculate the amount of grain necessary for an army, or to determine the quantity of money and of material required for the construction of the pyramids, multiplications were required.

The development of arithmetic therefore took place in a very slow tempo; it extends over the entire period from the invention of the art of writing to the Middle Kingdom, a stretch of many centuries. It is possible to indicate the successive phases of this development, but not to give their dates. It must have proceeded as follows:

From multiplication we come naturally to division which is nothing but inverse multiplication. But to carry out a division, fractions and operations on fractions are needed; thus the division problem led to a further development in the calculation with fractions. Next came the recognition of simple relations among the natural fractions. Additional rules resulted from halving. From $\overline{3} = \overline{2} + \overline{6}$, rules for the duplication of the unit fractions $\overline{3n}$ were derived, by division by 3, 5, etc. A decisive step was the discovery that every unit fraction $\overline{n}$ can be duplicated by division of 2 by n; this led to the first third of the $(2 : n)$-table.

As the calculations, which were now accessible, became more and more complicated, the need for a method of checking was recognized, a procedure to compare different sums of unit-fractions; in particular it became essential to be able to complement such sums to 1. This problem was solved by the introduction of auxiliary numbers. Their use made it possible to carry out every division, to solve every aha-problem, no matter in how sophisticated a manner it was formulated.

It is possible to understand this entire development without taking recourse to the

Hypothesis of an advanced science,

of which our texts would then merely represent the "first phase", as does Gillain [1]) for example.

We know absolutely nothing of such an advanced science; it would be without value for the explanation of arithmetic, because we can account for all the essential points without it.

Under the influence of Greek writers, the levels of Egyptian geometry and astronomy are frequently placed too high. Now that we know a little more about Egyptian astronomy, it becomes very clear that, even in the Roman period, when astrology flourished, Egyptian astronomy lagged far behind Greek and Babylonian astronomy. [2])

What about geometry however?

The geometry of the Egyptians.

We are going to show that Egyptian geometry is not a science in the Greek sense of the word, but merely *applied arithmetic*. While the other arithmetical problems are concerned with the calculation of wages, the quality of bread or of beer, etc., the geometric problems ask for the determination of areas and volumes. In all these cases, the calculator has to know the rules on which the calculations depend. But a systematic derivation of these rules occurs nowhere; it can not occur anywhere, because frequently, as for instance in the case of the area of a circle, only approximate formulas are used.

The analogy between the geometrical problems and the beer- and bread-exercises comes out clearly especially in the calculation of

The inclination of oblique planes.

Rhind No. 36: "example for the calculation of a pyramid. 360 is the side of the base, 250 the height, tell me the inclination (śkd)." The word śkd might be pronounced as "saykad".

Calculation: half of 360 $= 180$.

$180 : 250 = \bar{2} + \bar{5} + \bar{50}$ of a cubit. This quotient is now multiplied by 7, because a cubit is 7 hand's breadths. Result: the inclination is $5 + \bar{25}$ hands.

It becomes clear from the calculation that the saykad, or inclination, is the num-

[1]) O. Gillain, *L'arithmétique au Moyen Empire*, Brussels, 1927.

[2]) O. Neugebauer and A. Volten, *Quellen und Studien*, B 4, p. 383; O. Neugebauer, Trans. Amer. Philos. Soc. *32* (1942), p. 209; B. L. v. d. Waerden, Proc. Royal Acad. Amsterdam *50* (1947), pp. 536 and 782.

ber of hand's breadths by which the inclined plane departs from the vertical for a rise of one cubit. This means that the saykad of an inclined plane is the exact analogue of the pesu of a loaf of bread, or of beer. In all these problems the difficulty does not lie in the geometry, but in the calculation.

Areas

of triangles, rectangles and trapezoids are determined by use of the correct formulas. The base of the triangle is halved, "in order to make the triangle square" and is then multiplied by the height. In a similar way, the sum of the parallel sides of a trapezoid is halved and then multiplied by the height. Moreover there are reductions to other area-units.

In the deed of gift of the temple of Horus in Edfu appear the areas of a large number of triangles and quadrangles. [1] The quadrangles are treated as follows: half the sum of two opposite sides was multiplied by half the sum of the other two sides. This formula is obviously incorrect; it gives the correct result only if the quadrangle is approximately a rectangle.

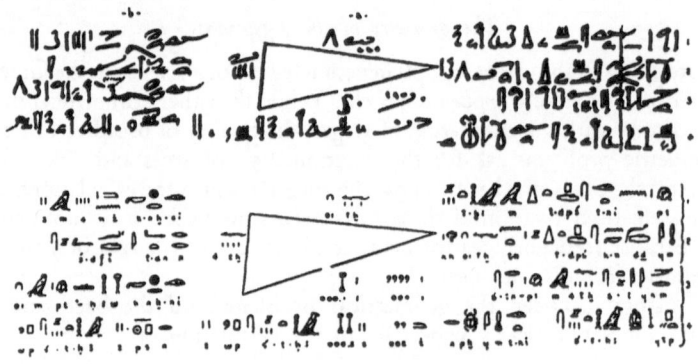

Fig. 1. Rhind 51: Area of a triangle
From A. B. Chace, The Rhind Mathematical Papyrus II.

To determine the area of a circle, the Egyptians square 8/9 of the diameter. This corresponds to a very good approximation

$$\pi \sim 4 \cdot (8/9)^2 = 3.1605 \ldots$$

It is a great accomplishment of the Egyptians to have obtained such a good approximation. The Babylonians, who had reached a much higher stage of mathematical development, always used $\pi = 3$. This is also the value given by Vitruvius; and it is found again in the Chinese literature. The Jews even held this as a

[1] Lepsius, Abh. preuss. Akad. Wiss., Berlin, 1855; philos.-hist. Abteilung, p. 69.

sacred number, authorized by the Bible (1 Kings VII, 23). It is true that Rabbi
Nehemiah, the writer of the Mishnat ha-Middot [1] had the courage, about 150 A.D.
to give the value $3^1/_7$ and to offer a different explanation of the Bible-text (V4),
but the Talmud restored the canonical value 3.

The genius of the Egyptians would have been wonderful and indeed incomprehensible, if they had succeeded in obtaining the correct formula for the

Area of the hemisphere,

as was thought to have been the case for some years, on the authority of Struve,
who edited and published the Moscow papyrus. In Struve's translation, problem
10 of this papyrus [2] is as follows:

> Form der Berechnung eines Korbes, wenn man dir nennt einen Korb mit einer Mündung
> zu 4½ in Erhaltung. O lass du mich wissen seine (Ober)fläche!
> Berechne du ¹/₉ von 9, weil ja der Korb die Hälfte eines Eies ist. Es entsteht 1. Berechne
> du den Rest als 8. Berechne du ¹/₉ von 8. Es entsteht 3̄ + 6̄ + 1̄8̄. Berechne du den Rest
> von dieser 8 nach diesen 3̄ + 6̄ + 1̄8̄. Es entsteht 7 + 9̄. Rechne du mit 7 + 9̄, 4½ mal.
> Es entsteht 32. Siehe, es ist seine (Ober)fläche. Du hast richtig gefunden.

Expressed in modern symbols, the diameter being taken as $x = 4\frac{1}{2}$, the calculation proceeds according to the formula

$$\Omega = (1 - 1/9)(1 - 1/9)2x \cdot x$$

and since $(1 - 1/9)^2$ is the Egyptian value for $\pi/4$, this would indeed give the
correct formula for the area of a sphere, viz.

$$2\Omega = \pi x^2.$$

But disappointment followed close upon amazement. According to Peet, [3] the
basket can also be taken to be a half-cylinder; this reduced this astounding accomplishment, which would have antedated Archimedes by more than a thousand
years, to something quite ordinary. Peet begins by straightening out the gram-
matical structure, which Struve had not made quite clear; he inserts the words
"of 4½" and translates as follows:

"when you are told a basket (of 4½) in diameter by 4½ in depth, then tell me
the area".

This creates a totally different state of affairs. Instead of one number $x = 4\frac{1}{2}$,
there are now given two numbers x and y, both of which have the value 4½, and
the formula which is used, becomes

$$\Omega = 2x \cdot (1 - 1/9)^2 \cdot y.$$

[1] S. Gandz, *The Mishnat ha-Middot, Quellen und Studien* A 2, p. 48.
[2] W. W. Struve, *Mathematischer Papyrus des Museums in Moskau, Quellen und Studien* A 1 (1930), p. 157.
[3] T. E. Peet, *Journal of Egyptian Archaeology*, 17, p. 134.

If we interpret now y as the height of a half cylinder and x as the diameter of its circular base, we obtain the correct formula for the lateral area

$$\Omega = (\pi x/2)y = 2x \cdot \pi/4 \cdot y.$$

Neugebauer has given a somewhat different interpretation. He takes the "basket" to be one of those dome-like barns, which we know from Egyptian illustrations and looks upon the calculation as an approximation. The further development of this idea can be found in Neugebauer's beautiful book. [1]

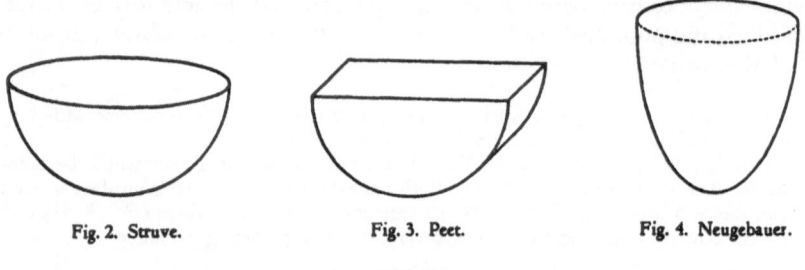

Fig. 2. Struve.   Fig. 3. Peet.   Fig. 4. Neugebauer.

Volumes

of cubes, of beams and of cylinders were of course determined by multiplying the area of the base by the height. The principal difficulty in those things arises from the relations between one another of different units of measure for volumes and for quantities of grain; indeed most of the problems are concerned with grain barns.

An outstanding accomplishment of the Egyptian mathematics is found however in the entirely correct calculation of the volume of the frustrum of a pyramid with square base, as found in the Moscow papyrus (Plate 5a), by means of the formula

(1) $$V = (a^2 + ab + b^2) \cdot h/3,$$

where h is the height and a and b the sides of the lower and upper base.

It is not to be supposed that such a formula can be found empirically. It must have been obtained on the basis of a theoretical argument; how? By dividing the frustrum into 4 parts. viz. a rectangular parallelopiped, two prisms and a pyramid (see Fig. 5), one finds, the volume of a pyramid being assumed as known, the formula

(2) $$V = b^2 h + b(a - b)h + (a - b)^2 \cdot h/3.$$

Neugebauer suspected that (1) came from (2) by means of an algebraic transformation. But can one justify the assumption that the Egyptians were able to make such an algebraic transformation? They were able to calculate with concrete numbers, but not with general quantities. This leads us to wonder whether in

[1] O. Neugebauer, *Vorgriechische Mathematik*, p. 136.

this case Egyptian arithmetic was influenced by Babylonian algebra. Or should we suppose that (1) was obtained from (2) by a geometric argument? One might imagine the following deduction:

For convenience, let us assume that one of the edges is perpendicular to the base. The two prisms of Fig. 5 are changed to rectangular blocks of half the height; the pyramid is also transformed into such a block, but having 1/3 of its original height (Fig. 6). Then the upper third of the first of these blocks is removed and placed on top of the second one (Fig. 7). The solid that is obtained in this way, can be divided into 3 horizontal layers, each of which has the height $h/3$; the lower one of these layers has a base equal to a^2, the middle one has a base ab and the upper one a base b^2.

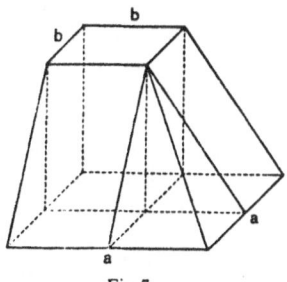

Fig. 5.

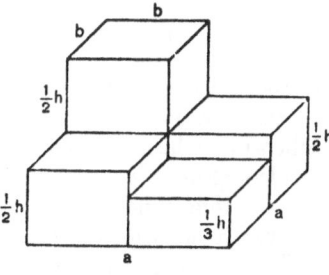

Fig. 6.

This derivation of the formula does not transcend the level of Egyptian mathematics. But, I certainly do not want to tell a fairy tale, and I definitely do not assert that the Egyptians actually proceeded in this manner. There are indeed other possibilities. For example, Cassina[1] has suggested another derivation of the formula for the special case (and this is indeed the only case dealt with in the papyrus) in which the area of the upper base is one half of that of the lower base. Moreover, we should not a priori eliminate a possible effect of Babylonian algebra.

Whichever one of these hypotheses is adopted, we must suppose that the Egyptians knew how to determine the volume of a pyramid.

Having reached the end of our study of Egyptian mathematics, we return to the question:

What could the Greeks learn from the Egyptians?

Looking at Egyptian mathematics as a whole, one cannot escape a feeling of disappointment at the general mathematical level, however much one may appreciate particular accomplishments. In the papyri one does not find a trace of "the construction of lines with proofs", in which Democritus is said to surpass even the Egyptian rope stretchers; there are only rules for calculation without any motivation.

[1] M. Cassina, *Periodico di Matematica* (4a seria), *22* (1942), pp. 1–29.

It is certain that from the Egyptians, the Greeks learned their multiplication and their computations with unit-fractions, which they then developed further;

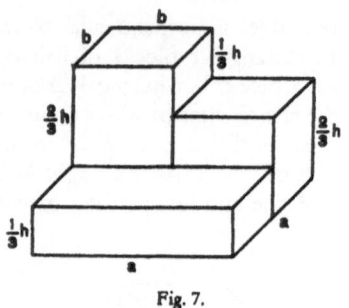

Fig. 7.

the Akhmen papyrus of the Hellenistic period [1] gives evidence of this. But calculation is not the same as mathematics.

The Greeks may also have taken from the Egyptians the rules for the determination of areas and volumes. But for the Greeks such rules did not constitute mathematics; they merely led them to ask; how does one prove this?

One might suppose, as Struve does, that the Egyptians knew a great deal more than is shown in the texts which we have and that the Greeks were familiar with such additional material. Against this assumption there are two arguments: in the first place, the general character of the mathematics which a people has at its command, remains the same, whether one considers elementary texts or more advanced ones. Literal calculation, decimal fractions, differential calculus and coordinate geometry, which are characteristic of Western mathematics, could be abstracted by a competent historian from a handbook for engineers, just as well as from a volume of the Mathematische Annalen. But, characteristic of Egyptian mathematics are the elaborate calculations with fractions, which can not serve as a basis for higher algebra, and the treatment of geometry as applied arithmetic.

In the second place, we shall show that Babylonian mathematics could very well supply, and indeed did supply, a basis for Greek mathematics. Since we know then the actual historical foundations on which the fine edifice of Greek mathematics was erected, we do not need to set up hypotheses concerning a lost Egyptian higher mathematics. We would not be able to demonstrate the validity of such hypotheses, and they would in no way serve our purpose.

[1] See K. Vogel, *Griechische Logistik*, Sitzungsberichte der bayerischen Akademie, München, 1936, pp. 357—472.

CHAPTER II

NUMBER SYSTEMS, DIGITS AND
THE ART OF COMPUTING

In this chapter we shall give a brief survey of the number systems and the number notations in the principal cultural periods, and of the related arithmetical techniques. We shall see that these notations and these techniques are of very great importance for the development of mathematics; not, of course, in the sense that a good number system leads automatically to a high development in mathematics, but rather that a good notation and a convenient manipulation of the four fundamental operations are necessary conditions for the development of mathematics. Without mastery of these fundamental operations, mathematics can not get beyond a certain low level.

This shows itself most clearly in algebra. The Babylonians had, as we shall see, an excellent sexagesimal notation for whole numbers and fractions, which enabled them to calculate with fractions as easily as with integers. This made possible their highly developed algebra. They knew how to solve systems of linear and quadratic equations in two or more unknowns! This would have been impossible for the Egyptians, because their complicated fraction-technique made a difficult problem out of every division, every subtraction of fractions. How would they be able to determine the square root of a sum of unit-fractions like $\bar{2} + \overline{10} + \overline{25}$? The Babylonians however wrote this sum as a sexagesimal fraction:

$$0; 38, 24, \text{ i.e. } 38/60 + 24/60^2,$$

and looked for the square root in a table:

$$\sqrt{0; 38, 24} = 0; 48.$$

If the number was not a perfect square, they would simply use an approximate value.

The sexagesimal system

was taken over by the semitic Babylonians from their predecessors, the Sumerians. This remarkable cultural group, which had also invented the cuneiform script, dominated southern Mesopotamia during the third millenium B.C. The Sumerian language is neither Indo-Germanic nor Semitic, but of a totally different type, not inflective but agglutinate i.e. combining. The stems of words, usually monosyllabic, remain unchanged, but other invariant syllables are placed in front or behind to determine the grammatic function of the word. The oldest Sumerian cuneiform texts date from the first dynasty of Ur, which flourished around 3000 B.C., or perhaps a few centuries later.

The Sumerian civilization was taken over by the semitic Akkadians, who dwelt farther north. In the course of time the Semites became more and more dominant, and about 1700 Hammurabi, the great lawgiver and ruler of the first Babylonian dynasty could call himself "king of Sumer and Akkad." [1]

In the Sumerian-Babylonian notation, numbers under 60 are written in the ordinary decimal notation; the simple vertical wedge has the value 1, the wedge with two ends the value 10. Both signs were produced by pressing a sharpened stylus into a clay tablet (see Fig. 8).

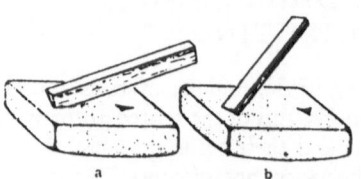

Fig. 8. Stylus for cuneiform script
From O. Neugebauer, Vorgriechische Mathematik

Thus the symbol ⟨⟨⟨ 𝐖 represents 35, the symbol 𝐖 , later abbreviated

to ▼▼▼ , the number 9.

There is nothing special in all this; it corresponds to the Egyptian system or to the familiar Roman numerals. But now comes the remarkable thing; the number 60 is again represented by the symbol for 1, and with the multiples of 60 we start counting anew. Thus the symbol for 10 can also mean 10×60, the symbol for 1 also 60×60 or even an arbitrarily high power of 60. But, more than this, fractions are also written in the sexagesimal system. The symbol for 1 can also mean $1/60$ or $1/60^2$, the symbol for 10 also $10/60$ or $10/60^2$. The fractions $\frac{1}{2} = 30/60$, $\frac{1}{3} = 20/60$ and $\frac{1}{5} = 12/60$ are then represented by the symbols

for 30, 20 and 12 respectively: ⟨⟨⟨ ⟨⟨ ⟨▼▼ .

Thus we see that the value of a symbol depends, as in our modern number system, upon the position of the symbol in the number; the higher powers of 60 are placed at the beginning, the lower powers at the end; hence the word

positional notation. Thus in ▼⟨⟨ 𝐖 , the first vertical wedge is worth 60

times as much as each of the four small ones. The value of the symbol can be, for example,

$$60 + 24 = 84 \text{ or } 1 + 24/60 = 84/60.$$

Following Neugebauer, we shall not put the sexagesimal numbers that occur in the texts in the decimal system, but we shall transcribe them sexagesimally and

---

[1] For Babylonian cultural history see, in the first place B. Meissner, *Babylonien und Assyrien*, I and II (Heidelberg 1921—25). Next E. Chiera, *They wrote on clay*, Chicago, 1938. For cuneiform script, see Ch. Fossey, *Manuel d'Assyriologie* I, Paris, 1904, and B. Meissner, *Die Keilschrift* (Sammlung Göschen), 1922. For the chronology, see B. L. v. d. Waerden, *Jaarboek, Ex Oriente Lux 10* (1948), p. 414 and the literature cited there.

separate the different powers of 60 by commas. The number just presented will therefore be written as 1, 24. And 1, 3, 30 will mean

$$1 . 60^2 + 3 . 60 + 30,$$

or this number multiplied or divided by a power of 60. When the context shows that we have to multiply or divide by a definite power of 60, we shall supply one or more zeros at the end (e.g. 1, 3, 30, 0), or we shall use a semi-colon, to serve the purpose of the modern decimal point, e.g.

$$1, 3; 30 = 63\frac{1}{2} \text{ and } 1; 3, 30 = 1^7/_{120}.$$

The systematic positional notation has enormous advantages in the technique of computation. Compare, e.g., a multiplication in the modern notation with such a calculation in Roman numerals! To compute 243×65, we treat the tens and the hundreds as if they were units moved one or two places to the left. But for the Romans, $CC \times LX$ is something very different from $II \times VI$. We can limit ourselves to the multiplication tables from 1×1 to 9×9, and, likewise, the Babylonian only requires tables from 1×1 to 59×59. In practice these numbers are of course split into tens and units, but this is of minor importance for us now. With these tables he can also multiply sexagesimal fractions, as if they were whole numbers; and he can ignore the entire muddle of calculating with common fractions. Do we not calculate with decimal fractions as if they were whole numbers, and then put the decimal point in the proper place in the result?

Babylonian positional notation also had disadvantages. It is true that in practice the lack of distinction between the symbols for 1 and for 60 is not serious, because the order of magnitude is usually known from the context (when we see the number 30 on a dress in a shop-window, we know very well that it does not stand for 30 cents!); but in theoretical problems it can be very unpleasant. Still more so is the fact that the notation does not enable us to distinguish between 1, 0, 30 and 1, 30, because there is no cypher. To overcome this drawback, a separate sign was introduced later on for the empty place between two digits; for example

$$\text{𒁹 𒌋 𒐘} = 1, 0, 4 = 3604.$$

The Greek astronomer Ptolemy (150 A.D.), who made all his computations in the sexagesimal system, uses the symbol o for zero, also at the end of a number. This gave the "finishing touch" to the sexagesimal positional system; in this way it became almost equivalent to our decimal system. It is true that Ptolemy wrote the whole numbers in the decimal system and only the fractions sexagesimally; but this is less important because he hardly ever needs large integers. The enormous superiority of the sexagesimal fractions in computation was responsible for their use by the astronomers, and hence for our minutes and seconds.

How did the sexagesimal system originate?

Originally the Sumerians did not have a systematic positional system for all powers of 60 and their multiples. They had the following symbols:

Oldest Sumerian period (before 3000 B.C.)

D   O   D   Ⓓ   O

  1   10   60   600   3600

The symbols for 1, 10 and 60 were made with the cylindrical lower end of a round stylus; for 1, the stylus was held obliquely, for 10 perpendicularly. The symbol for 600 is a combination of those for 10 and 60. The large circle for šár 3600, which terminated the number system, was scratched in with a sharp stylus.

Later on the symbols were separated into wedges which were pressed into clay by means of the sharpened stylus. The following sets of symbols are found, side by side, in the

Later Sumerian period (about 2000 B.C.)

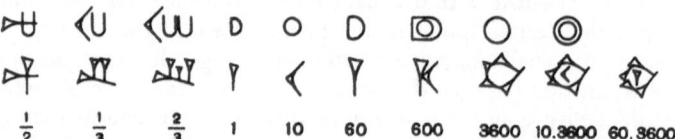

$\frac{1}{2}$ $\frac{1}{3}$ $\frac{2}{3}$ 1   10   60   600   3600   10.3600   60.3600

The last of these symbols was called šár-gal, i.e. big šár. The symbol for 60 is sometimes a little larger than that for 1. Apparently 60 was looked upon as the "big unit". On this supposition, it is also understandable why 10 · 60 was the next step in the scale and why 60 · 60, the šár terminated the old number system.

The fact that 1 and 60 are represented by the same symbol, is the fundamental principle of the positional notation. By simplification of the symbols (3600 = 1, etc.), and by a systematic continuation of the powers of 60 in both directions, one obtains quite naturally the consistent sexagesimal notation of the Babylonian texts.

Three questions remain:

1. What led to the choice of 60 as the next stage after 1 and 10?
2. Why was 60 considered as the "big unit" and represented by the symbol for 1?
3. What led to the representation of fractions in the sexagesimal system by taking 1/60 as the "small unit"?

The second and third questions are more important than the first, because it is not the magnitude of the basis of the number system which matters, but rather the principle of the positional notation and its extension to fractions. In a certain sense, the choice of 60 is a historical accident; 10, which appears everywhere (probably because we all have 10 fingers), is followed by 20 among the Celts (remember the French numerals quatre-vingt and quatre-vingt-dix, and the English three score for 60), by 100 among the Egyptians, perhaps by 50 among the Teutonic peoples, and by 60 among the Sumerians. Perhaps, as Neugebauer [1] assumes, a part in the choice of 60 was played by an old normalization of measures, in which it was important that one half and one third of the larger measure be simple multiples of the smaller one; we really do not know. But, as has already been said, this question is the least important of the three.

Neugebauer answers the second question by the very plausible statement that the source of 60 as the "big unit" must be looked for in metrology. Weights, and also monetary units, were arranged sexagesimally:

$$1 \text{ talent } = 60 \text{ mana,}$$
$$1 \text{ mana } = 60 \text{ bushel.}$$

As is well known, these monetary units were adopted by the Greeks. Nothing is more natural, when one monetary unit is sixty times another, than denoting the larger unit again by 1 and designating its multiples by the ordinary symbols for 1, 2, . . ., 10, ... In our own terminology, one-twenty is frequently used for one dollar and twenty cents.

The next step of great importance consists in carrying over this notation to non-denominate numbers.

Frankness compels us to say that the master of modern assyriologists, F. Thureau-Dangin, does not consider this hypothesis valid. [2] According to him, the development went in the opposite direction: the use of 60 as a big unit was carried over from pure numbers to the monetary units, not inversely. But what answer does he give then to our second question? None at all.

But, when it comes to our third question, Thureau-Dangin also thinks that the origin of sexagesimal fractions must be looked for in metrology. In this domain, and only there, the relation of a larger unit to a smaller one, is strictly reciprocal: one is 60 times the other, and this one is 1/60 of the first. In the domain of pure numbers, one would not be led to consider 1/60 as a new unit, but in the monetary realm one would. Analogously, one has the Roman custom of expressing the common fractions $\frac{1}{2}$, $\frac{1}{3}$ and $\frac{1}{6}$ in terms of the new unit $\frac{1}{12}$, the uncia. Again, the uncia was originally a weight, viz. $\frac{1}{12}$ of the as.

For a further discussion of ancient normalizations of units, the reader is referred to Neugebauer's article, mentioned above, and to Chapter 3, par. 4 of his excellent book, Vorgriechische Mathematik.

[1] O. Neugebauer, *Abhandlungen, Gesellschaft der Wissenschaften*, Göttingen, *13* (1927), p. 1.
[2] F. Thureau-Dangin, *Esquisse d'une histoire du système sexagésimal*, Paris, 1932, p. 11.

Sumerian technique of computation.

The most ancient Sumerian texts, from which the Sumerian number system was deduced, dating from the time of Shulgi (about 2000), were tables of inverses (1/x) and multiplication tables. The latter tables appear singly or in combination. A separate table contains the multiples of a single number. Two examples:

Table of 7 and of 16,40.

(cuneiform)		
7 a-rá 1	7	a-rá 1	16, 40
a-rá 2	14	a-rá 2	33, 20
a-rá 3	21	a-rá 3	50
. . .		. . .	
a-rá 19	2, 13	a-rá 19	5, 16, 40
a-rá 20	2, 20	a-rá 20	5, 33, 20
a-rá 30	3, 30	a-rá 30	8, 20
a-rá 40	4, 40	a-rá 40	11, 6, 40
a-rá 50	5, 50	a-rá 50	13, 53, 20

Obviously, a-rá means "times".

Frequently, several of these small tables are combined with a table of inverses and a table of squares, to form a large combination-table. Before we can understand the arrangement of such a combination table, we must first have a look at the short tables of inverses.

There are large tables of inverses from the times of the Seleucids (311—1 B.C.), probably designed for the use of astronomers. One of them begins as follows:

$$1 : 1 \qquad\qquad\qquad\qquad = 1$$
$$1 : 1, 0, 16, 53, 53, 20 = 59, 43, 10, 50, 52, 48$$
$$1 : 1, 0, 40, 53, 20 \quad = 59, 19, 34, 13, 7, 30$$
$$1 : 1, 0, 45 \qquad\qquad = 59, 15, 33, 20$$
$$1 : 1, 1, 2, 6, 33, 45 = 58, 58, 56, 38, 24.$$

In this way it continues throughout 7 pages of Neugebauer's "Mathematische Keilschriftentexte"[1] up to

$$1 : 3 \qquad\qquad\qquad = 20.$$

---

[1] O. Neugebauer, *Quellen und Studien*, A·3, Berlin, 1935.

The older tables of inverses are not so extensive. They usually contain the reciprocals of those integers between 0 and 81, which contain only factors 2, 3 and 5 and which have therefore reciprocals that can be expressed as finite sexagesimal fractions.

Normal table of inverses

1 :	2 = 30			16	3, 45		45	1, 20
	3	20		18	3, 20		48	1, 15
	4	15		20	3		50	1, 12
	5	12		24	2, 30		54	1, 6, 40
	6	10		25	2, 24		1	1
	8	7	30	27	2, 13, 20		1, 4	56, 15
	9	6, 40		30,	2		1, 12	50
	10	6		32	1, 52, 30		1, 15	48
	12	5		36	1, 40		1, 20	45
	15	4		40	1, 30		1, 21	44, 26, 40

The corresponding multiplication tables contain not only tables for the multiples of 2, 3, 4, 5, 6, 7, 8, 9, 10, 20, 30, 40, 50 which are to be expected in any ordinary multiplication table, but also tables for the multiples of several other numbers of two and three digits, in all of the following numbers:

50	24	12	6, 40	2, 30
48	22, 30	10	6	2, 24
45	20	9	5	2, 15
44, 26, 40	18	8, 20	4, 30	2
40	16, 40	8	4·	1, 40
36	16	7, 30	3, 45	1, 30
30	15	7, 12	3, 20	1, 20
25	12, 30	7	3	1, 15

What determined the choice of these numbers? Most of them occur also in the normal table of inverses, and the others, except the number 7, are inverses of simple numbers. This also accounts for the arrangement in order of decreasing magnitude; it is indeed the order of the reciprocals in the table of inverses.

Thus we see that the multiplication table served not only for ordinary multiplications $a \cdot b$, but especially for multiplications of the form $a \cdot b^{-1}$, i.e. for divisions $a : b$.

The matter can also be stated as follows: the combined tables of inverses and of multiplications is used to multiply numbers, but also to represent common fractions as sexagesimal fractions. For example, to write $\frac{3}{8}$ in sexagesimal form, we first locate $1 : 8 = 0;7,30$ in the table of inverse and then the product of this result by 3 in the multiplication table: $0;22,30$.

The mathematical texts fully confirm this interpretation. Whenever a division $a : b$ is to be carried out in these texts, one is told (not in terms of a general formula, but for definitely specified numbers): calculate the reciprocal b^{-1} and multiply it by a.

It appears therefore that the Sumerian-Babylonian calculation tables were arranged in a very useful manner. A systematic use of the advantages of the positional notation avoids all messing about with fractions; the four rational operations can be carried out rapidly without further thought.

Whenever a division did not go without remainder, approximations were used. An ancient Babylonian text, YBC 10529[1], gives approximate values for the reciprocals of all numbers from 40 or 50 to 80 in the following manner:

$$
\begin{aligned}
1 : 59 \quad &= 1, 1, 1 \\
1 : \ 1 \quad &= 1 \\
1 : \ 1, 1 &= 59, 0, 59 \\
1 : \ 1, 2 &= 58, 3, 52 \text{ etc.}
\end{aligned}
$$

Squares, square roots and cube roots.

From a short table of squares, such as supplements most multiplication tables:

$$
\begin{aligned}
&1 \text{ a-rá } 1 \quad \quad 1 \\
&2 \text{ a-rá } 2 \quad \quad 4 \\
&3 \text{ a-rá } 3 \quad \quad 9 \\
&\quad \ldots \ldots \\
&59 \text{ a-rá } 59 \quad 58, 1
\end{aligned}
$$

one can of course immediately derive a table of square roots:

$$
\begin{aligned}
1-e \quad &1 \quad \text{ib-si (i.e.: of 1 is 1 the root)} \\
4-e \quad &2 \quad \text{ib-si} \\
9-e \quad &3 \quad \text{ib-si.}
\end{aligned}
$$

In Babylonian mathematics these short tables are used in the solution of quadratic equations. In the same way, tables for cube roots,

$$
\begin{aligned}
1-e \quad &1 \quad \text{ba-si} \\
8-e \quad &2 \quad \text{ba-si} \\
27-e \quad &3 \quad \text{ba-si, etc.}
\end{aligned}
$$

are used for the solution of pure cubic equations

$$x^3 = a.$$

The word ba-si however does not only mean cube root, but more generally "root of an equation". Indeed there are also tables for the equation

$$x^2(x + 1) = a.$$

in which this word occurs as well:

$$
\begin{aligned}
2-e \quad &1 \quad \text{ba-si} \\
12-e \quad &2 \quad \text{ba-si} \\
36-e \quad &3 \quad \text{ba-si, etc.}
\end{aligned}
$$

[1] YBC stands for Yale Babylonian Collection. The text is reproduced in O. Neugebauer and A. Sachs, *Mathematical Cuneiform Texts*, New Haven, Conn., 1945, p. 16.

PLATE 5

PL. 5a. Two Columns of the Moscow Papyrus, containing the computation of the volume of a truncated pyramid with sides of 2 and 4 cubits and a height of 6 cubits. Agove, the hieratic text, below, the hieroglyphic transcription, executed by J. J. Perepelkın, both reproduced from W.W. Struve, *Math. Papyrus des Museums in Moskau,* Quellen und Studien A1. The text says: (1) Add together this 16 (2) with this 8 and this 4. (3) You get 28. Compute (4)$^1/_3$ of 6. You get 2. Mul- (5) tiply 28 by 2. You get 56. (6) Behold: it is 56. You have found right.

The hieroglyphic text ıs read from right to left. The upper length 2 with its square 4 is written above the drawing, the lower length 4 below, the height 6 and the volume 56 within, and the multiplication of 28 by 2 to the left of the drawing.

PL. 5b. Sumerian clay tablet with number signs, Berlin Museum (VAT 12593). The tablet was found in Fara, the ancient Shurupak, in Southern Mesopotamia. According to Neugebauer (*Vorgriechische Mathematik* p. 51) the text was probably written in the beginning of the third millennium B.C. The columns should be read from bottom to top. Later texts were written in lines from left to right. The middle column shows the numbers 1, 2, 3, 4, 5, – 7, 8, 9. At the top of the last column we see the numbers 10 and 20 (flat circular impressions). See p. 40.

PLATE 6

PL. 6a. Marble counting (1.50 by 0.75 m), found on the island of Salamis, Athens National Museum, probably from the 4th century B.C. This may not have been actually used as a counting board. At the bottom the old number signs are found (see p. 46) and some signs for special monetray units.

PL. 6b. Nabu, god of Babylonian science. Limestone statue (height 1.65 m), consecrated under Adadnirari III (810–782 B.C.), British Museum (D. 377).

Moreover the same word ba-si is found in VAT [1] 8521, accompanied by 1-lal, i.e. "less 1", with the meaning: root of the equation

$$x^2(x-1) = a.$$

It is worthy of notice that the numbers $n^2(n+1)$ and $n^2(n-1)$ occur here, the same which turn up in later Greek arithmetic as "paramekepipedoi" numbers.[2]

To determine the square root of a number which is not a perfect square, approximation was used. The following method is also used frequently by the Greeks: Suppose that a is a first approximation for $\sqrt{2}$. Then, if a is too small, $2/a$ will be too large, and vice versa. A better approximation is then obtained by taking the arithmetic mean

$$\tfrac{1}{2}(a + 2/a).$$

For instance, if one starts with $a = 1\tfrac{1}{2}$, one finds $2/a = \tfrac{4}{3}$, so that the closer approximation is

$$1^5/_{12} = 1; 25.$$

This approximation frequently occurs in Babylonian texts. A repetition of the process gives, as the average of $a = 1;25$ and $2 : a = 1;24,42,21 \ldots$ the very close approximation

$$\sqrt{2} = 1; 24, 51, 10.$$

This is found in the recently discovered cuneiform text YBC 7289.[3] (Plate 8b)

Application of the same method of the arithmetical mean to the root of the sum $a^2 + b$, gives the general approximation-formula

$$\sqrt{a^2 + b} \sim a + b/2a,$$

which is also found in Babylonian texts.

The approximations obtained in this way are always too large, as is seen by squaring $a + b/2a$. The Greeks (Archimedes, Heron) knew also approximations for $\sqrt{2}$ from below, and even continued-fraction approximations ("lateral- and diagonal-numbers"); to these we shall return at a later point.

The Greek notation for numbers,

as compared with the excellent Babylonian notation, was really a retrogression. In most remote antiquity, they had a notation, which resembles the well-known Roman numerals:

[1] VAT stands for Vorder-Asiatische Textsammlung Berliner Museum. For this text see F. Thureau-Dangin, *Textes math. babyl.*, Leiden, 1938, p. 123.
[2] O. Becker, *Quellen und Studien*, B 4, p. 181.
[3] Neugebauer-Sachs, *Mathematical Cuneiform Texts*, New Haven, 1945, p. 43. See Plate 8b.

The letters Π, Δ, H, X, M are of course the initial letters of the Greek words for 5, 10, 100, 1000 and 10 000.

Later on, a briefer, alphabetic notation was introduced:

1 – 9 $\alpha, \beta, \gamma, \delta, \varepsilon, \varsigma, \zeta, \eta, \vartheta$ (6 = ς = Vau)

10 – 90 $\iota, \kappa, \lambda, \mu, \nu, \xi, o, \pi, \varsigma$ (90 = ς = Koppa)

100 – 900 $\varrho, \sigma, \tau, \upsilon, \varphi, \chi, \psi, \omega, \mathcal{m}$ (900 = $\mathcal{m}$ = Sampi)

1000 – 9000· ${}_{,}\alpha, {}_{,}\beta$, etc.   (accent at lower left)

To distinguish numbers from words, an accent was added at the end, or a dash was placed over them, such as

$$\overline{{}_{,}\alpha\tau\varepsilon} \text{ or } {}_{,}\alpha\tau\varepsilon' = 1305.$$

Numbers beyond the myriad $M = 10^4$ were designated by use of the symbol M, e.g.:

$$\overset{\varkappa\varepsilon}{M} \mu\gamma' = 250\,043$$

In place of $\overset{\varkappa\varepsilon}{M}$ one could also write $\overset{\cdot\cdot}{\varkappa\varepsilon}$. For higher powers of M, Archimedes and Apollonius used still different notations.

The use of letters for specified numbers was not advantageous for the development of algebra. It did not leave the letters available for indeterminates or for unknowns, as in our algebra. Until the time of Plato's friend Archytas (390 B.C.), letters *were* used for indeterminates; in Archytas, $\Gamma\Delta$ represented for instance the sum of the numbers Γ and Δ. If supplemented by a sign for multiplication, a minus sign and a symbol for fractions, this system might have provided an effective notation for theoretical arithmetic. But even Euclid (300 B.C.) had already abandoned this simple notation for sums, probably, I think, to avoid confusion with the alphabetical number symbols. When Euclid wants to add two numbers, he represents them by means of line segments AB and $B\Gamma$, and denotes the sum by $A\Gamma$.

For purposes of calculation, the Greek number symbols were about equally troublesome. In our schools we only have to learn the multiplication tables from 1×1 to 9×9, but in the 14th-century Greek arithmetic of Nikolaos Rhabdas [1], one finds tables in which first 1 is multiplied by all the 37 numbers

$$\alpha, \beta, \ldots, \iota, \varkappa, \ldots, \varrho, \sigma, \ldots, {}_{,}\alpha, {}_{,}\beta, \ldots, \overset{\cdot\cdot}{\alpha} = 10^4,$$

then 2 by these same numbers, starting with 2, and so on up to $10^4 \cdot 10^4 = 10^8$.

---

[1] The letters of Rhabdas were edited by P. Tannery, *Mémoires scientifiques*, IV, p. 61–198. Undoubtedly they are based on much older Greek sources. Compare K. Vogel, *Griechische Logistik*, Sitzungsber. Bayer. Akad., München (Math.-Naturwiss. Abt.), *1936*, p. 357.

It is all arranged in three columns, the first two contain the factors, the third the product, e.g.

$$\varrho \, \nu \, \ddot{\delta} \text{ i.e. } 100 \times 400 = 40000.$$

To avoid this unmanageable table and to require only the use of the ordinary multiplication tables from 1×1 to 9×9, one would have to replace factors like 20, 200 or 2000 by their "Pythmen" or radix, in this case 2, then multiply these radices and finally determine the power of 10, by which this product has to be multiplied. In the "Sand-reckomer", Archimedes gives for this purpose a rule which is equivalent to our formula

$$a^p \cdot a^q = a^{p+q}.$$

For instance, to use this rule for the calculation of σ' times τ' (200 times 300), one determines first the radices 2 and 3 for σ' and τ' respectively, finds their product 6, then, by the rule of Archimedes, $10^2 \cdot 10^2 = 10^4$, so that the product turns out to be 6 myriads. [1] All this is very cumbersome!

Numbers represented by a single letter could be multiplied in this manner. A number consisting of more letters was split and written as a sum of numbers represented by a single letter. The example 265×265 is found in Eutocius:

		i.e.	265		
$\tau\grave{\alpha}\ \delta\grave{\epsilon}\ \overline{\sigma\xi\epsilon}$					
$\overset{}{\dot{\epsilon}\pi\iota\ \overline{\sigma\xi\epsilon}}$		times	265		
$\overset{\delta}{M}\ \overset{\alpha}{M}\ ,\beta\ ,\bar{a}$		40 000	12 000	1 000	
$\overset{-\alpha}{M}\ ,\beta\ ,\overline{\gamma\chi\tau}$			12 000	3 600	300
$,\overline{\alpha\tau\varkappa\epsilon}$			1 000	300	25
$\delta\mu o\tilde{\nu}$	$\overset{\zeta}{M}\ \overline{\sigma\varkappa\epsilon}$	together	70 225.		

This method was called "Greek multiplication". There was known also "Egyptian multiplication', by means of continued doubling, familiar to us from Chapter I.

Written computations were obviously rather complicated; moreover paper was expensive, and therefore one took recourse in practice to

Counting boards and counting pebbles.

The pebbles, called Psephoi, were aligned in parallel rows on a counting board (or perhaps on an ordinary table on which lines had been drawn). The pebbles in the last row had the value 1, those in the next to the last row 10, etc. There were separate rows for special monetary units, such as the obols.

Remarks, scattered in the works of various writers, indicate how general was the use of the "Abacus", the counting board, in antiquity. In Polybius (V 26) we

[1] See the examples of Pappus in Vogel, Sitzungsber. Munchen, 1936, p. 393.

find: "those in the courts of kings . . . are in truth exactly like counters on a counting board. For these, at the will of the reckoner are now worth a copper and now worth a talent". It is also significant that the common verb for "to calculate" is Psephizein derived from the word Psephos the counting pebble.

Among the Romans we find, besides the large counting board, patterned after the Salamis counting board, (Plate 6a) also the small hand-abacus with a restricted number of pebbles, or rather beads, which can be pushed back and forth along wires. Many peoples have adopted this convenient instrument; even to-day Chinese and Russian merchants follow the Roman example in using the abacus for calculating. Up to the late Middle Ages, the counting board was in general use in Western Europe. It was not until modern numbers were introduced that the abacus gradually disappeared from the scene, because it was no longer needed.

Calculation with fractions.

A scholium, i.e. a marginal note, in Plato's Charmides, describes in the following words the most important aspects of arithmetic or "logistics": "The so-called Greek and Egyptian methods for multiplying and dividing, and the combining and splitting of fractions."

Fig. 9. The chancellor of the treasury on the Darius-vase (plate 7) adds on the counting board the taxes that have been received.
From Menninger, Zahlwort und Ziffer.

"Combining" apparently refers to the addition of fractions, which Rhabdas performs, exactly as we do it, by reducing to a common denominator. For example, to add 3 and $\frac{1}{3}$ and $\frac{1}{14}$ and $\frac{1}{42}$, he changes $\frac{1}{3}$ to fourteen forty-seconds, $\frac{1}{14}$ to three forty-seconds; added to $\frac{1}{42}$, this gives $14 + 3 + 1 = 18$ forty-seconds, or 3 sevenths; combined with 3 units, this leads to 39 sevenths.

What is meant by the splitting of fractions, can be learned from the Akhmîm papyrus[1], in which the fractions m/n are split into unit fractions for $n = 3, 4, \ldots, 20$ and for various values of m, for example:

"The seventeenth part of 3 is $\frac{1}{12} + \frac{1}{17} + \frac{1}{51} + \frac{1}{68}$."

We get to know the notation for fractions from the works of the later mathe-

[1] J. Baillet, Le Papyrus Mathématique d'Akhmîm, Mémoires de la Mission archéologique française au Caire 9, fasc. 1, Paris 1892.

maticians (Archimedes, Heron, Diophantus) and, most reliably, from papyri like those already mentioned, because these are originals and not later copies, in which the copyists may have altered the notation. For $\frac{1}{3}$, the word $\tau\grave{o}\ \tau\varrho\acute{\iota}\tau o\nu$ could be written out in full, or it might be abbreviated to $\tau\grave{o}\ \gamma''$, or, still shorter, to γ', γ'' or something similar. Frequently the denominator was placed above the numerator: $\frac{\epsilon}{\gamma} = \frac{3}{5}$. According to Vogel, this notation came into use as early as the time of Archimedes (3rd century B.C.). A papyrus from the 1st century A.D. [1] has also the inverted form, the "Indian notation" $\frac{\gamma}{\epsilon} = \frac{3}{5}$, which led to our notation for fractions.

From this we see that the Greeks did not restrict themselves to unit-fractions, according to the Egyptian example. Why should they? They were not bound to a frozen tradition. It is true that they frequently operated (especially in the later papyri which were subject to a strong Egyptian influence) with sequences of unit-fractions, such as $\frac{2}{3} + \frac{1}{10} + \frac{1}{30}$; but they could also replace this by $\frac{4}{5}$. For example. Archimedes proves that the perimeter of the circle lies between $3\frac{10}{71}$ times and $3\frac{1}{7}$ times the diameter, and his contemporary Eratosthenes uses the fraction $\frac{11}{83}$ to designate the inclination of the ecliptic.

Official Greek mathematics before Archimedes does not have any fractions at all. This was not however because they were not known, but rather because one did not wish to know them. For, according to Plato, the unit was indivisible and, in Plato's own words, "the experts in this study" were absolutely opposed to dividing the unit (The Republic, 525E). Fractions were scorned and left to the merchants; for, so it was said, visible things are divisible, but not mathematical units, Instead of operating with fractions, they operated with ratios of integers.

Nevertheless, traces are found of an ancient technique of fractions. In Book VII of Euclid's Elements, which, as we shall see later, was produced before 400 B.C., we find the following definitions [2]:

Def. 3. A number is a part of a number, the less of the greater, when it measures the greater.

Def. 4. But parts when it does not measure it.

"Part" means here n-th part, where n is an integer. "Parts" means a number of n-th parts, e.g. 3 fifths. These definitions introduce therefore arbitrary fractions. One application is found in the definition of proportionality.

Def. 20. Numbers are proportional when the first is the same multiple, or the same part, or the same parts, of the second that the third is of the fourth.

Then follows an explanation of the reduction of ratios to lowest terms or, translated into the terminology of fractions, how fractions are reduced by determining the GCD of numerator and denominator and dividing it into both. Furthermore, we find in VII 34—36 a development of the properties of the LCM. The determ-

---

[1] H. Gerstinger und K. Vogel, *Eine stereometrische Aufgabensammlung im Papyrus Vindobon.* 19996, *Mitteilungen Papyrussammlung Erzherzog Rainer*, Neue Serie, 1. Folge, Wien 1932, p. 11.
[2] All references to Euclid's Elements, here and in the sequel, follow the translation given by T. L. Heath, *The thirteen books of Euclid's Elements*, second edition, Cambridge, 1926.

ination of the LCM is important for reducing fractions to the lowest common denominator.

The terminology of the ratios of numbers in the Pythagorean theory of harmony also recalls the fact that these ratios are originally fractions. The ratio 4 : 3, which corresponds to the musical interval of the fourth, is called the "epitriton", which means "one third additional" ($1\frac{1}{3}$), and the whole tone, to which corresponds the ratio 9 : 8 is called the "epogdo-on", i.e. $1\frac{1}{8}$.

The most ancient occurrence of fractions is in Homer's Iliad, Book K, 253: "Two parts of the night are past, the third part remains".

We conclude from this that, from remotest antiquity, the Greeks have known fractions and that in the 5th century, at the latest, they had mastered the operations on fractions; reduction to lowest terms, to a common denominator, etc. For them therefore, in contrast with the Egyptians, difficulties in the calculation with fractions can not have been an obstacle in the way of the development of mathematics.

Sexagesimal fractions.

Common fractions were too awkward for astronomical calculations; that is why the Greek astronomers adopted the Babylonian sexagesimal fractions. It is not known who was the first to do this, because the superb "Almagest" [1] of Claudius Ptolemy has cast the works of his predecessor into oblivion. In the Almagest, the circle is divided, on the Babylonian pattern, into 360 degrees, each degree into 60 minutes and each minute into 60 seconds. For 47°42'40". Ptolemy writes $\mu\zeta\ \mu\beta'\ \mu''$. But other units are also divided in the same way; when Ptolemy considers a circle, he usually divides the diameter into 120 parts, and then he subdivides each part according to the same scheme. For zero he has the symbol o, an abbreviation of $o\upsilon\delta\acute{e}\nu$ = nothing. There was no danger of confusion with the number symbol o = 70, because numbers beyond 60 do not occur in counting minutes, seconds, etc.

Ptolemy is a virtuoso in computing with these sexagesimal fractions; he divides and multiplies them, extracts square roots, etc., without wasting a word on the technique of these calculations. For this reason we have to be grateful to his comm ntator Theon of Alexandria [2] for giving an example of a sexagesimal division, viz. of '$\alpha\varphi\iota\varepsilon\ \varkappa'\ \iota\varepsilon''$' by $\varkappa\varepsilon\ \iota\beta'\ \iota''$. In modern symbols:
1515°20'15" : 25°12'10" = 60°7'33".

One begins by es' mating the number of units in the quotient. The method for estimating is quite remarkable, indeed more effective than ours. According to our school books, we would try whether it goes 60 times, multiply 25°12'10" by 60 and see whether the result exceeds 1515°20'15", etc. Theon does not mul-

---

[1] The actual title of this astronomical standard treatise is Syntaxis mathematika. The word Almagest is an Arabic corruption of Megale Syntaxis or Megiste with the Arabic article Al.

[2]. À Rome, *Commentaires de Pappus et de Théon d'Alexandrie sur l'Almageste*, Tome 2, Roma 1936.

tiply by 60, but he divides by 60. That is to say, he divides the dividend by the estimated initial digit of the quotient. Division of 1515° by 60 gives a quotient in excess of 25°12′10″, but division by 61 results in an answer that is too small; thus we have to take 60. Now 60 · 25°, 60 · 12′ and 60 · 10″ are successively subtracted from 1515°20′15″; this leaves 190′15″.

In the same manner, we estimate the 7′ of the quotient. Dividing 190′15″ by 7′ gives more than 25°12′10″, but division by 8′ gives less, so that we have to take 7′. Does the reader see the advantage in this method of calculation? Our method would require the multiplication of the entire big number 25°12′10″ by 7 or 8, but Theon only requires that 190 be divided mentally by 7 or 8. Then 7′ . 25°, 7′ . 12′ and 7′ . 10″ are successively subtracted from 190′15″; a remainder of 829″50‴ is obtained, etc.

I assume that Hipparchus (150 B.C.), Ptolemy's predecessor, also used sexagesimal fractions for his calculations, since he was thoroughly familiar with Babylonian astronomy; he knew the Babylonian eclipse observations and lunar periods. Like Ptolemy, Hipparchus calculated tables of chords and I can not imagine that these were arranged in any other way than sexagesimally.

Sexagesimal fractions never disappeared entirely from astronomy, but they did vanish from other fields of exact science. Why? Because they were supplanted by the decimal fractions, introduced by the Dutch engineer and mathematician Simon Stevin in his work "De Thiende". This was the endpoint of a development which began with the introduction of the Hindu-Arabic numerals.

Hindu numerals.

Where do our Arabic numerals 0 1 2 3 4 5 6 7 8 9 come from? Which people invented our excellent decimal positional system?

Arabic and Persian arithmetic books are unanimous in ascribing the invention of the nine digits to the Hindus. Indeed, if we look at the forms of the digits in Fig. 10, taken from Menninger's excellent little book, we see at a glance that our numerals come from the West-Arabian ones, which in turn derive from the Hindu numerals.

But let us start at the beginning.

In prehistoric times, let us say between 2000 and 1400 B.C., the Aryans penetrated India, subjugated the population and introduced the caste system. We do not know whether the separation of the castes was as sharp then as it is now, but it is certain that the conquerors constituted the highest castes, those of the warriors and the brahmans, i.e. nobility and clergy.

These people did not pay much attention to science; they preferred the sword to the pen. But we observe that in the Buddhistic period, i.e. in the last six centuries B.C., they show an interest in numbers, particularly in terribly large numbers. The following scene occurs in the book Lalitavistara [1]. Prince Gautama

[1] See, e.g. Datta and Singh, History of Hindu mathematics I.

(Bhudda) asks the prince Dandapani for the hand of his daughter Gopa. He is
now required to compete with five other suitors in writing, wrestling, archery,

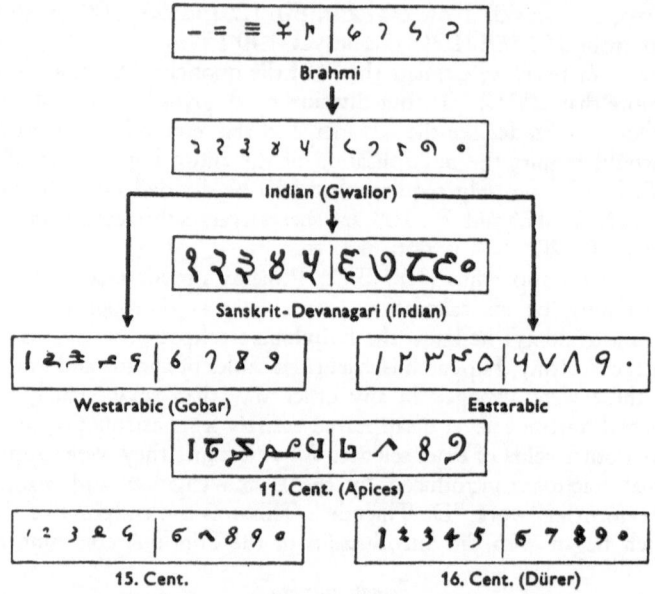

Fig. 10. Genealogy of our digits. From Menninger, Zahlwort und Ziffer. p. 329

running, swimming and arithmetic. Of course, he wins with flying colors. Now
the great mathematician Arjuna questions him;

"Oh, young man, do you know how the numbers beyond the *koṭi* continue by hundreds?"
"I know it."
"How then do the numbers beyond the *koṭi* continue by hundreds?"
"One hundred koṭis are called *ayuta*, one hundred ayutas *niyuta*, one hundred niyutas
kaṅkara, one hundred kaṅkaras *vivara* . . .

In this manner Buddha continues, through 23 stages. According to an arithmetic book, *koṭi* is a hundred times one hundred thousand (sata sata sahassa), so
that the largest number mentioned by Buddha is $10^7 \cdot 10^{46} = 10^{53}$. But in most
arithmetics, these same words ayuta and niyuta have other values, viz. 10^4 and
10^5.

But Buddha has not yet reached the end: This is only the first series, he says.
Beyond this there are 8 other series.

It is clear that these numerals were never used for actual counting or for calculations. They are pure fantasies which, like Indian towers, were constructed in
stages to dazzling heights.

Number symbols: Kharosti and Brahmi.

From the time of the great Buddhist king Asoka (3rd century B.C.) on, number symbols are also found in books. Different types of script were in use at that time, such as Kharosti and Brahmi, each having its own number symbols. The Kharosti are shown in fig. 11. There were special symbols for 1, 4, 10, 20; these were repeated as many times as necessary, as in Roman numerals. The steps 4 and 20 also appear elsewhere; thus they do not present anything of special interest.

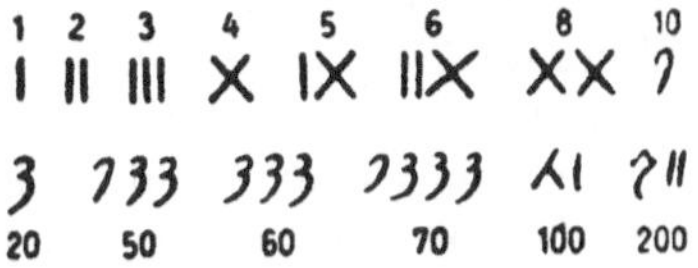

Fig. 11. Kharosti numbers.

Much greater interest attaches to the Brahmi numbers, shown in fig. 12, taken once again from Menninger. It is seen that a symbol for zero is still absent and that there is no positional system. The symbol for 60 is in no way related to that for 6. Nevertheless the system holds promise; it contains the germs of the later development. Each of the numbers less than 10 is represented by a single symbol, not by two or three as in the Kharosti system. The symbols themselves have a striking resemblance to the Arabic numerals; look at those for 6, 7, 8, 9!

Fig. 12 Brahmi numbers.

Some kind of "denominate positional system" begins to emerge in the numbers beyond 100; the symbol for 100 or for 1000 is combined with the numeral which indicates the number of hundreds or of thousands.

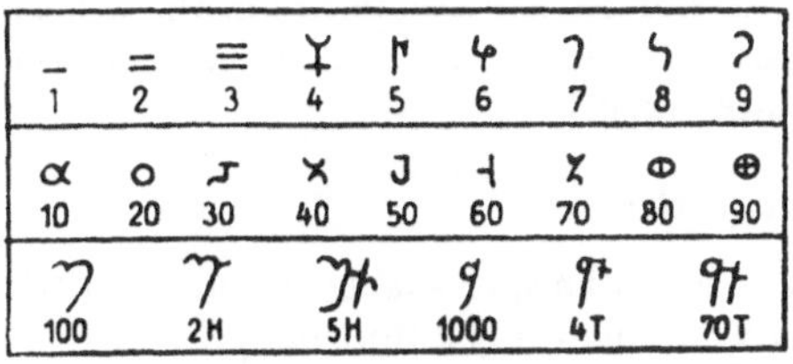

The invention of the positional system.

For all other peoples who have adopted the "Arabic" numerals, the Arabs themselves included, the written numerals appear in the general written language like strange ducks in a pond. For the Hindus alone, they are not strange; they are the same numerals from 1 to 9 inclusive, which already appear in the third century B.C. and which, later on, supplemented by a zero, were to be used to represent all numbers. As Datta and Singh justly observe, this alone suffices to show that the Hindus have to be recognized as the inventors of our positional system.

The date of the invention.

When did they invent it? This is a difficult question to answer! On a proclamation of gift of the year 595 A.D.[1], the year Samvat 346 is expressed by means

[1] *Epigraphica Indica* II, p. 19.

of the Brahmi digits 3, 4 and 6. But, for understandable reasons, such pro-
clamations were often falsified. Kaye [1], who lists 18 of them, declares that 16 of
them are not genuine. This seems to me like a very high percentage. In their
History of Hindu mathematics, Datta and Singh list 30 inscriptions on which
appear dates and other numbers, written in the decimal system, dating from the
years 595, 646, 674, etc., up to 972.

Professor Gonda, the Sanskritist of the University of Utrecht, sees no reason
to consider these inscriptions as not genuine. Even though there may be some falsi-
fications among them, we can still conclude that in the seventh century, at the
latest, the nine digits and the zero were generally known. The invention itself must
have taken place before 600, for with such matters, it is usual that at first they
are used only by scholars, and that they become known in wider circles only
very gradually. This is what happened in Arabia and in Eutrope.

The principle of the positional system, that is to say, the principle that the same
digit can, according to its position, have the values four, or forty, or four hundred,
was certainly known among Hindu astronomers and computers in the 6th century. To prove this statement, we must take a closer look at the number notation
which these astronomers and computers used.

Poetic numbers.

Have you ever tried to write a table of logarithms, or a table of sines in a rhyth-
mical and rhymed form? Try it, and you will find that it does not go very easily.
Numerals are too rigid, too prosaic, they have no synonyms, or almost none.
That is why the Hindu astronomers, who wrote in verse form, substituted other
words for the numerals, In place of 1, e.g., they wrote śaši, moon, because there
is only one moon. For 2, they wrote "eyes", "arms" or "wings"; for 3, they
used, for instance, "fire", because mythology knows 3 fires, or "brothers", because Rama had 3 brothers, etc. It is as if we were to say "muses" for 9, or
"graces" for 3; every one would at once know what is meant.

For a large number, they began with the units, followed by the tens, etc., in
the opposite direction therefore from the one used at present. Zeros were also
mentioned; the number 1021 might be written as follows:

$$\acute{s}a\check{s}i-paksa-kha-eka$$
$$\text{moon}-\text{wings}-\text{hole}-\text{one}$$
$$1 \qquad 2 \qquad 0 \qquad 1$$

In this manner the Indian astronomers learned an entire table of sines by heart
in verse form.

The most ancient work known to us, in which such a table of sines occurs in
verse, is the famous Sûrya-Siddhânta [2] which has remained the standard work on

[1] G. R. Kaye, *Notes on Indian mathematics*, Journal of the Asiatic Society of Bengal, *1907*, III, p. 482.
[2] A translation with excellent commentary is given by Burgess and Whitney, *Journal of the American Oriental Society 6*
(1860), p. 141.

Hindu astronomy until the present day. This work existed as early as the first half of the sixth century, when Varāha-Mihira wrote his compendium in which he quotes from 5 astronomical handbooks (Siddhāntas). Among these is also the Puliśa-Siddhânta, which impresses one as more primitive than the Sûrya-Siddhânta, but in which word-numbers already occur (judging from a quotation by Bhattotpala). [1] These also appear in the Agni-purâna, a religious compendium, which is ascribed to the first centuries of our era by Pargiter [2], but, at least in part, to a later date by others.

Conclusion: We do not know exactly when the word-numbers were first used, but they were certainly in circulation about 500 A.D.

Āryabhaṭa and his syllable-numbers.

About this time lived also Āryabhaṭa, the most ancient astronomer and computer whose date we can establish. [3] For, according to his own statement, he reached the age of 23 in 499. He knew the table of sines, but he used a different notation for numbers, not a positional one. As an example, we give in Āryabhaṭa's notation the number 57 75 33 36 (the number of revolutions of the moon in 432,000 years):

cayagiyiṅuśuchlṛ
6 3 3 3 5 7 7 5

The syllables which contain an *a* denote the units and tens, those with *i* the hundreds and thousands, etc. The 25 consonants from *k* to *m* have the values 1—25, e.g. $c = 6$, $g = 3$, $ṅ = 5$, $ch = 7$. The remaining 8 constants from *y* to *h* have the values 30, 40, . . ., 100. The vowels *a*, *i*, *u*, *ṛ* etc. indicate the powers of 100. A zero is not needed in this system; missing places are simply omitted.

Bhâskara I, a pupil of Āryabhaṭa, introduced an improved system, which *is* positional and *has* a zero; it has the further advantage of leaving the poet greater freedom in the choice of syllables and thus enabling him better to meet metrical requirements. According to Datta and Singh, this Bhâskara lived around 520. Like Āryabhaṭa, he begins with the units, followed by the tens, etc., but his system is actually positional; the same syllables which can designate 4, can also have the value 40 or 400.

All this makes it highly probable that around 500, the astronomers and computers introduced the positional system. Originally, they began with the units. The first to reverse the order (as far as we know) was Jinabhadra Gani, who lived about 537, according to Datta and Singh.

---

[1] According to Datta and Singh (*History of Hindu mathematics* I), the quotation is from the original Puliśa-Siddhânta, but Thibaut (*Astronomie*, Grundriss Indo-Ar. Phil.) says that the two Puliśa-Siddhântas, to which Bhattotpala had access, were totally different from the original Puliśa-Siddhânta, upon which Varāha-Mihira could draw.

[2] Pargiter, *Journal Royal Asiatic Society*, 1902, p. 254.

[3] See W. E. Clark, *the Āryabhatiya of Āryabhaṭa*, Chicago, 1930.

Where does the zero come from?

The zero is the most important digit. It is a stroke of genius, to make something out of nothing by giving it a name and inventing a symbol for it. "It is like coining the Nirvana into dynamos", says Halsted. The Babylonian sexagesimal system was imperfect because the zero was lacking; there was no difference between 60 and 1 or $1/_{60}$. In a later period there was a symbol for an absent digit in the interior of a number symbol, but not for its absence at the end.

It was the Greek astronomers who completed the system by adding a *o* for zero, an abbreviation for *οὐδέν* = nothing. The Neo-Pythagorean Iamblichus also knew the Zero, as was pointed out by Freudenthal in his masterful inaugural address. [1]

Is there any connection between the Babylonian system with the Greek round zero and the Indian decimal system with its identical round zero?

Freudenthal thinks there is. He points out that the Hindus became acquainted with Greek astronomy during the same period, from 200 to 600, when the decimal positional system came into use in India. Their standard astronomical work, the *Sûrya Siddhânta*, abounds in Greek terms. *Kendra*, center or distance from center, is of course derived from *κέντρον*, *lipta* (minute) from *λεπτόν*, etc. More significantly, the theory of the *Sûrya Siddhânta* is to a large extent based on the Greek theory of the epicycles. [2]

Along with Greek astronomy and trigonometry, the Indian astronomers quite naturally became acquainted with the sexagesimal positional system and the zero.

According to Freudenthal, there is something else which indicates a foreign influence. The versified numbers, with which we have already become acquainted, appeared in the order units, tens, etc., which connects most naturally with the language. At the time the digital notation is introduced, this order is suddenly reversed. Could this not also be a result of the Western influence? The Babylonians and the Greeks always started with the largest units.

To these arguments of Freudenthal may be added the fact that the Hindu arithmetic books write fractions just as the later Greek papyri (such as Vindobon,

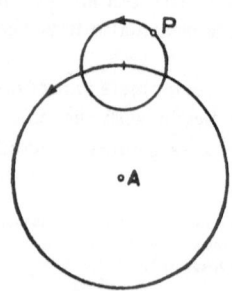

19996, see p. 49), viz. the numerator above the denominator, without a fraction line.

Freudenthal's hypothesis reduces therefore to the following: Before becoming subject to the Greek influence, the Hindus had a versified, positional system,

Fig. 13. Epicycle.

[1] H. Freudenthal, *5000 years of international science*, note 27, Groningen (Noordhoff), 1946.

[2] An epicycle is a small circle whose center describes a larger circle about the earth as a center. If a planet (or the sun) describes an epicycle, it has two simultaneous circular motions. On this account it appears, as seen from the earth, as if it had an irregular orbit, sometimes advancing, then again retrogressing. The great Greek astronomers Apollonius and Ptolemy explained all planetary motions systematically in terms of epicycles and eccentric circles.

arranged decimally and starting with the lowest units. They had the digits 1—9 and similar symbols for 10, 20, . . . Along with Greek astronomy, the Hindus became acquainted with the sexagesimal system and the zero. They amalgamated this positional system with their own; to their own Brahmin digits 1—9, they adjoined the Greek o and they adopted the Greek-Babylonian order.

It is quite possible that things went in this way. This detracts in no way from the honor due to the Hindus; it is they who developed the most perfect notation for numbers, known to us.

The rational operations with integers and fractions, as they are taught in our schools, are found in exactly the same forms in the Hindu arithmetic books. With the numbers themselves, they reached us by way of the Arabs.

The triumph of the Hindu numerals.

When in 622 Mohammed fled to Medina, to return to Mecca triumphantly shortly thereafter, there began not only the Mohammedan era, but also an entirely new period in the history of our culture. United and strong in the faith of one God, the Arabs conquered in less than a century the richest part of the world from the Indus to the Pyrenees. The prophet dies in 632, the Byzantine army is defeated in 635, the Persian army in 637. But the ancient cultures of the subjected peoples are not destroyed; on the contrary, they are absorbed in the florescent civilization of Islam. In 635 the Caliph Omar moves his seat to rich Damascus, a center of Graeco-Roman and of Semitic culture. The law gave religious freedom to Jews and to Christians, because they believed in one God. The pagans had to be converted. Arabic, the sacred language, became the world language of the Moslem empire. Tolerance, wise government, concern for safety and for welfare, the nurture of the arts and the sciences, these are the characteristics of the Arab dominion.

In the year 145 after the Hegira (766 of our era), the Caliph Al-Mansûr erects the fabled city of Baghdad not far from the ruins of Seleucia and of Babylon. Princely stipends attract Jewish, Syrian and Persian scholars and artists. Two of the greatest of Arab astronomers and mathematicians, Tabit ben Qurra and al-Battânî, even came from the pagan sect of Sabians, who worshipped the stars. It is not surprising therefore that much of Babylonian algebra, astronomy and astrology has been preserved in Arab literature.

Barely 10 years after the founding of Baghdad, a Hindu brought to the Caliph Al-Mansûr an astronomical treatise, a Siddhânta, which was immediately translated into Arabic and became very popular under the title "Sindhind". But the works of the great Greek scientists were also zealously translated. Euclid, Ptolemy and Aristotle became the great authorities for geometry, astronomy and philosophy.

Al-Ma'mun, a grandson of Al-Mansûr, established in Baghdad an academy, an astronomical observatory and a library. In this library, *Muhammed ben Musa*, called *Alkhwarizmi*, wrote the first Arabic book on algebra. Later on his name

was corrupted to Algorithmus, and even at present the word algorithm is used
to designate a process of calculation. The same al-Khwarizmi wrote a small work
on Indian calculation. The original is lost, but in the 12th century, the English
monk Adelhard of Bath (or some one else) translated it into Latin. It is by means
of this Latin opusculum, which is extant, that the Western world became acquainted with the Hindu-Arabic numerals.

Concerning the zero, al-Khwarizmi writes as follows:

> When (in subtraction) nothing is left over, then write the little circle, so that the place
> does not remain empty. The little circle has to occupy the position, because otherwise there
> would be fewer places, so that the second might be mistaken for the first.

Al-Khwarizmi was not the first in Arabic culture who knew the Hindu numerals
and Hindu arithmetic. The Syrian bishop Severus Sêbokht, who lived in the
first century after the Hegira, speaks with deep admiration of the arithmetic of
the Hindus and of the nine digits, by means of which they carried out all calculations. He judges Hindu astronomy to be "superior to that of the Greeks and the
Babylonians", and he is scornful of conceited people "who think, because they
speak Greek, that they have attained the extreme limits of science" and who ignore
"that there are others who know something". An extreme example of overvalu-
ation of Hindu astronomy, which was, after all, nothing but a distillation of Greek
astronomy! Indeed, from the introduction to his Algebra, it becomes clear that
Greek science was beyond the understanding of al-Khwarizmi.

Al-Khwarizmi already knew two forms of the digits. We distinguish the West-
Arabic or Gobar digits, and the East-Arabic (see Fig. 10.)

Our digits derive from the Gobar digits which were used in Moorish Spain.
The East-Arabic digits are still in use to-day in Turkey, Arabia and Egypt; they
are there called "Indian digits". It is clear that both were derived from the Brahmi-
digits.

Let us see now, how the Hindu-Arabic numerals reached the North.

Fig. 14. Abacus of Gerbert.

The abacus of Gerbert.

Gerbert, the later Pope Sylvester II, was born in
Auvergne in 940. Of modest descent, his connections
and his gifts enabled him to rise to the highest ecclesia-
tical dignity. He travelled much, In Spain he became
acquainted with the Gobar numerals. In Rheims he
taught mathematics and he wrote a small book on calculating with the abacus. He invented a new type of
abacus, carefully described by one of his pupils (see
Fig. 14). In place of the pebbles, he used "Apices",
his own invention, discs, carrying numbers which have a remarkable appearance

and look like fantastic cabbalistic signs. But when one takes a good look at them (see Fig. 10) one recognizes the West-Arabic Gobar numerals! The zero is lacking; the abacus does not require it.

It is the first appearance of the 9 digits on this side of the Pyrenees. But they had little success; the times were not ripe for them. Indeed on the abacus, one operates much more easily with pebbles or counters. It was a mistake on Gerbert's part to use the Arabic numerals on the abacus; their real destiny was to make the abacus superfluous. They were guest performers on the abacus, but soon disappeared from the stage.

In the 12th century, the nine digits and the zero appear once more, this time in the Latin translation of al-Khwarizmi's arithmetic. This book had been studied industriously in Spain ever since the 10th century. For that matter, it was by way of Spain, that we first became acquainted with Euclid and Aristotle; all these Greek classics were first translated into Latin from the Arabic.

In 1143, there appeared in Germany an extract from the Algorithm book from which an extract is reproduced in Fig. 15. Going from top to bottom, one sees the digits from 1 to 9 with their multiples. The captions at the head of the columns are the digits from 2 to 9 which serve as multipliers.

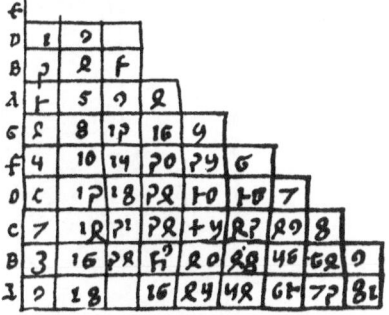

Fig. 15. Medieval multiplication table with Arabic numerals.

The new digits were used especially in the monasteries for the determination of Easter. The people did not want to have anything to do with them; they preferred to calculate with the familiar Roman numerals, just as the old Jewess, so masterfully portrayed in Keller's Der Grüne Heinrich (I, p. 6). As late as 1514, the reckoning master Ködel published an arithmetic, printed entirely in "German", i.e. Roman numerals.

"Ich habe disz rechenbüchlein, dem Leyen zu gutt und nutz (dem die Ziffernzahle am ersten zu lernen schwer) durch die gewöhnlich teutsch Zal geordnet."

The little book went through many editions.

It was particularly the zero which seemed queer to people. Some one in the 15 th century spoke of "a symbol that merely causes trouble and lack of clarity". In Latin the zero is called "Nulla figura" — no sign. Nevertheless it is a cypher, indeed the cypher par excellence, for the word cypher is derived from the Latin cifra = zero (the Arabic alṣifr = the empty). The zero is nothing, yet it decuples the number in front of it. This it was impossible to comprehend. "Algorismuscifra" was a word of obloquy in France, like the word „zany" with us and "Vocativus" with Fritz Reuter.

But, in the end, the Hindu numerals were bound to win out. Here, as in every domain, Italy led the way. In 1202, there appeared an excellent arithmetic, the Liber Abaci of Leonardo of Pisa, known as Fibonacci. Leonardo employs systematically and exclusively the "figurae Indorum" which he had learned as a boy from a Moorish teacher. He considers all other methods of calculation, in-

Fig. 16. Pythagoras and Boethius.
From the "Marguerita philosophica" of Gregor Reisch.

cluding the abacus of Gerbert, as inferior in comparison with the "modus Indorum".

Leonardo's book scored a great success. The large merchants realized very soon the great advantages of the new numbers, and they used them in their book-keeping.

But in 1299, the city of Florence issued an ordinance which prohibited the writing of numbers in columns, as well as the use of Hindu numerals. Why? The answer is supplied by a Venetian treatise on book-keeping; it is so easy to change a 0 to a 6 or a 9. It was not so easy to falsify Roman numerals. Even in

PLATE 7

PL. 7. Large volutecrater (height 1.30 m) from Canosa, in the National Museum in Naples, third century B.C. South Italian-Greek art, somewhat showy in character and not attaining the pure beauty of the Attic vases of the 5th century. In the central band, the painting represents the council of war of King Darius, before his famous expedition of 490 B.C. against the Greeks. Darius sits in the midst of his counsillors. The person who stands on a platform, addressing the king, warns him – as may be seen from his gesture – against the dangers of the expedition. A bodyguard stands behind the king. Below, the king's treasurer is seen at his counting board (see also Fig. 9 in the text). To his left and right, violently agitated persons, representing the tributary provinces, too heavily taxed.

(Photo Alinari)

PLATE 8

PL. 8a. Plimpton 322, Plimpton Library, Columbia University, New York. Old-Babylonian cuneiform text, containing a list of right-angled triangle with rational sides, h, b and d. On the left, several columns are broken off. The first preserved column gives the ratios d^2/h^2, in decreasing order. The heading of this column is unintelligible. The next 2 columns give "width" b and "diagonal" d as whole numbers. The last column contains only the running integers from 1 to 15. In line 11 the numbers b and d have a common factor 15, in line 15 a common factor 2; in all other cases b end d have no factor in common. See pp. 78–80.

PL. 8b. YBC 7289, Cuneiform text from the Yale Babylonian Collection. A square with its diagonals. The side is 30 (the number is written above the left upper side). On the diagonal the ratio 1,24,51,10 is written; under the diagonal its length 42,25,35. See p. 45.

1594 an Antwerp canon warns merchants not to use numerals in contracts or in drafts.

Finally, during the 15th century, along with all such Italian expressions as conto, giro, saldo, lombard and bankrupt, the number symbols reach us from Italy. In 1494, the mayor of Frankfurt cautions the clerks to use these numerals sparingly, i.e. not to use them too much. In 1482, the first German arithmetic of Ulrich Wagner appeared in Nürnberg. It was followed by many others; the most famous one is that of Adam Riese. All these books for popular use teach "calculation on lines and with numerals", i.e. the use of the abacus and the use of Arabic numerals.

In these days, there were violent controversies between the "abacists", the defenders of the old abacus, and the "Algorithmici". A picture in an arithmetic of 1504 represents this quarrel in a very amusing way: On the right sits Pythagoras, taken to be its inventor, in front of his abacus; the left shows Boethius in front of a disc with Arabic numerals, which he was supposed to have invented. Dame Arithmetica plays the role of the judge. Pythagoras' sour looks suggest that he is the loser.

It took a few more centuries, but the cause of the Hindu numerals is now definitely settled. Who would believe that, 500 years ago, practically everybody, north of the Alps, still calculated with the abacus and used Roman numerals?

CHAPTER III

BABYLONIAN MATHEMATICS

Chronological Summary.

General History	History of Civilization	History of Science
3000 Sumerian city states	Cuneiform writing High level of culture	Sexagesimal system
2800–1800 Semitification		Tables for dividing and multiplying
First Babylonian dynasty 1700 Hammurabi	Cultural flowering Legislatio, administration of justice	Phenomenal flowering of algebra and geometry. Observations of Venus.
1500–1250 Babylon under the rule of the Cassites	Series of astrological omens: "Enuma Anu Enlil"	Primitive astronomical calculations. Observations of heliacal rising of fixed stars.
747, Nabonassar, king of Babylon	Beginning of the astronomical "Era Nabonassar"	Dated observations of eclipses in Babylon.
729 The Assyrian Tiglatpilesar II ascends the throne of Babylon as Pulu.		
722 Sargon II 700 Sanherib	Assyrian royal palaces Court astrologers	Astronomical compendia: I·NAM·GIS·HAR and mul
650 Assurbanipal	Library of Assurbanipal	APIN of Babylonian origin, copied in Assyria about 700
612 Destruction of Niniveh. End of the Assyrian empire New Babylonian empire of the Chaldeans 580 Nebukadnezar II	New period of flowering of arts and sciences	Observation of moon and planets.
540 Cyrus, founder of the Persian empire 500 Darius	Babylonian religion not affected Calendar periods	Increased accuracy of observations of the Zodiac. Periods of planets
333 Alexander the Great 311 Beginning of the era of the Seleucids 247 Beginning of the Arsacid era	Hellenism Birth-horoscopes	Flowering of astronomy. Lunar and planetary tables. Revival of algebra. Extensive calculation tables.

Babylonian Algebra.

It seems best to begin with a characteristic old-Babylonian cuneiform text, instead of indulging in extended preliminary considerations. For indeed, the only things we know about what went on in the minds of our mathematical colleagues of Hammurabi's day, are the things we can dig out of the texts themselves.

All the text translations have been taken from the monumental work of O. Neugebauer, Mathematische Keilschrifttexte (Quellen und Studien, A 3, Berlin 1935), supplemented by Thureau-Dangin, Textes mathématiques babyloniens, Leiden 1938, and by Neugebauer-Sachs. Mathematical Cuneiform Texts, New Haven 1935.

First example. [1]

Text AO 8862, from Senkereh, old Babylonian (i.e. from the Hammurabi dynasty), begins as follows:

> Length, width. I have multiplied length and width, thus obtaining the area. Then I added to the area, the excess of the length over the width: 3,3 (i.e. 183 was the result). Moreover, I have added length and width: 27. Required length, width and area.
>
> (given:) 27 and 3,3, the sums
> (result:) 15 length 3,0 area.
>   12 width
>
> One follows this method:
> $$27 + 3,3 = 3,30$$
> $$2 + 27 = 29.$$
> Take one half of 29 (this gives 14;30).
> $$14;30 \times 14;30 = 3,30;15$$
> $$3,30;15 - 3,30 = 0;15.$$
> The square root of 0;15 is 0;30.
> $$14;30 + 0;30 = 15 \text{ length}$$
> $$14;30 - 0;30 = 14 \text{ width.}$$
> Subtract 2, which has been added to 27, from 14, the width. 12 is the actual width. I have multiplied 15 length by 12 width.
> $$15 \times 12 = 3,0 \text{ area.}$$
> $$15 - 12 = 3$$
> $$3,0 + 3 = 3,3.$$

Interpretation.

The first lines formulate the problem: 2 equations with 2 unknowns, each represented by a symbol, uś and sag, length and width. The Sumerian symbols are dealt with as our algebraic symbols x and y; they possess the same advantages of remaining unchanged in declension. We can therefore safely put the problem in the form of 2 algebraic equations:

(1)
$$xy + x - y = 183$$
$$x + y = 27.$$

[1] Neugebauer, MKT (*Mathematische Keilschrifttexte*) I, p. 113.

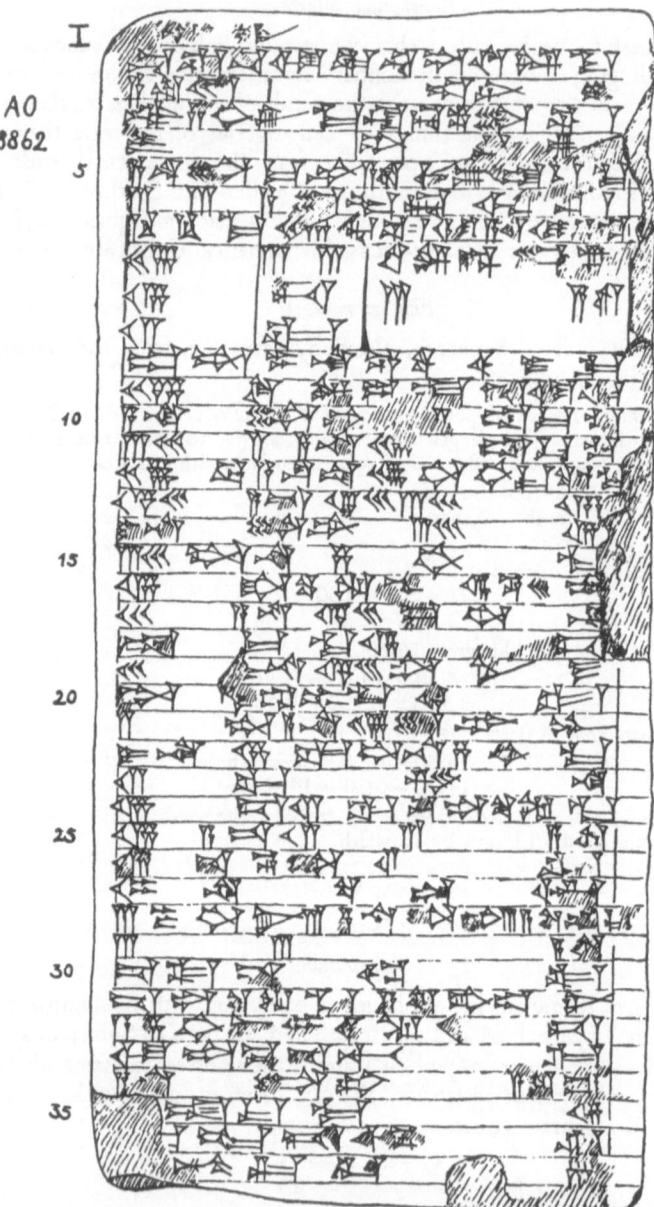

Fig. 17.   AO 8862, copied by Neugebauer, MKT II Table 35. The reader will have little difficulty in finding all
the numbers, which occur in the following translation, in lines 6 to 29, starting with 3,3 in line 6.

The last 4 lines of the text merely verify that the resulting numbers $x = 15$ and $y = 12$ indeed satisfy (1).

The preceding sentence indicates that the author, wishing to simplify the problem, has introduced a new variable y', in place of the actual width y:

$$y' = y + 2, \text{ i.e. } y = y' - 2.$$

This change does indeed simplify the problem; the equations in x and y' are:

(2) $$\begin{aligned} xy' &= 183 + 27 = 210, \\ x + y' &= 27 + 2 = 29. \end{aligned}$$

The first of the equations (2) is found immediately by adding the two equations (1). The transformation from (1) to (2) is indicated very succintly in the text by means of the two short lines:

$$\begin{aligned} 27 + 3,3 &= 3,30 \\ 2 + 27 &= 29. \end{aligned}$$

Then follows the solution of the simplified system (2). This follows a fixed recipe, which appears again and again in other texts. In modern algebraic symbolism, this recipe may be described as follows:

The solution of the system of equations

(3) $$\begin{aligned} xy' &= P \\ x + y' &= a \end{aligned}$$

is

(4) $$\begin{aligned} x &= \tfrac{1}{2}a + w \\ y' &= \tfrac{1}{2}a - w \\ w &= \sqrt{(\tfrac{1}{2}a)^2 - P}. \end{aligned}$$

What the Babylonians do, step by step, in numbers, amounts indeed to application of the formulas (4). But they do not give these formulas; they merely give one example after another, each of which illustrates the same method of calculation.

Second example (MKT I, p. 280).

Another old-Babylonian text (VAT 6598) contains, besides a system of the form (3), an analogous system, in which the difference is given instead of the sum:

(5) $$\begin{aligned} xy &= P \\ x - y &= d. \end{aligned}$$

The solution is as follows:

(6) $$\begin{aligned} x &= w + d/2 \\ y &= w - d/2 \\ w &= \sqrt{(d/2)^2 + P}. \end{aligned}$$

Where did the Babylonians get this method of solution? We can only guess at the answer. But we can say what they did not do: they did not use the Arabian method of determining one of the unknowns from one of the equations (5) and substituting in the other. The Arabians used to reduce every algebraic problem to one equation with one unknown. As we shall see in the sequel, the Babylonians also knew how to eliminate by substitution, but in the cases we have just discussed, they did not do this; if they had, they would have determined y from $a - x$ or from $x - d$, after having found x by solving a quadratic equation. But instead, in case (4), they found x in the form $\frac{1}{2}a + w$, and y as $\frac{1}{2}a - w$.

As Gandz has observed, Diophantus, a late Greek writer on Arithmetica, usually proceeds in the same way. Frequently his methods have more in common with Babylonian algebra, than with the processes of the classical Greek writers. When Diophantus wants to determine two unknowns x and y, of which the sum a or the difference d is given, he very often sets

$$\begin{cases} x = \frac{1}{2}a + z \\ y = \frac{1}{2}a - z \end{cases} \quad \text{or} \quad \begin{cases} x = z + \frac{1}{2}d \\ y = z - \frac{1}{2}d, \end{cases}$$

and then expresses in terms of z all further conditions which x and y have to satisfy.

In our cuneiform texts we find the same method applied, not only to quadratic equations, but also to linear equations. Let us take, e.g., the very remarkable old-Babylonian text VAT 8389:

Third example (MKT I, p. 323).

Per bùr (surface unit) I have harvested 4 gur of grain. From a second bùr I have harvested 3 gur of grain. The yield of the first field was 8,20 more than that of the second. The areas of the two fields were together 30,0. How large were the fields?

For a full understanding of the calculation which follows, one has to know that the combined area of the two fields (30,0) is measured in SAR (1 SAR = 12 yds. square), and the difference in yields in terms of sila, and that

$$1 \text{ bur } = 30,0 \text{ SAR},$$
$$1 \text{ gur } = 5,0 \text{ sila}.$$

In the terminology of Thureau-Dangin, the SAR and the sila are the "scholar's units", in terms of which mathematical calculations are carried out, while bùr and gur are larger, practically more convenient, units.

The first field yields 4 gur = 20,0 sila per 1 bùr = 30,0 SAR, and the second field 3 gur = 15,0 sila per 30,0 SAR. Call the unknown areas (expressed in SAR) x and y. Then we have to solve two equations with 2 unknowns:

$$(7) \qquad \frac{20,0}{30,0}x - \frac{15,0}{30,0}y = 8,20$$

$$x \quad + \quad y = 30,0$$

The Babylonians were fully able to solve the second equation for x, to substitute in the first and then to solve for y. Indeed this is what they did in another problem of the same text, in which $x - y$ was given instead of $x + y$. But in the present problem they followed a different path. They begin by dividing the total area into two equal parts:

Divide 30,0, the sum of the areas, into two parts: 15,0. Thus take 15,0 and again 15,0.

Then they calculate what the yield would be, if each of the fields had an area of 15,0.

Everything is elaborately worked out in great detail; the reciprocal of 30,0 is multiplied by 20,0 and we find "the wrong yield of grain" 0;40, i.e. the yield of the first field for 1 SAR. Hence the yield of a field of 15,0 SAR is

$$0,40 \times 15,0 = 10,0.$$

"Keep this in mind", the text says. In the same manner one finds for the second field the "wrong yield of grain" of 0;30 for 1 SAR, and hence a yield of 7,30 for 15,0 SAR.

It is concluded that, if each of the fields had an area of 15,0 SAR, the difference in yield would be

$$10,00 - 7,30 = 2,30.$$

But it is given that the difference is 8,20. "Subtract", the text says:

$$8,20 - 2,30 = 5,50.$$

"Keep 5,50 in mind", says the text and then continues the calculation as follows:

$$0;40 + 0;30 = 1;10.$$

I don't know the reciprocal of 1;10. What must I multiply by 1;10 to obtain 5,50? Take 5,0, since $5,0 \times 1;10 = 5,50$.

Subtract this 5,0 from one of the areas of 15,0 and add it to the other. The first is 20,0, the second 10,0. So 20,0 is the area of the first field, 10,0 that of the other.

An elementary school teacher might explain the procedure to the children as follows:

If each of the fields had an area of 15,0 SAR, the difference in yield would be 2,30. It has to be 8,20, so that 5,50 has to be added. For every unit of area, added to the first field and subtracted from the second, the first would produce 0;40 more and the second 0;30 less, so that the difference would be increased each time by $0;40 + 0;30 = 1;10$. This has to be taken 5,0 times to obtain exactly 5,50. Hence the first area must be $15,0 + 5,0 = 20,0$ and the second 15,0—5,0 $= 10,0$.

We do not know whether the Babylonians reasoned exactly as our teacher; but I believe that their thought process is expressed better by this primitive argument than by the elaborate algebraic transformation, which Neugebauer gives on page 334.

At least we see that, just as Diophantus, when the Babylonians were given that

$x + y = 2h$, they set

$$y = h + w, \qquad y = h - w$$

and then tried to determine w. If, e.g., the product P is given as well as the sum, then the value of w is determined from the equation

$$xy = (h + w)(h - w) = h^2 - w^2 = P,$$

so that

$$w^2 = h^2 - P.$$

In this ways they could therefore derive formula (4), as soon as the formula

(8) $$(h - w)(h + w) = h^2 - w^2$$

was known to them. The derivation of (6) is entirely analogous.

Fourth example (MKT I, p. 154).

It becomes clear also from other texts that the special product (8) was known. For example, BM (British Museum) 85 194 requires the construction of the profile of a dike in the form of an isosceles trapezoid, of which are given the base a, the inclination

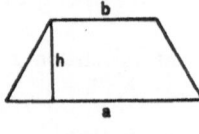

$$\beta = \frac{a - b}{2h}$$

and the area $S = \dfrac{a + b}{2} h.$

Fig. 18.

After multiplying 2β by $2S$, one obtains

$$4\beta S = (a - b)(a + b) = a^2 - b^2,$$

which can be solved for b^2

(9) $$b^2 = a^2 - 4\beta S.$$

The text actually calculates b^2 by use of (9). Another problem in the same text takes b as given and computes a from

(10) $$a^2 = b^2 + 4\beta S.$$

All of this would be totally inexplicable, unless the special product (8) were known.

Fifth example (MKT III, p. 8, no. 14).

The formulas

(11) $$(a + b)^2 = a^2 + 2ab + b^2, \text{ and}$$

(12) $$(a - b)^2 = a^2 - 2ab + b^2$$

must also have been known to the Babylonians. For the old-Babylonian text BM 13901 contains the following problem:

> I have added the areas of my two squares: 25,25. (The side of) the second square is $^2/_3$ of that of the first plus 5 GAR.

That is to say:

(13)
$$x^2 + y^2 = 25,25$$
$$y = (^2/_3)x + 5$$

In order to substitute the value of y, obtained from the second equation, in the first equation, formula (11) has to be used:

$$(0;40x + 5)^2 = 0;40^2x^2 + 2 \cdot 0;40 \cdot 5x + 5^2.$$

This leads to a quadratic equation

(14)
$$ax^2 + 2bx = c$$

for x, in which

$$a = 1 + 0;40^2 = 1;26,40, \quad b = 5 \times 0;40 = 3;20, \quad c = 25,25 - 5^2 = 25,0.$$

The text first calculates the 3 coefficients a, b, c; then the quadratic equation (14) is solved by use of the correct formula:

(15)
$$x = a^{-1}(\sqrt{ac + b^2} - b)$$

and finally $y = (^2/_3)x + 5$ is determined. It follows that the method of elimination, described above, was used and that the formula (11) was known.

Quadratic equations (MKT III, p. 6)

The beginning of the same text exhibits simpler examples for the solution of quadratic equations, such as

    1. $x^2 + x = 0;45$,
    2. $x^2 - x = 14,30$,
    6. $x^2 + (2/3)x = 0;35$,
    7. $11x^2 + 7x = 6;15$.

The following solution is given of 2:

> Take 1, the coefficient (of x). Divide 1 into two parts. $0;30 \times 0;30 = 0;15$, you add to 14,30 and 14,30;15 has the root 29;30. You add to 29;30 the 0;30 which you have multiplied by itself, and 30 is (the side of) the square.

How may the Babylonians have obtained the solution of the quadratic equations

(16)
$$x^2 \pm ax = b?$$

I believe it likely that they proceeded just like the Arabians did and like we do, viz. by changing the left side to a perfect square:

$$(x \pm \tfrac{1}{2}a)^2 = b + (\tfrac{1}{2}a)^2.$$

Another possibility, although a less probable one, would be to introduce a second unknown $y = x \pm a$, which would give us the difference a of x and y and the product

$$xy = x(x \pm a) = b,$$

thus reducing this case to (5).

Sixth example (MKT III, p. 9, no. 18)

Our texts also contain systems of equations with 3 or more unknowns; in the same text BM 13 901, we find:

$$x^2 + y^2 + z^2 = 23{,}20, \qquad x - y = 10, \qquad y - z = 10.$$

The method of solution is again the same; x and y are expressed in terms of z, and a quadratic equation is obtained for z. Such eliminations were obviously like rolling off a log for the old algebraists.

In other cases, elimination was not used; instead, the two unknowns were determined in parallel manner. E.g., the same text solves the system

(17) $$x^2 + y^2 = S = 21{,}40, \qquad x + y = a = 50$$

in the form

(18) $$x = \tfrac{1}{2}a + w, \qquad y = \tfrac{1}{2}a - w, \qquad w = \sqrt{\tfrac{1}{2}S - (\tfrac{1}{2}a)^2}$$

We have here the same idea that we met before: when $x + y = a$ is given, we set both x and y equal to $\tfrac{1}{2}a$, plus or minus a correction; this correction is then found from the other condition which x and y have to satisfy.

In the following problem (no. 9), $x - y$ is given instead of $x + y$:

(19) $$x^2 + y^2 = S = 21{,}40, \qquad x - y = d = 10.$$

Entirely analogously, the solution is:

(20) $$x = w + \tfrac{1}{2}d, \qquad y = w - \tfrac{1}{2}d, \qquad w = \sqrt{\tfrac{1}{2}S - (\tfrac{1}{2}d)^2}.$$

There are also texts which contain merely long series of problems without solutions. Sometimes, the solutions are supplied in other texts; the text G of Neugebauer-Sachs, Math. Cun. Texts (p. 66), contains, for example, 31 problems, of which the first 8 are solved in text H and the last 10 in text J. All this indicates that Babylonian algebra was taught systematically in the schools by the aid of sets of problems.

Seventh example (MKT I, p. 485)

Sometimes, the solution of such a Babylonian problem is far from easy, not even for us! Let us look for instance at YBC 4697 (Yale Babylonian Collection), problem D2:

(21) $$1/3 \cdot (x + y) - 0;1(x - y)^2 = 15, \qquad xy = 10{,}0.$$

Neugebauer thought at first that this would lead to a cubic equation for x or y. But, if one starts from the Babylonian idea that the unknowns are equal to one half of their sum plus or minus a correction, i.e. if we set

(22) $\qquad\qquad x = u + v, \qquad\qquad y = u - v,$

then (21) reduces to

(23) $\qquad\qquad 0;40u - 0;40v^2 = 15, \qquad u^2 - v^2 = 10,0.$

From these equations v^2 is readily eliminated, leaving a quadratic equation for u.

Eighth example (MKT I, p. 204)

But even cubic equations did not frighten the Babylonians. As an example, we quote problem 22 of the text BM 85 200, believed by Neugebauer (p. 193) to belong to a later date than the texts which have been discussed thus far. In this problem, the pure cubic

$$12x^3 = 1;30$$

is solved, by using a table, to show that the cube root of

$$1/12 \cdot 1;30 = 0;7,30$$

is $0;30$; the use of a table is, of course, as simple for cube roots as for square roots. The next problem, no. 23, leads to a mixed cuibc

$$x^2(12x + 1) = 1;45.$$

By multiplying both sides by 12^2, the author obtains

$$(12x)^2(12x + 1) = 4,12,$$

and from this he obtains, without any further ado, $12x = 6$. Where did he get this result? From a table, of course. Indeed we know that tables existed, which gave the "roots" n for numbers of the form $n^2(n + 1)$.

By means of their tables, the Babylonians were therefore able to solve mixed cubics of the form

$$x^2(\mu x + 1) = V,$$

as readily as pure quadratics or pure cubics.

Geometrical proofs of algebraic formulas?

How may the Babylonians have obtained formulas like

$$(a - b)(a + b) = a^2 - b^2,$$
$$(a + b)^2 = a^2 + 2ab + b^2$$
$$(a - b)^2 = a^2 - 2ab + b^2?$$

We don't know. Perhaps they derived them by the use of diagrams, such as are found in Euclid and in Arabic writers:

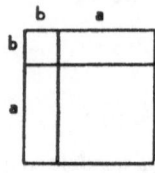

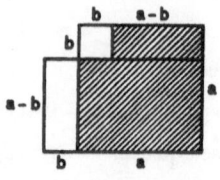

Fig. 19.

$(a + b)^2 = a^2 + 2ab + b^2.$

Fig. 20.

$a^2 - b^2 = (a - b)(a + b).$

It is certain that they interpreted a product as the area of a rectangle, a square as the area of a square; this becomes clear from their own terminology.

But we must guard against being led astray by the geometric terminology. The thought processes of the Babylonians were chiefly algebraic. It is true that they illustrated unknown numbers by means of lines and areas, but they always remained numbers. This is shown at once in the first example, in which the area xy and the segment $x - y$ are calmly added, geometrically nonsensical. Neither did the Babylonians hesitate to multiply two areas, something the very careful Euclid would never do.

Even in problems which were formulated in geometrical terms, the question which the Babylonians asked, was always to calculate something, never to construct or to prove something. Through the geometric exterior, the algebraic kernel is always visible. Just let us look at the

Ninth example (MKT I, p. 342).

In the old-Babylonian text VAT 8512, from Larsa, the following problem is proposed:

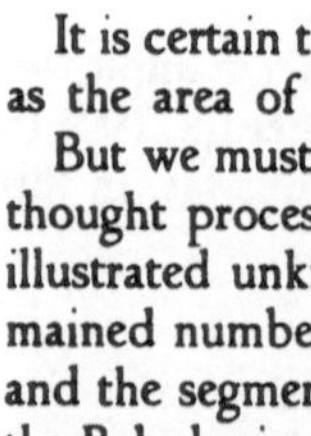

A triangle, whose base b is given equal to 30, is divided into two parts by a line parallel to the base, viz. the trapezoid F_1 of height y_1 and the triangle F_2 of height y_2. It is given that

$$F_1 - F_2 = \Delta = 7.0$$
$$y_1 - y_2 = \delta = 20.$$

Fig. 21.

Hence we have three unknowns: the dividing line x and the heights y_1 and y_2; between them, we have the relations:

(1) $\tfrac{1}{2}y_1(x + b) - \tfrac{1}{2}y_2 x = \Delta,$

(2) $y_2 - y_1 = \delta$

From the diagram we see furthermore, that the proportionality

(3) $y_2 : y_1 = x : (b - x)$

must hold.

The solution proceeds as follows: First x is calculated by

(4) $$x = \sqrt{\tfrac{1}{2}\{(\Delta/\delta + b)^2 + (\Delta/\delta)^2\}} - \Delta/\delta.$$

Next, the given base b is determined anew by subtracting Δ/δ from $\Delta/\delta+b$. The heights y_1 and y_2 are found from

(5) $$y_1 = (b-x) \cdot \Delta/(\tfrac{1}{2}b^2 - x^2),$$

(6) $$y_2 = y_1 + \delta.$$

The formulas for x, y_1 and y_2 are algebraically correct, but probably they were not found by pure algebra. A satisfactory explanation of the solution (4) was found by P. Huber (Isis 46, p. 104).

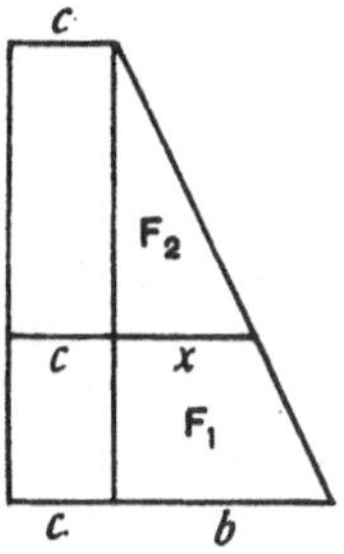

Suppose we add a rectangle having the same height $y_1 + y_2$ to the triangle, thus obtaining a trapezoid with parallel sides c and $a = c + b$. The line x, prolonged, divides the trapezoid into two parts. The difference of the areas of these parts is

$$(F_1 - F_2) - c(y_2 - y_1) = \Delta - c\delta.$$

If we want to make this difference zero, we have to put $c = \Delta/\delta$. Now the Babylonians knew that the line z parallel to the sides a and c, which halves the area of a trapezoid, is given by the formula

$$z^2 = \tfrac{1}{2}(a^2 + c^2)$$

(see MKT I, p. 131). Hence we have

$$x = z - c = \sqrt{\tfrac{1}{2}(a^2 + c^2)} - c$$

which explains formula (4). The idea of this solution is geometrical, not algebraic.

Fig. 22

A lesson-text (MKT II, p. 39)

It is to be regretted, that almost all the texts merely contain problems and solutions, but no derivations. The solutions are given in the form of recipes, without saying how they were obtained. But these recipes must have been derived in some way and the teachers must certainly have told their pupils how they could solve equations, and how they could express one unknown in terms of the others.

An example of such a derivation was found by H. Freudenthal in a text which had remained unintelligible to Neugebauer and Thureau-Dangin. [1] A.O. 6770, old-Babylonian from Uruk is a "lesson-text". If in the translation of the first line we follow Neugebauer (MKT II, p. 39), but in lines 2—8 Thureau-Dangin, we

[1] I owe thanks to my friend Professor H. Freudenthal for his permission to publish his interpretation here.

obtain the following meaning for these lines:

> Length and width as much as area; let them be equal.
> You in your procedure,
> The product you take twice.
> From this you subtract 1.
> You form the reciprocal.
> With the product that you have taken
> You multiply and
> The width it gives you.

In the first line the problem is formulated:

(7) $x + y = xy.$

The following lines give the solution, entirely in abstract form, without numbers:

(8) $y = \dfrac{1}{x-1} \cdot x$

Dr. Bruins [1] rejects this interpretation on philological grounds, because *ù* in the first line, although it can mean "and", can not indicate addition, and because *iku* (a measure of area) is translated as if the text had aša (area). I can not judge how much importance is to be attributed to these philological objections. But we must recognize that formula (7) is mathematically equivalent to (8). And, if Thureau-Dangin's translation of lines 2—8 is correct, formula (8) is stated in the text. If this translation is taken as the point of departure, then (7) must also be accepted as giving the factually correct interpretation of the writer's intention. Whether he expresses himself in a grammatically correct manner or not, we can not get rid of the fact that the sum of the length and width, which the author had in mind, is equal to the area. It is usually so in mathematical texts, that the meaning of the words has to be determined in relation to the mathematical content.

There is another example of a general rule for calculation in Neugebauer's MKT III (p. 19). A free translation of BM 34 568 gives:

> Length, width and diagonal times length, width and diagonal take. The area times 2 take. The product from the (square of the length, width and) diagonal you subtract. What remains times a half take. (The reciprocal of) length, width and diagonal you must multiply by the half. The result is the diagonal.

The words in parentheses do not occur in the text; Neugebauer and Thureau-Dangin have added them in order to obtain the correct formula. One sees again that, in the interpretation of a text, the leading investigators give precedence to mathematical correctness. If *l*, *b*, *F* and *d* designate length, width, area and diagonal of the rectangle respectively, the formula is

$$d = \frac{\frac{1}{2}[(l+b+d)^2 - 2F]}{l+b+d}.$$

E. M. Bruins, *Nouvelles découvertes sur les mathématiques babyloniennes*, Conférences, Palais de Découverte, Univ. de Paris, Serie D, no. 11 (1951).

I am indebted to Dr. Bruins for calling my attention to this formula.

Babylonian Geometry.

We have already discussed proportionalities related to parallel lines (see Example 8).

Volumes and areas.

It has been mentioned above that the Babylonians knew how to calculate the area of a triangle and of a trapezoid.

They took the area of a circle of radius r to be equal to $3r^2$, the perimeter $6r$. It has been shown that the Egyptians had a better approximation for π.

The volumes of prisms and of cylinders were determined by multiplying the area of the base and the height.

Unfortunately we have no examples of cones and of pyramids, but we do have them for

Frustra of cones and of pyramids (MKT I, pp. 176 and 187)

In the old-Babylonian text BM 85 194 (Plate 9) (see also the fourth example), the volume of a frustrum of a cone is determined by use of the wrong formula

½ height × sum of the bases.

The same text also deals with a frustrum of a pyramid, whose height is given and whose bases are squares with sides $a = 10$ and $b = 7$. The first step is to show that

$$\left(\frac{a+b}{2}\right)^2 = 1 . 12;15$$

and that $a - b = 3$. From this, an operation, which is not entirely clear, leads to a number . . .,45, that might be interpreted perhaps as 0;45.

Neugebauer interprets 0;45 as representing $\frac{1}{3} \cdot \left(\frac{a-b}{2}\right)^2$. Thureau-Dangin considers it to be $\frac{1}{4} \cdot (a - b)$. By adding this 0;45 to 1,12; 15, one finds 1,13. This amount is multiplied by the height 18. The multiplication is wrong; instead of producing the result 21,54, the text gives 22,30. A rounding off like this occurs not infrequently.

According to Neugebauer, the calculation is based on

$$(1) \qquad V = \left\{\left(\frac{a+b}{2}\right)^2 + 1/3 \cdot \left(\frac{a-b}{2}\right)^2\right\} h,$$

which is a correct formula for the volume of a frustrum of a pyramid with square bases. According to Thureau-Dangin however, the nonsensical formula

$$(2) \qquad V = \left\{\left(\frac{a+b}{2}\right)^2 + \frac{a-b}{4}\right\} \cdot h$$

was used.

An objection to his own interpretation, raised by Neugebauer, is that the space in the text, which precedes the number 45, and which contains a few illegible symbols, is too small for the calculation of $\frac{1}{3} \cdot \left(\frac{a-b}{2}\right)^2$. Another difficulty arises rom the fact, that in two other texts, which are closely related to BM 85 194, viz. BM 85 196 and BM 85 210, the wrong formula

(3) $$V = \frac{1}{2}(a^2 + b^2)h$$

is used. Both difficulties disappear if we suppose that 0;45 is an error of calculation and should be replaced by

$$\left(\frac{a-b}{2}\right)^2 = 2;15.$$

This would mean that the work is based on the formula

(4) $$V = \left\{ \left(\frac{a+b}{2}\right)^2 + \left(\frac{a-b}{2}\right)^2 \right\} \cdot h,$$

which is indeed wrong, but which agrees with (3).

The "Theorem of Pythagoras" (MKT II, p. 53)

The old-Babylonian text BM 85 196 contains the following nice problem (no. 9):

A patû (beam?) of length 0;30 (stands against a wall). The upper end has slipped down a distance 0;6. How far did the lower end move?

The problem amounts to the consideration of a right triangle, of which are given the hypotenuse equal to $d = 0;30$ and one leg equal to $0;30 — 0;6 = 0;24$. The other side is determined quite properly, using "Pythagoras" and found to be

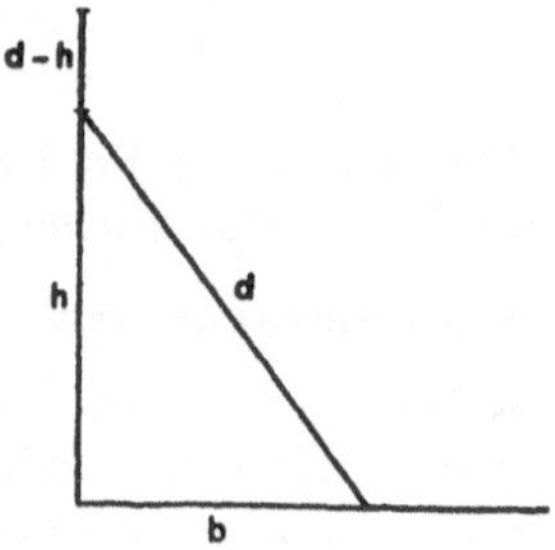

$$b = \sqrt{d^2 - h^2}.$$

Fig. 23.

The faithfulness with which the Babylonians preserved, throughout one and one-half millenia, the tradition of the theorem of Pythagoras, is shown by a text from the era of the Seleucids[1], BM 34 568, in which occurs, among a number of other small problems concerning lengths, widths and diagonals of rectangles, the following little exercise (MKT III, p. 22):

A reed stands against a wall. If I go down 3 yards (at the top), the (lower) end slides away 9 yards. How long is the reed, how high the wall?

[1] i.e. after 310 B.C. The Seleucids were the successor of Alexander the Great.

PLATE 9

PL. 9. BM 85,194, Old Babylonian cuneiform text. This side of the tablet contains 16 problems with solutions. There are problems on dams, walls, wells, water-clocks and excavations. The fourth problem, illustrated by a drawing, is concerned with a circular wall. A part of this problem is treated on p. 68, fourth example. The 14th problem is concerned with a frustrated cone. The volume of the frustrum is determined by multiplying the height by half the sum of the upper and lower areas (see p. 75). *(British Museum)*

PLATE 10

PL. 10. The "seven sages" (*not* Plato teaching geometry). A mosaic from Torre Annunziata near Pompeii, in the National Museum in Naples; first century B.C. From the mosaic one sees how the "seven sages" were honored in Roman times as the representatives of wisdom and science. The distribution of names among the seven figures is uncertain. The third from the left, pointing to the celestial sphere, is identified by some as Thales. The figure to the left of the sun-dial could be the inventor Anaximander (or Solon?).

(Photo Alinari)

The diagram is exactly the same. This time it is given that $b = 9$ and $d \dot{-} h = 3$, while h and d are required. The solution is

$$d = \frac{\frac{1}{2}(9^2 + 3^2)}{3} = 15$$

$$h = \sqrt{d^2 - b^2} = 12$$

The other problems of the same text give d and h, or $d + h$ and b, or again $d + h$ and $d + b$, etc. The last and most complicated problem gives $d + h + b$ and dh. As always, the real difficulty in these problems is algebraic in character. The only geometric property, used over and over again, is the "Theorem of Pythagoras".

It would be wrong to suppose that in all these problems the ratios of width, height and diagonal are $3 : 4 : 5$. One finds other ratios, such as

$$5 : 12 : 13, \qquad 8 : 15 : 17, \qquad 20 : 21 : 29.$$

These examples of right triangles with rational sides, except the last, are also found in the writings of Heron of Alexandria.

Where did the Babylonians get such examples? This question leads us to the domain of the

Babylonian Theory of Numbers.

The Babylonians were not only virtuosos in algebra, they had also gone far in Arithmetic. At the end of his "Mathematische Keilschrift-Texte", Neugebauer expresses the hope that we might get to know quite a great deal of elementary number theory, and he conjectured that we would more properly have to call "Babylonian" many things which the Greek tradition had brought down to us as "Pythagorean".

This hope and this conjecture have been brilliantly realized by the later discovery of the text "Plimpton 322". But let us first summarize what other texts have revealed about Babylonian Theory of Numbers.

Progressions (MKT I, p. 99)

The summing of arithmetic progressions was like rolling off a log for our Babylonian calculators. This is shown by many little problems about the distribution of a sum of money among a number of brothers, according to an arithmetic progression.

In text AO 6484, we also find the summation of a geometrical progression whose ratio is 2:

$$1 + 2 + 4 + \ldots + 2^9 = 2^9 + (2^9 - 1).$$

The same text computes the sum of the squares of the integers from 1 to 10, according to the formula

$$1^2 + 2^2 + \ldots + n^2 = (1 \cdot 1/3 + n \cdot 2/3)(1 + 2 + \ldots + n).$$

The text carries a late date, but nevertheless it resembles the old-Babylonian texts very closely. We still find the solutions carried out step by step for concrete numerical cases. Two problems are concerned with the Pythagorean Theorem, four with systems of equations of the type

$$xy = 1, \qquad x + y = a.$$

These texts and other similar ones indicate, that the tradition of Babylonian algebra was carried on uninterruptedly from the days of the flowering of old-Babylonian science well into the Hellenistic period.

Plimpton 322: Right triangles with rational sides.

We come now to the very remarkable old-Babylonian text "Plimpton 322", discovered and published by Neugebauer and Sachs in their Mathematical Cuneiform Texts (1945).

This text is the right-hand part of a larger tablet with several columns. The last column contains nothing but the numbers 1, 2, . . ., 15. The two preceding columns refer, according to the legend at the head of the columns, to the "width" and the "diagonal"; these numbers do indeed satisfy the relation

$$d^2 - b^2 = h^2,$$

in which the height h is always an integer, having only factors 2, 3 and 5. Undoubtedly these heights were found in an earlier column. The numbers are the following:

h	b	d	number
2, 0	1,59	2,49	1
57,36	56, 7	1,20,25	2
1,20, 0	1,16,41	1,50,49	3
3,45, 0	3,31,49	5, 9, 1	4
1,12	1, 5	1,37	5
6, 0	5,19	8, 1	6
45, 0	38,11	59, 1	7
16. 0	13,19	20,49	8
10, 0	8, 1	12,49	9
1,48, 0	1,22,41	2,16, 1	10
1, 0	45	1,15	11
40, 0	27,59	48,49	12
4, 0	2,41	4,49	13
45, 0	29,31	53,49	14
1,30	56	1,46	15

Where did these numbers come from and by which principle were they ordered? The preceding column throws some light on these questions. For it contains the squares d^2/h^2.

These decrease fairly steadily from nearly 2 to slightly more than $^4/_3$. Subtracting 1, we obtain

$$d^2/h^2 - 1 = (d^2 - h^2)/h^2 = b^2/h^2.$$

It is a plausible conjecture that the column preceding this one, contained the ratios

$$\beta = b/h \text{ and } \delta = d/h.$$

This conjecture can also be formulated by saying that, first, h was taken to be equal to 1. The problem would then be, to construct right triangles, whose sides 1, β, δ are measured by rational numbers, and such that the relation

$$\delta^2 - \beta^2 = 1$$

is satisfied. This condition can be written in the form

$$(\delta + \beta)(\delta - \beta) = 1.$$

This means that the sum $\delta + \beta$ and the difference $\delta - \beta$ are reciprocals. Setting

we find

$$\delta + \beta = \alpha = p/q,$$
$$\delta - \beta = \alpha^{-1} = q/p,$$

so that

$$\delta = \frac{\alpha + \alpha^{-1}}{2} = \frac{p^2 + q^2}{2pq},$$

$$\beta = \frac{\alpha - \alpha^{-1}}{2} = \frac{p^2 - q^2}{2pq}.$$

If now one sets

$$h = 2pq,$$

so as to obtain integers, one finds immediately

$$b = p^2 - q^2, \qquad d = p^2 + q^2.$$

These are the well-known modern formulas for right triangles with integral sides, also used frequently by Diophantus.

The way in which the numbers β and δ were derived here, is a typically Babylonian procedure. For it reduces to the problem of constructing right triangles, of which are given one leg, taken as 1, and the sum of the other two sides, $\beta + \delta = \alpha$. But, we already know that this problem was formulated and solved in the text BM 34 568 (see the discussion of the "Pythagorean Theorem" above).

We find support for our hypothesis that the numbers h, b, d were obtained from 1, β, δ in line 11, on which the numbers 1, β, δ have been preserved. If multiplied by $2pq = 4$, the simpler numbers $h = 4$, $b = 3$, $d = 5$ would have been obtained.

P. Huber has discussed, in L'Enseignement mathématique 3 (1957), p. 19, the errors in Plimpton 322. He succeeded in explaining the errors by assuming that β was calculated from δ by the formula

$$\beta = \sqrt{\delta^2 - 1}.$$

This hypothesis also explains why a column for $\delta^2 = d^2/h^2$ preceeds the columns for b and d.

Problems of this character are very closely related to Greek arithmetic, especially to the part that is traditionally ascribed to Pythagoras and the Pythagoreans. We shall not discuss

Applied mathematics

such as compound interest, problems related to excavations and constructions, to alloys, etc. The technical terms which are involved are rather difficult to understand; moreover, these matters are of little importance for the history of mathematics.

Summary.

I. Algebra and Arithmetic.

The standard forms, which the Babylonians could solve with ease and to which they tried to reduce the solution of all algebraic equations, are the following:

A. *Equations with one unknown.* [1]

(A1)	$ax = b$
(A2)	$x^2 = a$
(A3)	$x^2 + ax = b$
(A4)	$x^2 - ax = b$
(A5)	$x^3 = a$
(A6)	$x^2(x + 1) = a$

B. *Systems of equations with 2 unknowns.*

(B1)	$x + y = a,$	$xy = b$
(B2)	$x - y = a,$	$xy = b$
(B3)	$x + y = a,$	$x^2 + y^2 = b$
(B4)	$x - y = a,$	$x^2 + y^2 = b$

Furthermore, the following formulas were known:

(C1)	$(a + b)^2 = a^2 + 2ab + b^2$
(C2)	$(a + b)(a - b) = a^2 - b^2$
(R1)	$1 + 2 + 4 + \ldots + 2^n = 2^n + (2^n - 1)$
(R2)	$1^2 + 2^2 + 3^2 + \ldots + n^2 = (1/3 + 2/3 \cdot n)(1 + 2 + 3 + \ldots + n);$

and also the summation of arithmetical progressions.

Pythagorean numbers $x^2 + y^2 = z^2$ were found from the formulas

(R3) $\qquad\qquad z/y = \frac{1}{2}(\alpha + \alpha^{-1}), \qquad x/y = \sqrt{(z/y)^2 - 1}$

Compare S. Gandz, Osiris 3 (1938), p. 405 on the absence of the third form of the quadratic equation $x^2 + b = ax$.

II. Geometry.

The Theorem of Pythagoras.
A formula for the line z halving the area of a trapezoid.
The area of a triangle and of a trapezoid.
The area of a circle $3r^2$ (poor approximation).
Perimeter of a circle $6r$ (poor approximation).
Volume of a prism and of a cylinder.
Volume of a frustrum of a cone $\frac{1}{2}(3R^2 + 3r^2)h$ (wrong)
Volume of a frustrum of a pyramid with square bases

(M1) $\frac{1}{2}(a^2 + b^2) \cdot h$ (wrong)

It is possible that in one case the correct formula

(M2) $\left\{ \left(\frac{a+b}{2}\right)^2 + \frac{1}{3}\left(\frac{a-b}{2}\right)^2 \right\} \cdot h$

was used; more probably it was the formula

(M3) $\left\{ \left(\frac{a+b}{2}\right)^2 + \left(\frac{a-b}{2}\right)^2 \right\} \cdot h,$

which is equivalent to (M1).

CHAPTER IV

THE AGE OF THALES AND PYTHAGORAS

Chronological Summary

General History	Philosophers and Historians	Mathematicians and Astronomers
610 B.C. The beginning of the New Babylonian empire 540 B.C. The beginning of the Persian empire	Milesian school: Thales Anaximander Anaximenes	585 B.C. Thales 550 B.C. Pythagoras Anaximander
500 B.C. Ionian revolt 480 B.C. Persian wars 450 B.C. Pericles 420 B.C. Peloponnesian war	Heraclitus The Eleatics 450 B.C. Anaxagoras Herodotus 430 B.C. Atomists 410 B.C. Thucydides	500 B.C. Hippasus 500 B.C.–350 B.C. Pythagoreans Anaxagoras Oenopides 430 B.C. Hippocrates Democritus Theodorus
370 B.C. Epaminondas	Socrates †399 Plato Heraclides of Pontus Aristotle Eudemus	390 B.C. Archytas Theaetetus 370 B.C. Eudoxus Callippus Hicetas 350 B.C. Menaechmus Dinostratus Autolycus
333 B.C. Alexander the Great Hellenism	Stoics	300 B.C. Euclid 280 B.C. Aristarchus 250 B.C. Archimedes 240 B.C. Eratosthenes Nicomedes 210 B.C. Apollonius 150 B.C. Hipparchus
60 B.C. Julius Caesar 1 A.D. Augustus 400 A.D. Migration	Neo-Pythagoreans Neo-Platonists: Proclus	60 A.D. Heron 100 A.D. Menelaus 150 A.D. Ptolemy 250 A.D. Diophantus 320 A.D. Pappus

Hellas and the Orient.

In his posthumous dialogue Epinomis, Plato sketches very strikingly the relation of the Greeks to the old cultures of the Orient, as follows: ". . . whatever Greeks acquire from foreigners is finally turned by them into something nobler" (987 E).

This is also applicable to the exact sciences. The Greeks themselves declare unanimously that they found in Egypt and in Babylon the material for their geometry and their astronomy. Thales and Pythagoras, Democritus and Eudoxus, all of them are reported to have travelled to Egypt and to Babylonia. Even if one does not accept these reports of travels as historical facts, but looks upon them rather as an anecdotal way of saying that Oriental elements were recognized in their theories, even then they prove enough. Only a few of the modern philologists absolutely refuse to recognize that the Greeks have taken anything essential from the Orient. As if the Hellenes were so narrow-minded as not to recognize the elements of value in an alien culture!

It is certainly not accidental that the Ionians were the first torch-bearers of Greek civilization. Did they not dwell near the boundaries of the great oriental empires, had they not, for many years, been subjects of the kings of Lydia and Persia? They had abundant opportunity to get well acquainted with oriental culture.

The close ties which linked Ionia and Asia Minor, politically and economically, become evident to any one who has looked through the first book of the Histories of Herodotus. And the relations between Hellas and Egypt are within easy reach. Numerous Greeks lived in the Nile delta. The Greek city of Naucratis, founded during the reign of Psammetichus (663—609) even received under Amasis (569—525) a trade monopoly for the whole of Egypt.

Less obvious, but nevertheless unmistakable, are the connections with the Assyrian empire. In 709, Sargon II, who expanded this empire to Syria, received presents from 7 city-kings on Cyprus. His successor Sanherib defeated the Ionians after they had landed in Cilicia. Later on, they succeeded however in establishing a trade-center there.

It was a serious blow to Ionian commerce, when Gyges, the founder of the Lydian empire (680—652), put a barrier across the Ionians' access to the hinterland. They attempted to find another route for commercial communication with Babylon. In the middle of the 7th century, they established the commercial towns of Sinope and Trapezus, on the Black Sea, at the terminals of old trade routes with Mesopotamia. Moreover, after the later kings of Lydia, Alyattes and Croesus, had annexed most of the Ionian coastal towns, the route through Asia Minor became again accessible to the Ionians.

In the mean time, important political changes had taken place. The powerful military empire of the Assyrians, which had oppressed the eastern peoples for so many years (consult the Bible!), had crumbled. In alliance with the Medes,

Nabopolassar, king of the Chaldeans, had liberated Babylon from Assyrian domination. In 612 the Medes destroyed Nineveh. When they attempted to advance further to the West, Alyattes of Lydia marched to meet them with Ionian soldiers. The battle at the Halys was brought to a sudden stop by the solar eclipse of 585, which had been predicted by Thales of Milete. "Day turned into night", writes Herodotus. The soldiers were frightened to such an extent, that they were unwilling to continue the fight. A peace was concluded between the contending forces, in which Cilicia and Babylonia were included.

Since that time three great empires maintained a balance of power: Lydia, Media and the "New-Babylonian empire", founded by the kings of Chaldea. Cultural and commercial communications were opened in both directions; the entire hinterland became accessible to the Greeks. The powerful Chaldean Nebukadnezar, king of Babylon (604—562) was not only a great general, but he also nurtured the cultural development of his country. Greek noblemen, for instance the brother of the poet Alcaius, took part in battles under his leadership. A cuneiform text from Uruk, bearing the date 551, mentions copper and iron from Ionia.

A statement of Herodotus (II 109) shows that there was also a cultural exchange with Greece; the Gnomon and the Polos, and the 12 hours of the day came to Greece from Babylon. The Polos was probably a sundial in the form of a hemisphere. The Gnomon is also a sun-dial, consisting of a vertical bar, which throws its shadow on a horizontal disc. There is no doubt as to the correctness of the statement, as far as it relates to the Gnomon, for a cuneiform text exists which contains a table for the lengths of the shadows of the bar at the various times of day. [1] The statement concerning the 12 hours, in which the Greeks divided the day from sunrise to sunset, is also correct, for another text from Nineveh, gives a table for the duration of $^1/_{12}$, $^2/_{12}$, $^3/_{12}$, . . ., $^{12}/_{12}$ of the period of daylight for various times of the year, expressed in terms of astronomical time-units (bêru and uš). [2] The names of the signs of the Zodiac, with which the Greeks became acquainted around the year 550 through Cleostratus of Tenedos, are also derived from Babylon. [3]

The political equilibrium was disturbed again, when Cyrus subjected the entire Orient to Persian domination about the year 540. The Ionian cities, which had come to the aid of his opponent Croesus, had to pay heavy tributes (plate 11). Many Ionians left the country; the Phocaeans, for example, established the town of Elea, in Italy, which was destined to play such an important part in the history of philosophy. It was also at this time the Pythagoras migrated from Samos to Croton. The center of gravity of the world of mathematics and philosophy moved from Ionia to Italy.

[1] See E. F. Weidner, *Ein babylonisches Kompendium der Himmelskunde*, American Journal for Semitic Languages, **40** (1924), p. 198.
[2] See van der Waerden, Babylonian Astronomy III, Journal of Near Eastern Studies **10** (1951) p. 25.
[3] See Weidner, loc. cit., p. 192.

But it was not long before the Persian empire reestablished economic and cultural links with the Greeks. Ionian artisans and artists took part in the construction of the palace of Darius. [1] The sculptor Telephanes of Phocia worked for Darius and for Xerxes. The Greek physician Democedes from Croton and, later on, Histiaeus, the tyrant of Milete, lived at the court of Darius, although not entirely voluntarily.

The great Persian kings Cyrus and Darius (plate 12) were very tolerant; they did not interfere with the cultures and the religions of subject peoples (a reference to the Bible is again in order here). Babylonian stellar rituals continued to exist. The observation of the moon and the planets of the Babylonian priest-astronomers, were continued systematically during the Persian regime. Without these carefully dated observations, the later flowering of Babylonian theoretical astronomy, during the era of the Seleucids, the successors of Alexander the Great [2], would have been impossible. The Greeks also showed much interest in these observations; Callisthenes, a pupil of Aristotle, who accompanied Alexander the Great to Babylon, sent his uncle Aristotle, upon his request, Babylonian observations. [3] Hypsicles, a Greek astronomer of the 3rd century calculates the times of rising and setting of the signs in the Babylonian manner, because Greek geometry of the sphere was not yet able to solve this problem. [4] And in his Isagoge, Geminus discusses a method of the Chaldeans for the calculation of the velocity of the moon. Hipparchus (150 B.C.) makes use of Babylonian observations and periods of the moon, which Ptolemy could use 300 years later, practically without corrections. [2]

From all this we see that even during the period of flowering of their own astronomy, the Greeks were glad to learn from the Babylonians in any respect in which the latter had advanced beyond them. Might this not be applicable as well to the initial period of Greek mathematics, when the Babylonians were already in possession of a highly developed algebra and geometry, which the Greeks had not yet acquired?

The natural point at which fruitful contacts between East and West could take place at the beginning of the 6th century, was the flourishing commercial town of Milete on the coast of Asia Minor, the most important center of Ionian culture. And the first Ionian natural philosopher, the first mathematician and also astronomer, is

Thales of Milete.

Thales was the first of the "seven sages" (plates 10 and 12). In general, these sages were not scholars, but statesmen, lawgivers and moralists. Sayings, such as the celebrated Delphic "Know thyself" were ascribed to them. It was they who,

---

[1] F. W. König, *Der Burgbau zu Susa*, Mitteilungen Vorderasiatische Gesellschaft, 35 (1930) 1.

[2] Compare F. X. Kugler, *Babylonische Mondrechnung* (1900) and *Sternkunde und Sterndienst in Babel* I; O. Neugebauer, *Quellen und Studien* B 4 (1938), p. 193 and p. 407; also A. Pannekoek and B. L. v. d. Waerden in *Eudemus* I (1940).

[3] Simplicii in Arist. *De Caelo Comment.* II 12, p. 506 (Heiberg).

[4] Hypsicles. Anaphoricus, ed. Manitius, Programm Gymnasium Heil. Kreuz, Dresden 1888, Compare O. Neugebauer, *Trans. Am. Phil. Soc.*, 32 (1942), p. 251.

after the collapse of the old feudal system, laid the foundations for the new
political organization of the Greek cities; it is true that this was still aristocratic,
but it was no longer feudal.

But the wisdom of Thales had a more universal, a more philosophical character.
He asked for the origin of things. It is well known that he taught that everything
originates in water.

Plato relates that Thales fell into a well while looking at the stars, and that a
comely Thracian slave-girl laughed at him, saying: "he wanted to know what
happens in the heavens, but he did not observe what was in front of his own feet".

It would be a mistake to draw from this anecdote the conclusion that Thales
was a scholar, alien to this world, some sort of absent-minded professor. On the
contrary, he was at the very center of the intense Ionian life of his time. It is reported that he made a great deal of money in an oil speculation and that he had a
new river bed made to facilitate the crossing of Croesus' army. But he advised
his fellow citizens against an alliance with Croesus, and later on he attempted to
establish a federation of Ionian towns with Teos as the capital.

Another example of his practical wisdom is furnished by his advice to naviga-
tors to depend upon Ursa Minor rather than upon Ursa Major. Thus he was not
only a theoretician and a philosopher, but also a statesman and a man of practical
judgment, a "sage" in the complete ancient sense.

Prediction of a solar eclipse.

Herodotus reports (see p. 84) that, during the battle on the Halys, day was
suddenly turned into night and that Thales had predicted this event to the Delians
for that year. According to Diogenes Laertius, Xenophanes voiced his admiration
of Thales for this prediction. Thus, besides Herodotus, we have the older witness
Xenophanes for this accomplishment. At present it is generally agreed that this
event refers to the solar eclipse of 585 B.C.

How was it possible for Thales, who, according to all our sources, is the first
Greek astronomer, to predict a solar eclipse? Such a feat requires the experience of
more than forty years, no matter how one proceeds. It is not possible for one man
alone to gather this experience. But Thales had no Greek predecessors. The
conclusion is inescapable that he must have drawn upon the experience of Oriental
astronomers.

This points in the first place to the astronomers of Mesopotamia. For we know
from the letters of Assyrian court astrologers of about 700 that these had predicted
solar and lunar eclipses (with varying success) [1].

The prediction of Thales fits in very well with this series of predictions. It is
a fairy tale of modern times that, in making it, Thales had used the "Saros", the
18-year period which had been known to the Babylonians since about 400 B.C.

---

[1] R. C. Thompson, *The reports of the magicians and astrologers of Nineveh and Babylon* (1900).

I prefer to think that Thales, as well as the ancient Babylonians, start from the approximate relation:

51 draconitic lunar periods = 47 synodic months.

According to this relation, the possibility for the repetition of a lunar eclipse exists 47 months after a total lunar eclipse, while the chance of a solar eclipse occurs 23½ months after a total lunar eclipse. Indeed, a considerable lunar eclipse could be seen 23½ months before the eclipse of Thales. [1] Whatever may actually have happened, the prediction of Thales indicates that he was acquainted with Babylonian astronomy.

The geometry of Thales.

Did Thales also know Babylonian mathematics?

The following information concerning Thales is given by Proclus, the commentator of the first book of Euclid's Elements,[2] who obtained it from the History of Mathematics of Eudemus, itself unfortunately lost:

1. He was the first to prove that a circle is divided into two equal parts by its diameter (Proclus, p. 275).

2. Besides several other theorems, he had obtained the equality of the base angles in an isosceles triangle; in ancient fashion, he called these angles not equal, but similar (Proclus, p. 341).

3. According to Eudemus, he discovered that when two straight lines intersect, angles are equal, (Proclus, p. 374).

4. The congruence proposition concerning two triangles, in which a side and two angles are equal, was ascribed by Eudemus to Thales, with the remark that, in order to demonstrate the validity of his method for determining the distance between two ships at sea, Thales had to make use of this congruence theorem (Proclus, p. 409).

How might Thales have determined the distance between ships at sea? According to Tannery, the most ancient method that has been brought down to us is that of the Roman surveyor Marcus Junius Nipsius, who gives the following, indeed very primitive, rule:

In order to find the distance from A to the inaccessible point B, one erects in the plane a perpendicular AC to AB, of arbitrary length and determines its midpoint D. In C one constructs a line CE perpendicular to CA, in a direction opposite to that of AB, and one extends it to a point E, collinear with D and B. Then CE has the same length as AB.

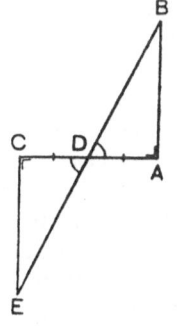

Fig. 24.

The congruence theorem (4), referred to by Eudemus, is indeed used in the

[1] See my paper Voraussage von Finsternissen, Ber. sächs. Akad. Wiss., Leipzig 92 (1940).
[2] Proklus Diadochus, Euklidkommentar, translated by P. L. Schönberger, edited by M. Steck, Halle 1945.

proof of this rule, as well as proposition (3), concerning the equality of vertical angles, also known to Thales. It is therefore possible that this was Thales' method.

Pamphile, as reported by Diogenes Laërtius, says that Thales was the first to construct a circle about a right triangle and that, in honor of this discovery, he sacrificed a bull. Hence, the proposition that an angle inscribed in a semicircle is a right angle, is ascribed to Thales. On the other hand, this proposition is related to certain calculations concerning chords and their apothems, which occur in Babylonian mathematics; but this relation is of course in no way sufficient to prove that Thales knew Babylonian mathematics. The matter must be looked at from a broader point of view, and the statements of Proclus and of Pamphile must be studied more closely.

The absolute accuracy of statements (1) and (4) has been drawn into question, even though they come from the best source. It is thought that the structure of old Thales' mathematics can not have been so strictly logical, that he would undertake to *prove* such obvious things as the equality of the parts into which a diameter divides a circle. Heath, the eminent English historian of Greek mathematics, observes in this connection, that even in Euclid this proposition is not proved.[1] It has been held that the statements of Eudemus should be accounted for in this way, that he incorrectly assumed for the mathematics of Thales the same external structure as had been given to the mathematics of his own time (about 330—300), in which every proposition was derived by strictly logical steps from previous propositions, definitions and axioms. Thus Eudemus might, for instance, have reasoned as follows: Purely logically, the measurement of the distance between ships at sea, depends upon the congruence theorem mentioned in (4); therefore Thales must have known this theorem and have formulated it explicitly. It is thought that, in reality, Thales did perhaps apply this congruence proposition without being aware of it. Some people even believe that Thales did in no way prove his discoveries, but that he established them empirically, the very opposite of what Eudemus explicitly says in (1)!

The first objection to this view is that Eudemus not only knew the results of the mathematics of Thales, but also their external form, at least to a certain extent; he even knew the terminology which Thales used for the equality of angles. We are hardly justified therefore in simply ignoring his judgment that the geometry of Thales was constructed logically in the same way as that of the later mathematicians — and that evidently is his judgment because otherwise he would not have drawn the conclusion (4) — and certainly not his explicit statement that Thales had proved proposition (1). Only on the basis of better knowledge can one fairly correct an antique historian.

Furthermore, the entire criticism of the statements of Eudemus-Proclus, which has just occupied the stage, stands and falls with the view that Thales stands at

[1] T. Heath, *A History of Greek mathematics* I (Oxford), 1921, p. 131.

the beginning of ancient mathematics. The reasoning is as follows: since he was the first, he must have discovered the theorems empirically. But we know now that mathematics does not start with Thales, but, at least 1200 years earlier, in Babylon. This knocks out every resaon for refusing Thales credit for the proofs and for the strictly logical structure which Eudemus evidently attributes to him.

A closer look at the propositions which are ascribed to Thales, shows that, instead of belonging to the early discoveries of mathematics, they form part of a systematic, logical exposition of mathematics. At the start, in the first excitement of discovery, one is occupied with questions such as these: how do I calculate the area of a quadrangle, of a circle, the volume of a pyramid, the length of a chord; how do I divide a trapezoid into two equal parts by means of a line parallel to the bases? These are indeed the questions with which the Egyptian and the Babylonian texts are concerned. It is only later on that the question arises: How do I prove all of this?

This question becomes a central one, especially when the results of the old mathematics, logically unconnected, partly right and partly wrong, are communicated to a younger generation of keenly interested foreigners. At the time of Thales, the Egyptian and the Babylonian mathematics had long been dead wisdom. The rules for computing could be deciphered and shown to Thales, but the train of thought which underlay them, was no longer known. From the Babylonians he might hear that the area of a circle is $3r^2$, while the Egyptians asserted that it is $(^8/_9 . 2r)^2$. How was Thales to discriminate between the exact, the correct recipes for computation, and the approximate, the incorrect ones? Obviously, by proving them, by fitting them into a logically connected system! This is exactly what he did, according to Eudemus and it is exactly at the beginning of such a logical system, that one may expect to find such Irish bulls as: vertical angles are equal, the base angles of an isosceles triangle are equal, a diameter divides a circle into two equal parts, etc.

It follows that we have to abandon the traditional belief that the oldest Greek mathematicians discovered geometry entirely by themselves and that they owed hardly anything to older cultures, a belief which was tenable only as long as nothing was known about Babylonian mathematics. This in no way diminishes the stature of Thales; on the contrary, his genius receives only now the honor that is due to it, the honor of having developed a logical structure for geometry, of having introduced proof into geometry.

Indeed, what is characteristic and absolutely new in Greek mathematics, is the advance by means of demonstration from theorem to theorem. Evidently, Greek geometry has had this character from the beginning, and it is Thales to whom it is due.

The material, from which Greek mathematics was constructed, was not new; the disjecta membra can be dug out of the remnants of the old civilizations. But the style in which the edifice was erected, was new; it bears witness to the clear thinking of the Greeks, the thinking that does not tolerate obscurities, nor doubts as to the correctness of the conclusions that have been acquired.

From Thales to Euclid.

In his commentary on the first book of Euclid's Elements (Friedlein, pp. 65—68), the Neo-Platonist Proclus, who brought a late Indian summer to Plato's Academy, about 450 A.D., gives a rapid survey of the history of geometry from Thales (600 B.C.) to Euclid (300 B.C.):

> *Thales* traveled to Egypt and brought geometry to Hellas; he made many discoveries himself, in many other things he showed his successors the road to the principles. Sometimes he treated questions in a more general manner, sometimes in a more intuitive way. Following him *Mamercus*, brother of the poet Stesichorus, occupied himself with geometry; and Hippias of Elis says that he acquired reputation as a geometer.
>
> *Pythagoras*, who came after him, transformed this science into a free form of education; he examined this discipline from its first principles and he endeavoured to study the propositions, without concrete representation, by purely logical thinking. He also discovered the theory of irrationals (or of proportions) and the construction of the cosmic solids (i.e. of the regular polyhedra). After him *Anaxagoras* of Clazomenae dealt with many questions in geometry, and so did the somewhat younger *Oenopides* of Chios, also mentioned by Plato as famous mathematicians in the Rivals. Later on, *Hippocrates* of Chios, who discovered the quadrature of the lunules, and *Theodorus* of Cyrene, became famous geometers. Hippocrates was indeed the first of whom it is recorded that he compiled Elements.
>
> Through his enthusiastic interest, *Plato*, who came after them, aided a great deal in the development of geometry and of the other mathematical disciplines; it is indeed well known that he filled his writings with mathematical arguments and that he used every opportunity to arouse admiration for mathematics among those who were devoting themselves to philosophy. In his time lived also *Leodamas* of Thasos, *Archytas* of Taras and *Theaetetus* of Athens, who increased the number of theorems and arranged them in a more scientific system.
>
> Younger than Leodamas were *Neoclides* and his pupil *Leon*, who added many things to what was known before them. Thus Leon was able to develop Elements, better prepared as to the number of propositions and the use of proved propositions; and he could formulate restrictions as to the possibility or impossibility of solving a given problem.
>
> *Eudoxus* of Cnidos, a little younger than Leon, on terms of friendship with Plato's circle, enlarged for the first time the number of so-called general theorems, joined three more to the three mean proportionals and continued the researches on the section, begun by Plato, making use of analysis. *Amyclas* of Heraclea, one of Plato's friends, *Menaechmus*, a pupil of Eudoxus and a member of Plato's circle, and his brother *Dinostratus*, perfected geometry still further. *Theudius* of Magnesia was reputed to be excellent in mathematics, as well as in the other sciences, because he put together Elements admirably and succeeded in generalizing special propositions. And *Athenaeus* of Cyzicus, who lived in the same period, became famous in other parts of mathematics, but especially in geometry. These men assembled in the Academy and conducted their investigations in common.
>
> *Hermotimus* of Colophon continued the investigations, started by Eudoxus and Theaetetus; he discovered many of the propositions of the Elements and developed a part of the theory of geometrical loci. *Philippus* of Mende, one of Plato's pupils and led by him to an interest in mathematics, not only carried out researches in accordance with Plato's direc-

tions, but also undertook to do things which could, in his judgment, contribute to Plato's philosophy.

It is up to this point that the history of this science was carried by those who recorded events. Not much younger than these men is *Euclid* who composed the Elements, in which he collected many of the discoveries of Eudoxus, completed many of the results of Theaetetus, and supplied irrefutable proofs of the things which had not been proved strictly by his predecessors. This man lived in the days of Ptolemy I. For Archimedes, who came immediately after Ptolemy I, mentions Euclid. It is also reported that at one time, Ptolemy asked him whether there was not a shorter route through Geometry than by way of the Elements, and that he replied, that there is no royal road to geometry. Thus he was younger than the pupils of Plato, and older than Eratosthenes and Archimedes; for, as stated somewhere by Eratosthenes, these were contemporaries.

This "Catalogue of geometers" is obviously largely an extract from Eudemus' History of Mathematics. For who could have been meant, except Eudemus, by "those who recorded events", shortly after Philippus of Mende and before Euclid?

Eudemus is an excellent source. But unfortunately, the excerpt contains important additions which can not have been taken from Eudemus. For example, the passage at the end, about Euclid, does certainly not come from Eudemus. It is also striking that the entire catalogue bears the stamp of a Platonist, who is anxious to give fullest possible recognition to the merits of Plato and of his school. What is said about Plato and his pupil Philippus of Mende, obviously does not come from the shop of Eudemus, the pupil of Aristotle, who was not as fervent an admirer of Plato as the Neo-Platonist Proclus. On the other hand, Democritus is not mentioned at all, although he also played an important role in the development of geometry (he wrote several mathematical works and, according to Archimedes, he found the volume of the pyramid). This is probably due to the fact that Plato considered the influence of Democritus pernicious and would best have liked to burn his works. This indicates that, as a whole, the Catalogue serves a special interest.

The statements about Pythagoras must also be viewed in that light. It is possible that what Proclus (or whoever wrote the Catalogue before him) found in Eudemus concerning Pythagoras, did not satisfy him and that he went to other, more doubtful sources (e.g. to the Neo-Pythagorean Iamblichus, a fanciful and muddle-headed writer) for supplementary material, in order to be sure not to undervalue the famous and honored Pythagoras. In the absence of knowledge of the sources, the value of these statements is very doubtful.

In summary, we can say: What the Catalogue says about Pythagoras is unreliable, what it tells about Plato is not new; but the references to real mathematicians before Euclid, who were not also philosophers or legendary figures, must stem from Eudemus and therefore merits our confidence.

Unfortunately, not much is said about these mathematicians. We must therefore try to find better sources of information concerning the development of geometry after Thales.

Concerning the sixth century, we are restricted to scattered notices in various later writers. For the fifth century, there is moreover an important supplementary fragment of Eudemus, concerning the lunules of Hippocrates, which will be discussed when we come to Hippocrates of Chios. Relative to the fourth century, there is another fragment of Eudemus about the duplication of the cube by Archytas, and, furthermore, the mathematical passages in the writings of Plato and Aristotle.

Moreover, the Elements of Euclid are an extremely important source, especially for the mathematics of the Pythagoreans and of the contemporaries of Plato. It will be seen that the manner in which these Elements have been brought together from a variety of separate fragments, leads to important conclusions concerning the origin of these fragments. The reader who wishes to become acquainted with the special character of Greek mathematics, is stronlgy advised to read the Elements for himself, in Greek or in one of the excellent modern translations. [1]

The other sources will now be discussed in chronological order.

Pythagoras of Samos.

At the present time Pythagoras is thought of primarily as a mathematician. In Amsterdam a street is named after him, in the neighbourhood of Archimedes, Newton and Copernicus; his name immediately makes one think of the famous "Theorem of Pythagoras". (plate 13)

It was quite different in antiquity. Herodotus calls him "an important sophist", i.e. teacher of wisdom; and he relates that Zalmoxis, the saint of the Getae, who had, according to a legend, risen from the dead, had been a pupil of Pythagoras. He also tells that the Pythagoreans did not bury their dead in woollen clothing. [2] This looks more like religious ritual than like mathematics. The Pythagoreans, who were held up to ridicule on the stage, were presented as superstitious, as filthy vegetarians, [3] but not as mathematicians.

Pythagoras himself was looked upon by his contemporaries in the very first place as a religious prophet. The much-traveled poet-singer Xenophanes, for example, pokes fun at the Pythagorean doctrine of the transmigration of souls. He tells a story, in which Pythagoras, coming across a scene in which a little dog was being thrashed, said: "Stop the beating, for in this dog lives the soul of my friend; I recognize him by his voice". [4]

Pythagoras was also known as a performer of miracles. All kinds of wonderful tales concerning him were in circulation, as, e.g., that the calf of one of his legs was

[1] In English, with extensive commentaries: T. Heath, *The thirteen books of Euclid's Elements*, Cambridge, 1908 (2nd edition, 1926). In German: C. Thaer, *Ostwald's Klassiker der exakten Wissenschaften* 235, 236, 240, 241 and 243 (1933—37). The two volumes in Dutch of E. J. Dijksterhuis, *De Elementen van Euclides* (Hist. Bibl. ex. wet, I and III 1929—30) give a summary of the text and important commentaries.
[2] See e.g., Diels, *Fragmente der Vorsokratiker*, Pythagoras A 1—2.
[3] See, e.g. Diels, *Fragmente der Vorsokratiker*, Pythagoreische Schule E.
[4] H. Diels, *Fragmente der Vorsokratiker*, Xenophanes B 7.

PLATE 11

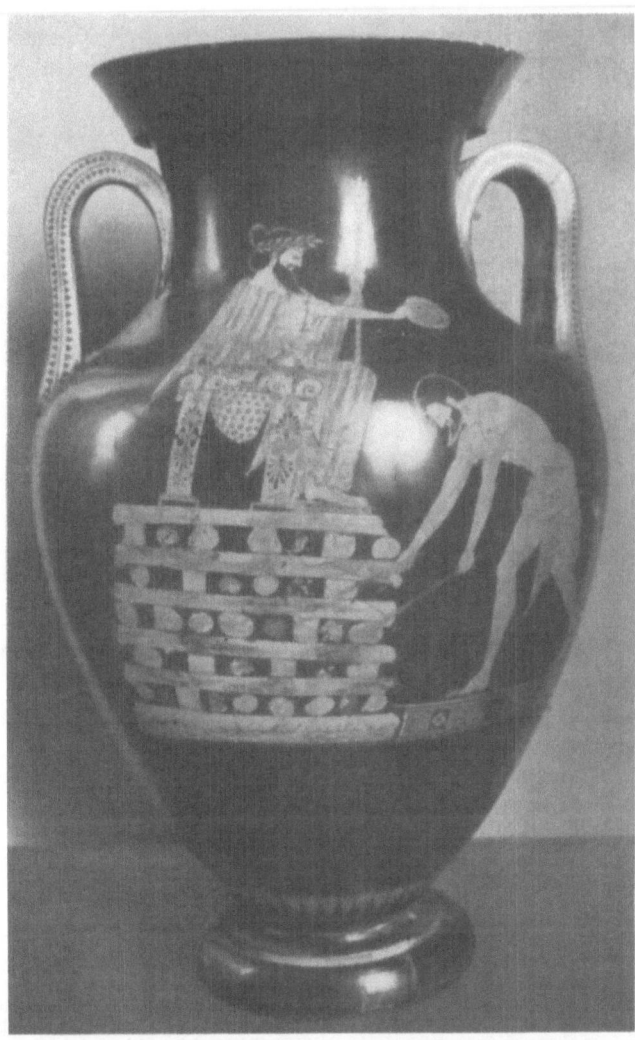

PL. 11. Croesus on the pyre. Attic Amphora, with figures in red, (500–475 B.C.), from Vulci, Italy; Louvre, Paris
The names of the king (Κροεσος, sic!) and of his slave (Ἐυθυμος: Cheerful) are written on the vase. The slave
holds two burning torches, with which he lights the pyre. While the light red-brown figures and pyre represent
the original surface, being surrounded by the shiny black background, and the inner drawing and the contours
have been indicated with black lines, the artist painted the flames blazing up from the pyre and also the burning
torches with thin paint over red and black. Croesus, the last king of Lydia (560–546 B.C.) was ruined by the
oncoming power of the Persians. According to Herodotus, his life was spared by Cyrus, the king of the Persians;
in one of his odes, Bacchilides relates that Croesus ended his own life. This is the version of the saga followed
by the painting on this vase. The technique of representing figures in red began about 530 B.C. (cf. Plates 7, 17).
The vase is a splendid specimen of the severe style which dominated Greek art during the first half of the 5th century B.C. The archaic profuseness of ornamentation has been abandoned, but the very decorative effect of the
ornaments and of the graceful folds recall the archaic art.

PLATE 12

Pl. 12a. King Darius of Persia as conqueror. Relief on the rock of Behistun (520/19 B.C.). In the center above the nine conquered liar-kings (the nine leaders of the revolt after the death of Cyrus) stand forth the symbol of Ahuramazda.

(British Museum)

Pl. 12b. Portrait of a philosopher, probably Thales of Milete (632–546 B.C.). Marble herma (Ny Carlsberg Glyptothek, Copenhagen). Roman copy from the 2nd century A.D. of a famous original, difficult to date. The name Thales is merely a plausible hypothesis. According to K. Schefold, the palm branch on the herma is an indication for Thales, because the poet Callimachus said of him: "The victory is to Thales". It can certainly not be said that we have here an authentic portrait of the philosopher. It is an "image" of the sage of which the style indicates that is is not earlier than the 4th century B.C.

of gold, and that he was seen at two places at the same time. When he crossed a small stream, the river rose out of its bed, greeted him and said: "Hail, Pythagoras." [1]

Was Pythagoras a mathematician, a philosopher, a prophet, a saint or a charlatan? He had some of the qualities of each of these. For his adorers he was the personification of the highest divine wisdom, but Heraclitus spoke of "a lot of knowledge without intellect". [2] He preached the immortality of the soul, he established a strict regime for his followers and founded a brotherhood of believers, the order of Pythagoreans, which later on spread from Croton to a number of Greek cities in Italy and which seems to have played an important role in the political life of these cities. [3] After a testing period and after rigorous selection, the initiates of this order were allowed to hear the voice of the Master behind a curtain; but only after some years, when their souls had been further purified by music and by living in purity in accordance with the regulations, were they allowed to see him. This purification and the initiation into the mysteries of harmony and of numbers, would enable the soul to approach the Divine and thus to escape the circular chain of rebirths. [4]

In this period there were many mystery-rites, which promised their followers eternal life. Orphic prophets roamed through Italy and Greece. The cult of Dionysus swept men and women into its wild ecstasy. A more tranquil way to attain immortality existed through initiation into the mysteries of Demeter and of Persephone in Eleusis.

All such initiations began with ritual purifications. After the soul had been freed from earthly blemishes, it could aspire to unity with the Divine. There was a rebirth in God, and thus eternal life was attained.

The Pythagoreans thus have purification and initiation in common with several other mystery-rites. Ascetic, monastic living, vegetarianism and common ownership of goods occur also in other sects. But, what distinguishes the Pythagoreans from all others, is the road along which they believe the elevation of the soul and the union with God to take place, namely by means of mathematics. Mathematics formed a part of their religion. Their doctrine proclaims that God has ordered the universe by means of numbers. God is unity, the world is plurality and it consists of contrasting elements. It is harmony which restores unity to the contrasting parts and which moulds them into a cosmos. Harmony is divine, it consists of numerical ratios. Whosoever acquires full understanding of this number-harmony, he becomes himself divine and immortal.

Music, harmony and numbers — these three are indissolubly united according

---

[1] For these legends, compare especially Is. Lévy, *Les sources de la légende de Pythagore*, Paris, 1926.

[2] Diels, *Fragmente der Vorsokratiker, Heraclitus* B 40.

[3] A. Delatte, *Essai sur la politique Pythagoricienne*, Liège 1922; K. v. Fritz, *Pythagorean Politics in Southern Italy*, New York, 1940.

[4] For the religious philosophy of Pythagoras and his followers, see P. Boyancé, *Le culte des muses chez les philosophes grecs*, Paris 1936.

to the doctrine of the Pythagoreans. All three are among the essential elements of the Pythagorean system of education and of its path for the elevation of the soul. According to Heraclides of Pontus, Pythagoras said that "Beatitude is the knowledge of the perfection of the numbers of the soul". Mathematics and number mysticism mingle fantastically in the Pythagorean doctrine. Nevertheless, it was from this mystical doctrine that the exact science of the later Pythagoreans developed.

The travels of Pythagoras. [1]

Mystery-rituals and number mysticism come from the orient. For this reason it is not surprising that, even in antiquity, it was believed that Pythagoras had made extensive journeys to practically all Oriental countries. The orator Isocrates, whom many later writers copy, says that he journeyed to Egypt. The story then relates that, in Egypt, Cambyses, the Persian conqueror, made him a prisoner and carried him to Babylon. Here the priests initiated him into the mysteries; according to Iamblichus, he remained there for seven years, during which time he learned from the Magi the theory of numbers, the theory of music and the other sciences. Iamblichus (Introductio in Nicomachi Arithm.) also reports that he learned from the Babylonians the "golden proportionality"

$$A : H = R : B,$$

in which H and R are, respectively, the harmonic and the arithmetic mean of A and B. [2] Older writers merely mention, that he came in contact with "Zaratas, the Chaldean" (i.e. Zarathustra) or that he received lessons from the Chaldeans.

Of how much value are these various statements? There was a time when everything was believed without any hesitation. This was followed by a hypercritical tendency, which pushed all these later tales aside as being entirely unreliable fantasies.

At the present time, since the work of Delatte and Rostagni, the late-antique tradition is again given greater credit. It has been recognized that these accounts rest, to a large extent, upon an old tradition recorded by Aristotle and his pupil Aristoxenus, as well as by historians like Dicaearchus and Timaeus of Tauromenion towards the close of the fourth century. At all accounts, the travel records show that, already in antiquity, Pythagorean and Oriental wisdom were considered as being related. [3]

---

[1] A good survey of the travel accounts, with indication of the sources, is found in Th. Höpfner, *Orientalische und griechische Philosophie, Beihefte zur alten Orient* 4 (1925).

[2] This proportionality plays an important role in the Pythagorean theory of music (see my article in *Hermes* 78). The two means may be defined (in modern notation) by

$$R = \frac{A + B}{2}, \qquad H = \frac{2AB}{A + B},$$

or (in classical form) by:

$$R - A = B - R, \qquad (H - A) : A = (B - H) : B.$$

It is easy to prove the golden proportionality $A : H = R : B$ by use of either of these definitions.

[3] Frequently Pythagoras is represented wearing an oriental turban.

In our further account we shall have occasion to point out a number of instances of the connections between Babylonian mathematics and the science of the Pythagoreans. I am convinced that it was Pythagoras himself who transmitted Babylonian scholarship; did he, as an Ionian, not stand much closer to the source of this old wisdom than the Pythagoreans in Italy?

Not only in pure mathematics, but also in the theory of music and in astronomy, Pythagoras must have learned quite a few things from the Babylonians. Did Heraclitus not speak of "a lot of knowledge without intelligence"? This contemptuous remark cannot refer to a logically constructed theory of numbers and a geometry such as we find in the writings of the later Pythagoreans. But, if Pythagoras gathered into one lump, all kinds of half-assimilated learning about the gods and the stars, about musical scales, sacred numbers and geometrical calculations, and proclaimed such an omnium-gatherum to his followers as divine wisdom in a prophetic manner, then Heraclitus' ridicule, as well as the veneration of mystics, such as Empedocles, become entirely understandable.

Pythagoras and the theory of harmony. [1]

When a string or a flute is shortened to half its length, the tone produced is an octave higher. Similarly, the ratios 3 : 2 and 4 : 3 correspond to the intervals of the fifth and the fourth. It was of eminent importance for the Pythagoreans to have learned that the most important consonant intervals could be obtained in this manner by ratios of the numbers 1, 2, 3, 4; it confirmed their general thesis "Everything is number", or "Everything is ordered by numbers". The numbers 1, 2, 3, 4 themselves constituted the famous "tetractys". A very old saying runs: "What is the oracle of Delphi? The tetractys! For it is the scale of the sirens".

Geometrically the tetractys was represented by the "perfect triangle", arithmetically by the "triangular number" $1 + 2 + 3 + 4 = 10$. According to Lucian, Pythagoras asked some one to count; when he had said 1, 2, 3, 4, Pythagoras interrupted him as follows: "Do you see? What you take to be 4, is 10, a perfect triangle, and our oath". For the Pythagoreans pledged themselves by oath to "him, who had entrusted to our soul the tetractys, the source and the root of eternal nature". All these sayings and these forms of oaths are characteristic of antiquity; it seems therefore necessary to ascribe to Pythagoras himself the Tetractys, the triangular numbers and the numerical ratios of the consonant intervals.

Fig. 25.

On his deathbed, Pythagoras urged his followers to practice on "the monochord". Gaudentius records the following history of this instrument: Pythagoras stretched a string over a straight edge and divided it into twelve parts. When he shortened the string from 12 to 6, or to 8 or 9, i.e. in the ratios 2 : 1, 3 : 2 or 4 : 3, he obtained tones which were an octave, a fifth or a fourth higher.

[1] See B. L. v. d. Waerden, *Die Harmonielehre der Pythagoreer*, Hermes 78 (1943).

The same 4 numbers 6, 8, 9, 12 are found again in practically all Pythagorean and Neo-Pythagorean writers on music. All these writers obtain the middle terms 9 and 8 as the arithmetic and the harmonic means of the extreme terms 12 and 6. Usually the number 12 was assigned to the highest note and 6 to the lowest note, these numbers being not proportional to the lengths of the strings, but inversely proportional. What was the empirical significance of these numbers? Apparently the Pythagoreans did not care very much whether they represented the lengths of the strings, or their tensions or velocities. The most important thing was that the correct ratios of the harmonic intervals appeared, such as $12 : 9 = 8 : 6$ for the fourth and $12 : 8 = 9 : 6$ for the fifth, as the Master had taught.

The tradition which credits Pythagoras with the calculation of the intervals of the diatonic scale, the whole tone $(9 : 8)$ and the major semi-tone or the "leimma" $(256: 243)$, also merits confidence, for these ratios can be obtained by successive division from those for the octave $(2 : 1)$, the fifth $(3 : 2)$ and the fourth $(4 : 3)$:

$$3/2 : 4/3 = 9/8,$$
$$4/3 : 9/8 = 32/27,$$
$$32/27 : 9/8 = 256/243.$$

Pythagoras and the Theory of Numbers.

As magic and number magic belong together, so do mysticism and number mysticism. Every magician utilizes the magic power of words and of numbers, every superstitious person knows sacred numbers, lucky numbers, etc. These things had of old played an important role among the Babylonians and the Magi, and among the Pythagoreans as well. For example, they called 10 a perfect number, they looked upon even and odd as the roots of all things; and, as Aristotle informs us[1], abstract concepts such as "justice" were identified with definite numbers. The even numbers were called feminine, the odd ones masculine, and the number 5, the sum of the first masculine and the first feminine number, was taken as a symbol of marriage, and so forth.

The following example, taken from Plutarch (Isis and Osiris 42) has some mathematical interest:

> The Pythagoreans also have a horror for the number 17. For 17 lies exactly halfway between 16, which is a square, and the number 18, which is the double of a square, these two being the only two numbers representing areas, for which the perimeter (of the rectangle) equals the area.

Interpretation: Let x and y be integers which measure the sides of a rectangle and let the area, the "plane-number" xy be equal to the perimeter

$$xy = 2x + 2y.$$

Then one can express the unknown y in terms of x, as is done in the Babylonian

[1] Aristotle, *Metaphysics* A 5; see also Diels, *Fragmente, Pythagoreische Schule* B 4.

lesson text AO 6770 (see p. 74):

$$y = \frac{2x}{x-2} = 2 + \frac{4}{x-2}.$$

In order that y be an integer, x—2 must be a divisor of 4, so that we must have

	$x - 2 = 1$,	$x = 3$,	$y = 6$	$xy = 18$,
or	$x - 2 = 2$,	$x = 4$,	$y = 4$,	$xy = 16$,
or	$x - 2 = 4$,	$x = 6$,	$y = 3$,	$xy = 18$.

This gives the two possibilities mentioned by Plutarch.

Neo-Pythagoreans, like Nicomachus of Gerasa (100 A.D.) and Iamblichus (300 A.D.) revel in this kind of number-mysticism. In his Arithmetic Theology, Iamblichus expatiates broadly on the mystical and divine significance of numbers. The purpose of the layman's Introduction to Arithmetic of Nicomachus [1], is to explain in a manner intelligible to every one, the wonderful and divine properties of numbers. In a very entertaining way he tells about triangular numbers, about square, rectangular and polygonal numbers, about gnomonic numbers and spatial numbers, about ratios and the distinction between multiple and epimoric [2] ratios, etc., about prime numbers, about geometric progressions, all illustrated by numerous examples, but never accompanied by proofs. He knew his public; he was well aware of the fact that his readers wanted to be initiated into the mystery of numbers, but that they would be bored by prosaic proofs, which would moreover deprive these things of a large part of their mystery.

Another source for the Pythagorean theory of numbers is found in the three arithmetical books of Euclid (Books 7, 8 and 9 of the Elements). But these are purely scientific, there is nothing left of the mystery, everything is carefully and neatly proved. Although Nicomachus lived four centuries after Euclid, he makes nevertheless a much more primitive impression. He is much closer to the original number-mysticism of Pythagoras and his school. It looks to me as if Pythagoras, the Prophet, clothed his wisdom in a mysterious oracular form, and that only much later, the Pythagoreans made the theory of numbers into an exact science.

For this reason we use Nicomachus as our source for the theory of numbers of Pythagoras and of his immediate followers. In the next chapter, in which we shall be concerned with the mathematics of the later Pythagoreans, we shall draw chiefly on Euclid.

Perfect numbers.

It was considered as something very remarkable by the Pythagoreans, when a number equals the sum of its proper divisors, such as

$$6 = 1 + 2 + 3.$$

---

[1] English translation with extensive commentary by Martin Luther d'Ooge, New York 1926.

[2] The word "epimoric" is introduced as the English equivalent of the Greek ἐπιμόριος (superparticularis). Two numbers, A and B, are in "epimoric ratio", if the diffference $B — A$ is a part of A and of B, i.e., if A and B are multiples of $B — A$. E.g., the ratio of $n + 1$ to n is an epimoric ratio. In the discussion of proportionalities in the next Chapter (see p. 110), we shall see that Archytas proved that every epimoric ratio can be reduced to the form $(n + 1) : n$.

They called such numbers "perfect". Nicomachus gives the four instances 6, 28, 496 and 8128. He also gives the following general rule, which is proved in Euclid (Elements IX 36): When the sum

$$1 + 2 + 2^2 + \ldots + 2^n = p$$

is a prime number, then $2^n p$ is a perfect number. For example, $1 + 2 + 4 = 7$ is a prime number, hence $4 \times 7 = 28$ is a perfect number.

The proof makes use of the formula for the sum of a geometric progression

$$1 + 2 + \ldots + 2^{n-1} = 2^n - 1,$$

which is already found in Pythagorean texts. It is therefore very probable that Pythagoras knew this formula.

The numbers mentioned by Nicomachus are $2(2^2 - 1)$, $2^2(2^3 - 1)$, $2^4(2^5 - 1)$ and $2^6(2^7 - 1)$. The next perfect number is $2^{12}(2^{13} - 1)$, the largest known one is $2^{126}(2^{127} - 1)$. For further details, see L. E. Dickson, History of the Theory of Numbers, I.

Similar interest is attached to

Amicable numbers

like 220 and 284, each of which equals the sum of the proper divisors of the other:

$$1 + 2 + 4 + 5 + 10 + 11 + 20 + 22 + 44 + 55 + 110 = 284,$$
$$1 + 2 + 4 + 71 + 142 \qquad\qquad\qquad\qquad = 220.$$

When Pythagoras was asked what a friend is, he said "a second I" and he mentioned the amicable numbers 284 and 220. The Babylonians also enjoyed such tricks based on the interchange of numbers.

In Nicomachus one finds many more links with Babylonian arithmetic than in Euclid. He pays especial attention, just as the Babylonians did, to the squares n^2 and the cubes n^3, and he considers, among the parallelopiped numbers, abc, especially those of the form $n^2(n \pm 1)$, of which the Babylonians had constructed tables.

This leads us to one of the favorite topics of Nicomachus:

Figurate numbers.

He knew triangular numbers, square numbers n^2, rectangular numbers $n(n+1)$,

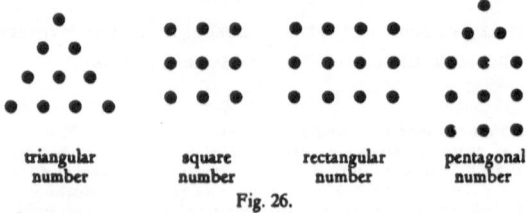

| triangular number | square number | rectangular number | pentagonal number |

Fig. 26.

pentagonal numbers, etc.

The marvellous properties of these numbers may be read in Nicomachus (in the translation of d'Ooge, see p. 97). They follow from the summation of simple arithmetical progressions such as

(1) $1 + 2 + 3 + \ldots + n = \frac{1}{2}n(n + 1)$ (triangular number),
(2) $1 + 3 + 5 + \ldots + (2n - 1) = n^2$ (square number),
(3) $2 + 4 + 6 + \ldots + 2n = n(n + 1)$ (rectangular number),
(4) $1 + 4 + 7 + \ldots + (3n - 2) = \frac{1}{2}n(3n - 1)$ (pentagonal number).

For example, to deduce (2) from a diagram, one observes that a square may be divided into a smaller square and a carpenter's square or "gnomon". By repeating this division, one concludes that the number of dots in a square is the sum of "gnomonic numbers", i.e. of odd numbers $1 + 3 + 5 + \ldots$

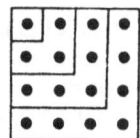

Fig. 27.

Less immediate, but nevertheless also a consequence of the formula for the sum of an arithmetic progression, is the following rule for the sums of one, two, three... successive odd numbers, found in Nicomachus:

$$1 = 1^3$$
$$3 + 5 = 2^3$$
$$7 + 9 + 11 = 3^3$$

From this, one obtains readily a rule for the sum of cubes, known to the Roman surveyors[1] (undoubtedly from a Greek source)

$$1^3 + 2^3 + 3^3 + \ldots + n^3 = \{\tfrac{1}{2}n(n + 1)\}^2.$$

An analogous formula for $1^2 + 2^2 + \ldots + n^2$ we have already encountered in a Babylonian text.

Pythagoras has been given credit for a rule for determining numerical solutions of the indeterminate equation

(5) $x^2 + y^2 = z^2,$

in the form

$$x = \tfrac{1}{2}(m^2 - 1), \qquad y = m, \qquad z = \tfrac{1}{2}(m^2 + 1).$$

where m is an odd integer (Proclus in Euclid I, p. 487).

It can be verified by means of a diagram that these values do indeed satisfy (5). To a square whose side is $x = \frac{1}{2}(m^2 - 1)$, we add a gnomon of width 1, thus obtaining a square of side $z = \frac{1}{2}(m^2 + 1)$. The area of the first square, of the gnomon and of the final square are equal exactly to x^2, y^2 and z^2, so that (5) follows. The special case $m = 3$ leads to the right triangle whose sides are 3, 4, 5, discovered by Pythagoras, according to Vitruvius. We have already observed that the Babylonians knew a more general formula for the determination of right triangles with integral sides.

It is hardly thinkable that all these coincidences between the Babylonian and the

[1] See M. Cantor, *Geschichte der Mathematik* I, p. 559.

Pythagorean mathematics are only a matter of chance. Apparently Pythagoras has taken from the Babylonians various remarkable things about numbers and about their mystical significance. As we will see in the next chapter, his disciples continued the investigation in a more systematic manner and built them into a logically consistent system.

Pythagoras and Geometry.

Who does not think, when he hears the name Pythagoras, of the famous theorem about the square of the hypothenuse? Alas, the proofs of the connection between the two are extremely doubtful. [1] In his commentary on this proposition (Euclid I 47), Proclus says, very indefinitely "If we listen to those who wish to recount ancient history, we may find some of them referring this theorem to Pythagoras, and saying that he sacrificed an ox in honor of the discovery". Plutarch quotes a distich: "When Pythagoras discovered his famous figure, for which he sacrificed a bull", and he says that the figure in question is either that of the square on the hypothenuse or that of the adaptation of areas. [2] In another place, the same Plutarch says however, that the bull was sacrificed in connection with the problem of constructing a figure of the same area as another figure and similar to a third one. [3] But Vitruvius is of the opinion that the bull fell victim to the discovery of the right triangle whose sides are 3, 4, 5.

But the entire story is an impossible one, because Pythagoras was strongly opposed to the killing and sacrificing of animals, of cattle especially.

It has been thought, that the theorem can not have been known in the days of Pythagoras, during the first stages of the development of geometry. But this objection loses force now that we have found it applied even 1200 years earlier in the cuneiform texts. It is quite possible that Pythagoras became acquainted with the theorem in Babylon. That is about all we can properly say about it.

Proclus' catalogue also ascribes to Pythagoras the construction of the regular polyhedra. But a scholium in the thirteenth book of the Elements says that the Pythagoreans knew only three regular polyhedra, viz. the cube, the tetrahedron and the dodecahedron, and that it was Theaetetus who first discovered the other two. Precisely because this scholium directly contradicts the tradition which used to ascribe to Pythagoras anything that came along, it is given greater credit nowadays than the catalogue. An Etruscan dodecahedron, made of soapstone, was found near Padua, dating from before 500 B.C. [4] It is therefore quite possible that Pythagoras was acquainted with the cube, the tetrahedron and the dodecahedron.

The faces of a dodecahedron are regular pentagons. The diagonals of such a pentagon form a star-pentagon.

[1] Collected by Allman, *Greek geometry from Thales to Euclid* (1889), p. 26.
[2] Non posse suaviter vivi sec. Epicurum IX.
[3] Plutarch, *Quaestiones Convivii*, VIII, Quaest. 2, 4.
[4] See F. Lindemann, *Sitzungsber. Bay. Akad. Wiss.* 26 (1897), pp. 625—768.

This figure, the symbol of health, served as a distinctive mark among the Pythagoreans. One of them, lying on his deathbed in a foreign country and unable to pay the man who had taken care of him until the end, advised this man to paint the star-pentagon on the outside wall of his house, so that any Pythagorean who might ever pass the house, would make inquiries. As a matter of fact, a Pythagorean did come past many years later, and the man was richly rewarded.

For the construction of the star-pentagon, the Pythagoreans could make use of the fact, that each of these 5 lines divides every other one in mean and extreme ratio, i.e. that the shorter piece $AK = a - x$

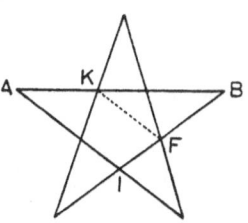

Fig. 28. Pentagramma.

is to the larger one $KB = x$ as the larger piece is to the whole segment $AB = a$. The correctness of this statement follows at once from the similar triangles AIB and KFB. The proportionality $BF : BK = BI : BA$, or

$$(a - x) : x = x : a,$$

leads to the quadratic equation

$$x^2 = a(a - x),$$

which the Pythagoreans knew well how to solve by their method of "adaptation of areas", to be discussed in the next chapter; we are already acquainted with the fact that the Babylonians also knew how to solve this equation. Again we do not know whether the Pythagoreans actually constructed the star-pentagon in such a way; they certainly had the knowledge for doing it.

The star-pentagon is also found on Babylonian drawings, another point of contact between Babylon and the most ancient Pythagorean mathematics.

Little more is to be said about Pythagorean geometry. We know still less about

The astronomy of the Pythagoreans.

According to Alexander Polyhistor [1], Pythagoras assumed the earth to be spherical in shape, placed at the center of the cosmos, and he knew the proper motions of sun, moon and planets to be opposite to the diurnal motion of the fixed stars.

By no means can we put unlimited faith in the compilations of Alexander, which he derives from certain "Pythagorean hypomnemata". They contain among other things, the statement that the numbers are produced from the unit and the indefinite duality. But Aristotle asserts explicitly, that this idea is Platonic and not Pythagorean. The Pythagoreans generated the numbers from the unit and the unlimited, and Plato replaced the unlimited by the indefinite duality of "large and

[1] See Diogenes Laërtius VIII 25. Compare J. E. Raven, *Pythagoreans and Eleatics* (Cambridge 1948), p. 160, concerning this much-discussed fragment.

small" (Aristotle, Metaphysics A 6). The compilation on which Polyhistor draws, was therefore written after Plato and by some one who was not able to discriminate between the doctrine of Pythagoras and that of Plato. Furthermore, as becomes clear from his terminology, the writer was also influenced by the ideas of the Stoa. Certainly not a reliable witness!

Nonetheless, the statement concerning the proper motions can be credited, particularly since Alcmaeon, a little younger than Pythagoras, with ideas closely related to those of Pythagoras, also speaks of the motion of the planets from West to East, contrary to that of the fixed stars. [1] Indeed, this view is already found among the Babylonians.

Summary.

Taking it all in all, we know something about the theory of music of Pythagoras, next to nothing about his theory of numbers, still less about his astronomy and, properly speaking, nothing at all about his geometry. A sorry result!

Are there no other sources of information for the development of mathematics in these ancient days? Is it not necessary that the architects, who built their marvelous temples in Ionia and in Southern Italy — the temple of Artemis in Ephesus was one of the seven wonders of the world — were also skilled in geometry?

Speaking frankly: we do not know. It is possible to build beautifully and on a large scale even without mathematics — witness the accomplishments of the Romans. The Roman architect Vitruvius describes very carefully how to build a hall of columns, but mathematics is not involved.

There is one case however in which we know something about the mathematical preparation of a Greek building construction, namely in the case of

The tunnel on Samos.

In about 530 Eupalinus constructed an aqueduct, at the request of the powerful tyrant Polycrates straight across the lime-stone of the mountain Castro on the island of Samos.

Herodotus (III 60) describes this work as follows:

> I have written thus at length of the Samians, because they are the makers of the three greatest works to be seen in any Greek land. First of these is the double-mouthed channel pierced for a hundred and fifty fathoms through the base of a high hill; the whole channel is seven furlongs long, eight feet high and eight feet wide; and throughout the whole of its length there runs another channel twenty cubits deep and three feet wide, wherethrough the water coming from an abundant spring is carried by its pipes to the city of Samos. The designer of this work was Eupalinus, son of Naustrophus, a Megarian . . .

When German archaeologists looked for antiquities on Samos in 1882, they found the tunnel, in a state of good preservation, exactly as described by Herodotus: 1 Kilometer in length, 7 feet high and wide, with a deep ditch con-

[1] H. Diels, *Fragmente, Alkmaion* A 4.

taining the pipes, with vertical vents for changing the air and for cleaning away rubble, and with niches, used by the working men to place their lamps. The ditch has a depth of 7 feet at the upper end and of 27 feet at the lower end; it was probably dug because the drop which had originally been planned, turned out to be too small. [1] (Plate 14a)

But now we come to the most important point: apparently the digging of the tunnel proceeded from both ends. The working men approaching from two directions, met at the center with an error of about 30 feet horizontally and 10 feet vertically. A splendid accomplishment!

When king Ezechias of Judaea had a similar aqueduct constructed through the rocks near Jerusalem about the year 700, his workers had to keep check on the direction in which the work was proceeding in a very primitive way, by means of vertical shafts from the top; the result was a zigzag tunnel, twice as long as the distance between the ends. [2]

Eupalinus did much better: his tunnel was essentially a straight line. How could this be accomplished?

The answer is found in the "Dioptra" of Heron of Alexandria (60 A.D.). Heron first describes a dioptra, i.e. a horizontal bar mounted so as to rotate, with two sights, which made it possible e.g. to sight a right angle in a plane. Then he proposed the following problem (no. 15): "To cut through a mountain $AB\Gamma\Delta$ in a straight line, the openings B and Δ of the tunnel being given. "He draws an arbitrary line BE in the plane, then, by means of the dioptra, the perpendicular EZ, next ZH perpendicular to EZ, and thus successively $H\Theta$, ΘK, $K\Lambda$. Now he moves the dioptra along the line $K\Lambda$, until the ponit Δ is sighted in a right

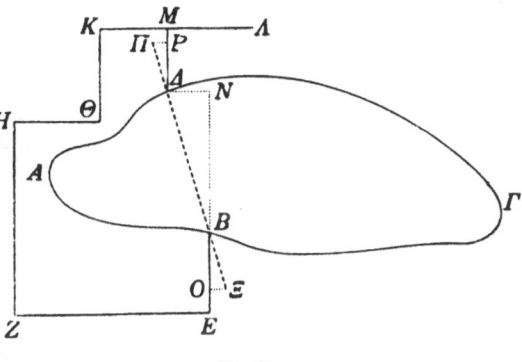

Fig. 29.

angle. If this takes place at M, then $M\Delta$ wil be perpendicular to $K\Lambda$.

Now drop the perpendicular ΔN from Δ on EB. Then ΔN can be determined from EZ, $H\Theta$ and KM; similarly BN can be found from BE, ZH, ΘK and $M\Delta$ by addition and subtraction. Consequently the ratio $BN : N\Delta$ is known; Heron gives for this ratio the value 5 : 1. Now, he constructs right triangles $BO\Xi$ and $\Delta P\Pi$ whose legs have the same ratio 5 : 1. The hypothenuses of these triangles

[1] See E. Fabricius, *Mitt. D. archäol. Inst. Athens 9* (1884), p. 165.
[2] See J. Bidez, *Eos, Platon et l'Orient*, Brussels 1945, p. 12.

then give the directions in which the digging must take place. "If the tunnel is dug in this manner", Heron concludes, "then the laborers will meet."

In no. 16, Heron also deals with the problem of making vertical shafts, to meet the tunnel, supposed to be rectilinear. Such shafts are indeed present on Samos.

Antique measuring instruments.

Eupalinus must also have been able to determine differences in altitude. It is quite likely that he did this like Heron and as we still do it: proceeding from point to point with vertical measuring rods and a horizontal sight. There were no telescopes; the sighting instrument must have been a dioptra. Heron obtains a level by means of communicating tubes.

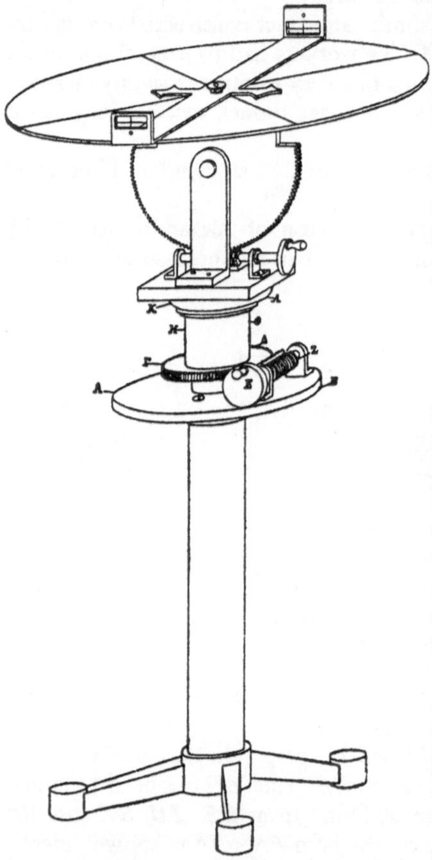

The reader may think that this instrument is too clever for the era of Eupalinus; we are frequently inclined to think of the ancient Greeks as being more primitive than they actually were. We must remember that Anaximander of Milete, the pupil of Thales, had a workshop in which, among other things, wooden celestial spheres were manufactured. Anaximander also was commissioned to place a gnomon on the market place in Sparta, i.e. a vertical sundial, supplied of course with all the cleverly constructed hourlines and monthlines, which belonged there according to Vitruvius. [1] This happened around 560, about thirty years before the construction of the tunnel on Samos.

A knowledge of these things makes it possible to form a clearer picture of scientific life in the sixth century than is possible merely on the basis of the vague reports concerning Thales and Pythagoras.

We can also better appreciate now the Proclus tradition, which says that Pythagoras "transformed mathematics into a free education". "Free education"

Fig. 30. Heron's dioptra, as reconstructed by Schöne from Heron's own description.

is to be understood in this connection as the kind of education which suits a free

[1] See Pauly-Wissowa, *Realenzyklopaedie des klassischen Altertums*, article *Horologium*.

man, in contrast with training for a trade. The Ionians cultivated mathematics not only for its inherent interest, but also for the sake of the practical applications. Surveyors and architects, such as Eupalinus, had to know something about geometry, and the training of an instrument maker in the workshop of Anaximander undoubtedly involved astronomy. But Pythagoras freed mathematics from these practical applications, The Pythagoreans pursued mathematics as a kind of religious contemplation, as a way to approach the eternal Truth.

CHAPTER V

THE GOLDEN AGE

The fifth century, the age of Pericles, the era of the widest unfurlment of power of the Athenian empire, and of its decline, is the golden age of Hellenic culture. It is the century during which the glorious temples of the Acropolis arose, when sculpture produced its ever admired, never equalled masterworks, the age also of the great tragic poets Aeschylus, Sophocles and Euripides, of the great historians Herodotus and Thucydides.

The fifth century begins with the unsuccessful revolt of the Ionian cities in Asia Minor against the Persian domination, which led to the great Persian wars, from which Athens and Sparta emerged as victors. Athens, which had been completely destroyed, was now splendidly rebuilt and Athenian naval power was established. And, democracy was introduced over the entire range of this power. Thus the aristocratic Pythagoreans were divested of their power and driven out of Italy (about 430). The contrasts between democratic Athens and the military aristocracy of Sparta became more and more pronounced, until at last the frightful Peloponnesian war brought about the end of the power of Athens and thus of the greatness of Hellas.

In philosophy, this century was the period of the Enlightenment (Aufklärung), of rationalism. The ancient religion and ethics were ridiculed by the intellectuals. The sophists taught that all truth and all values are only relative. Materialistic systems, such as that of the atomist Democritus, were at a premium in wide circles. Anaxagoras and Democritus held that the sun and the moon were not living beings, but inert rocks, tossed to and fro in the whirl of the atoms.

In contrast to this view stood the religious-mystical metaphysics of the Pythagoreans, which adhered for instance to the belief that the planets were animated, divine beings, carrying out perfect circular motions as a result of their perfection and their divine insight.

It was not until the fourth century, the century of Plato, that philosophy and mathematics reached their zenith. We can follow the growth which took place in the fifth century and which was preparatory to the period of flowering. For mathematics, this growth occurs chiefly in the Pythagorean school. At the beginning of the fifth century, the most interesting figure in this school is

Hippasus.

The Pythagorean doctrine had two aspects, a scientific one and a mystical-religious one. Pythagoras called himself a philosopher, not a sage therefore, but a seeker after wisdom, a lover of wisdom; but he was also a prophet whose teachings were looked upon as infallible, divine revelations.

This twofold aspect was bound to lead to a serious dilemma after his death.

Should the divine revelations of Pythagoras, the "ipse dixit" be looked upon as the source of all knowledge, concerning numbers, harmony and the stars? In this case there was little sense in attempting to perfect this absolute wisdom by one's own studies. Or should one search for the truth, relying upon one's own thoughts, in the manner of a genuine philosopher? In that case, reason should be ranged above revelation, as the supreme judge in all scientific questions.

And, there is something else. Any one who wants to advance in science can not look upon this discipline as a secret accessible only to the initiates. He has to take into account the results of others and should not conceal his own views. It is thus inevitable that he should come into conflict with the pledge of silence.

This conflict did indeed break out soon after the death of Pythagoras. Hippasus made bold to add several novelties to the doctrine of Pythagoras and to communicate his views to others. He constructed the circumscribed sphere of a dodecahedron; he developed the theory of the harmonic mean; to the three consonant intervals, octave, fifth and fourth, he added two more, the double octave and the fifth beyond the octave; he discussed the theory of musical ratios with the musician Lasos, who was led thereby to experiment with vases, empty and half-filled. These and other similar indiscretions caused a split: Hippasus was expelled. Later on he lost his life in a shipwreck, as a punishment for his sacrilege, according to his opponents. His followers called themselves "Mathematikoi", in opposition to the "Akousmatikoi", who strictly observed the sacred tenets and who faithfully transmitted the "Akousmata", the sacred sayings. The two sects entered into vicious disputes, as is customary among sects of one religion.

A century after Hippasus, there were various groups of Pythagoreans; their mutal relations and their relations to the two old sects are not entirely clear. One of these groups was driven out of Italy around 430 by the advancing democracy. Aristoxenes (a pupil of Aristotle, about 320) says that he had known the last remnant of this group, disciples of Philolaus and Eurytus. We are not certain of the truth of this; it would not be the only lie of which Aristoxenes is guilty. [1] Another group must have remained in Italy; for late in the 4th century, Pythagorean orgies there are still mentioned. A third group, which appears to carry on the tradition of the "Mathematikoi", is formed by the "so-called Pythagoreans in Italy", repeatedly referred to by Aristotle. One of the most important and most interesting figures among these is Archytas of Taras, friend and contemporary of Plato, whose work will be discussed in the next chapter. Around 350, mention is still made of the astronomers Ecphantus and Hicetas of Syracuse who taught that the earth revolves on its axis.

No names of Mathematikoi are known between Hippasus and Archytas. But, more important than such names is the development of mathematics itself in the school of the Pythagoreans; about this we do know something.

---

[1] Aristoxenos tells us in his Harmonics that his predecessors ignored the chromatic and diatonic scales; but we know from Ptolemy's Harmonics that Archytas calculated these scales.

The Mathemata of the Pythagoreans.

Four "mathemata", i.e. subjects of study, were known to the Pythagoreans:
Theory of Numbers (Arithmetica), Theory of Music (Harmonica), Geometry
(Geometria) and Astronomy (Astrologia). Astronomy will be discussed fully elsewhere. For the theory of music, I refer to my memoir in Hermes 78 (1943). We
begin therefore with

The theory of numbers.

Here and in the sequel, the word "numbers" always refers, according to Greek
usage, to *integral* positive numbers (quantities). [1] The theory of numbers is therefore the theory of natural numbers.

The oldest bit of the theory of numbers that we can reconstruct, is:

The theory of the even and the odd.

If we turn to the ninth book of Euclid's Elements, we find a sequence of propositions (21—34), which have no connection whatever with what precedes, and
which constitutes, according to O. Becker[2], together with IX 36, a piece of
archaic, and indeed typically Pythagorean mathematics. In abbreviated form,
they are:

21. A sum of even numbers is even.
22. A sum of an even number of odd numbers is even.
23. A sum of an odd number of odd numbers is odd.
24. Even less even is even.
25. Even less odd is odd.
26. Odd less odd is even.
27. Odd less even is odd.
28. Odd times even is even
29. Odd times odd is odd.
30. If an odd number measure an even number, it will also measure the half of
    it.
31. If an odd number be prime to any number, it will also be prime to the
    double of it.
32. A number that results from (repeated) duplication of 2, is exclusively
    even times even.
33. If a number has its half odd, it is even times odd only.
34. Every even number that does not belong among those mentioned in 32
    and 33, is even times even and also even times odd.

[1] The Greeks exclude even unity from the numbers because unity is not a quantity. This compels clumsy formulations, such as "if *a* is a number or 1 ...". We shall take no notice of these quibbles and we shall simply count
1 among the numbers.
[2] O. Becker, *Quellen und Studien*, B 4, p. 533.

PLATE 13

PL. 13. Bronze head of Pythagoras (?). Copy of a Greek original, probably from the end of the 4th century B.C., from the Villa dei Papiri in Herculaneum. Museo Nazionale, Naples. Aelianus relates that Pythagoras wore oriental dress; this would explain the turban. As Schefold points out, it is also possible that the head represents Archytas of Taras, the most important Pythagorean mathematician and musicologist (see p. 149). If this be the case we have probably a real, though somewhat idealized, portrait.

(Photo Alinari)

PLATE 14

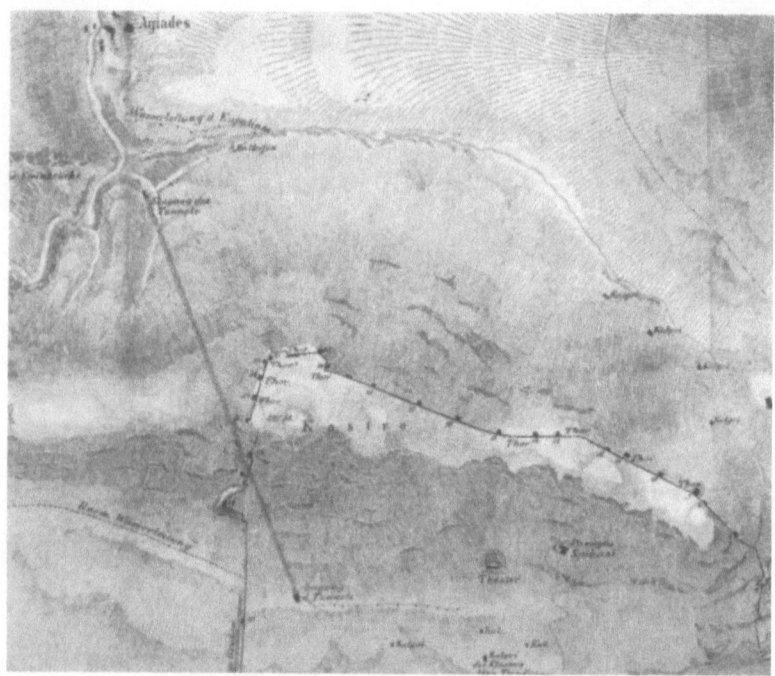

PL. 14a. Mt. Castro on the island of Samos with the conduit of Eupalinos. From E. Fabricius, Mitt. Deutsches Archäol. Institut, Athens, 9 (1884).

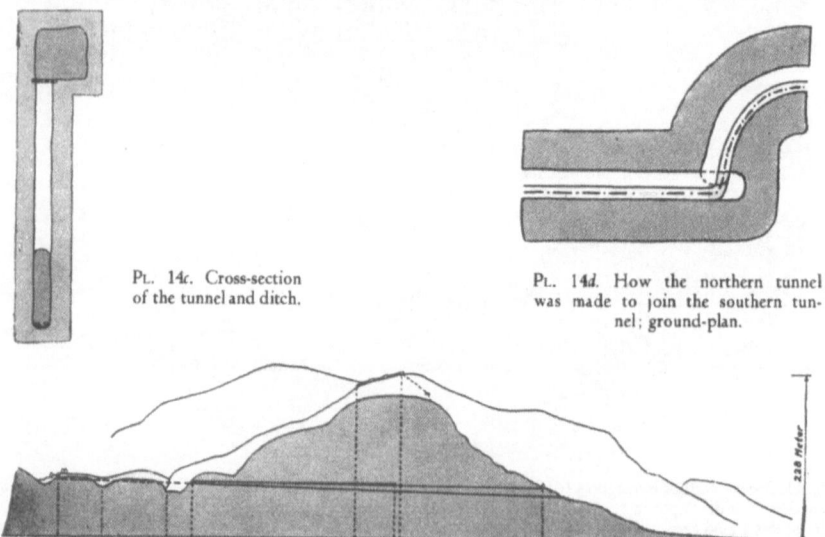

PL. 14c. Cross-section of the tunnel and ditch.

PL. 14d. How the northern tunnel was made to join the southern tunnel; ground-plan.

PL. 14b. Cross-section of Mt. Castro, the altitudes enlarged twofold. The upper line across the mountain represents the original tunnel, the lower one the ditch with greater fall.

The climax of the theory is proposition 36, which asserts that numbers of the form

$$2^n(1 + 2 + 2^2 + \ldots + 2^n)$$

are perfect, provided

$$p = 1 + 2 + \ldots + 2^n$$

is a prime number.

Becker has shown that this proposition can be derived from 21—34 alone. Euclid gives a somewhat different proof, probably to establish the connection with the Theory of Numbers developed in Books VII—IX, but there is no doubt that the original proof depended on the theory of the even and the odd. In particular, proposition 30 seems to have been introduced especially to make possible the proof that $2^n p$ has no other divisors than $1, 2, 2^2 \ldots 2^n$ and $p, 2p, \ldots, 2^n p$.

Propositions 32—34 are related to the classification of even numbers into even times even and even times odd, which is found, in a different form in the work of Nicomachus of Gerasa, and at which Plato hints here and there.

Plato always defines Arithmetica as "the theory of the even and the odd". For the Pythagoreans, even and odd are not only the fundamental concepts of arithmetic, but indeed the basic principles of all nature. Their point of view is expressed by Aristotle (Metaphysics A5) in the following words:

> The elements of number are the even and the odd . . . Unity consists of both . . . Number develops from Unity . . . Numbers constitute the entire heaven.
>
> Others of the same school say that there are 10 principles, which they group in pairs:
>
> | | | |
> |--------|---|-----------|
> | bounded | — | unbounded |
> | odd | — | even |
> | unity | — | plurality |
> | male | — | female, etc. |

It becomes clear that the antithesis even-odd plays a very fundamental role in Pythagorean metaphysics.

In a fragment of the old comic poet Epicharmus (around 500 B.C. or perhaps even earlier), we find an amusing allusion to the philosophy of the Pythagoreans. The following dialogue takes place:

> "When there is an even number present or, for all I care, an odd number, and some one wants to add a pebble or to take one away, do you think that the number remains unchanged?"
>
> "God forbid!"
>
> "And when some one wants to add some length to a cubit or wants to cut off a piece, will the measure continue as before?"
>
> "Of course not".
>
> "Well then, look at people. One grows, another one perhaps gets shorter, they are constantly subject to change. But whatever is changeable in character and does not remain the same, that is certainly different from what has changed. You and I are also different people from what we were yesterday, and we will be still different in future, so that by the same argument we are never the same."
>
> (Diels, Fragmente der Vorsokratiker, Epicharmus A 2).

To appreciate this properly, we must remember that this conversation takes place in a comedy, not in a philosophical discussion. Quite likely the first speaker owes the other some money and wants to prove by a philosophical argument, that it is not he who has borrowed the money, but a totally different person!

Epicharmus is poking fun therefore at the disputes of the philosophers of his day. But who are the philosophers he is thinking of? Obviously the Pythagoreans.[1] For, why should he speak in his first sentence of the antithesis even-odd? Epicharmus lived in Sicily, and the Pythagoreans played a big role throughout Southern Italy in this period; so the audience probably understood the allusion.

But it is time to return to the mathematics of the Pythagoreans!

The only place at which the theory of the even and the odd is applied in the Elements themselves, is the proof of the incommensurability of the side and the diagonal of a square, at the end of Book X; we will reproduce this proof in abbreviated form.

If the diagonal AC and the side AB of a square $ABCD$ are commensurable, let their ratio, reduced to lowest form be $m : n$. From $AC : AB = m : n$ follows $AC^2 : AB^2 = m^2 : n^2$; but, since $AC^2 = 2AB^2$, this leads to $m^2 = 2n^2$, so that m^2 is even. It follows that m is even; for, if m were odd, we would conclude from Proposition 29 that m^2 is also odd. Hence m is even; let one half of m be equal to h. Since m and n have no common factor, and m is even, it follows that n is odd. But, from $m = 2h$, we obtain $m^2 = 4h^2$ and hence $n^2 = 2h^2$. Therefore n^2 is even, so that a repetition of the earlier reasoning shows that n is also even. The conclusion would then be that the number n is both even and odd; this is impossible.

Aristotle repeatedly alludes to this proof. According to Plato, Theodorus of Cyrene (around 430) proved the irrationality of $\sqrt{3}$, $\sqrt{5}$, etc. to $\sqrt{17}$, (stated more accurately: of the sides of squares of areas 3, 5, ... 17), but the tacitly omitted $\sqrt{2}$. From this Zeuthen draws the conclusion, that the irrationality of $\sqrt{2}$, i.e. of the diagonal of the unit square, was known before his time. On the other hand, Pappus states that the theory of irrationals started in the school of Pythagoras, and their theory of the even and the odd gave the Pythagoreans the means to prove the irrationality of $\sqrt{2}$; it is therefore highly probable that the proof given in the preceding paragraph, came from them.

Proportions in numbers.

The systematic organization of the theory of the ratios of numbers and of divisibility, found in Book VII of the Elements, came later than the theory of the even and the odd. It is my judgment that this entire book should be attributed to the Pythagoreans before Archytas. To justify this conclusion, it is necessary to take first a closer look at the Theory of Numbers of Archytas himself.

[1] Compare A. Rostagni, *Il Verbo di Pitagora* (1924)

In the theory of music of Archytas, the following two propositions from the theory of numbers, which we shall call A and B, play a leading role:

A. *Between two numbers in epimoric ratio* [1] *there exists no mean proportional.*

B. *When the "combination" of a ratio with itself* [2] *is a "multiple ratio"* [3], *then the ratio itself is a multiple ratio.*

These two propositions are found, e.g., at the beginning of Euclid's "Sectio canonis". Boethius says that Archytas' proof of proposition A ran as follows: Let $A : B$ be the given epimoric ratio, and let C and $D + E$ be the smallest numbers in this ratio. [4] Then $D + E$ exceeds the number C by an amount D, which is a divisor of $D + E$ itself and also of C. I assert now that D must be unity. For, suppose that D is a number greater than 1 and a divisor of $D + E$; then D is also a divisor of E and hence of C. It follows that D is a common divisor of C and $D + E$; but this is impossible, because the smallest of the numbers whose ratio equals their ratio, are relatively prime (cf. Euclid's Elements VII 22). Therefore D is unity, i.e. $D + E$ exceeds C by unity, and hence a mean proportional between C and $D + E$ can not be found. Consequently, a mean proportional between the two original numbers A and B can not exist, since their ratio equals that of C and $D + E$.

This proof quotes almost word for word a proposition from Book VII, viz.:

VII 22. *The least numbers of those which have the same ratio with them are prime to one another.*

Furthermore, at the end of the proof a proposition from Book VIII is applied, viz.:

VIII 8. *If between two numbers there fall numbers in continued proportion with them* (i.e. if one has a geometric progression of which the given numbers are the extreme terms), *then, however many numbers fall between them in continued proportion, so many will also fall in continued proportion between the numbers which have the same ratio with the original numbers.*

The proof of Proposition A in Euclid's Sectio Canonis is essentially the same as that of Archytas. Proposition B is proved in the Sectio Canonis by appeal to another proposition from the Elements, viz.

VIII 7. *If there be as many numbers as we please in continued proportion, and the first measure the last, it will measure the second also.*

It becomes evident that Archytas' theory of music presupposes quite a bit of the theory of numbers. Tannery and Heath rightly conclude from this, that at the time of Archytas, i.e. around 400, there must have existed in the school of Pytha-

---

[1] See footnote 2 on p. 97 for the meaning of this term.

[2] The "combination" of the ratios $a:b$ and $b : c$ is the ratio $a . c$. The combination of two ratios can $a : b$ and $c : d$ can always be obtained by reducing them to the form $ac : bc$ and $bc : bd$. This operation corresponds to our multiplication of fractions.

[3] Two numbers are said to be in "multiple ratio" if one of them is a multiple of the other.

[4] For the sake of clarity I have written $D + E$. The Latin text simply writes DE. The sequel indicates that the equality of C and E is assumed.

goras some kind of "traité d'arithmétique", a "treatise of some kind on the Elements of Arithmetics", something similar to Books VII—IX of Euclid's Elements. I am now going to inquire more closely what must certainly have been contained in these Pythagorean Elements of Arithmetic.

Propositions VIII 7 and 8 both depend necessarily upon

VIII 3. If as many numbers as we please in continued proportion [1] *be the least of those which have the same ratio with them* [2], *the extremes of them are prime to one another.*

This proposition in turn depends on VII 27 and on

VIII 2. To find numbers in continued proportion, as many as may be prescribed, and the least that are in a given ratio.

The solution of this problem proceeds as follows: Let the given ratio, in its lowest terms, be $A : B$. Multiply A and B, each by themselves and by each other; we obtain then a geometric progression of 3 terms A^2, AB, B^2. Multiply these by A and the last also by B, thus getting a progression of 4 terms: A^3, A^2B, AB^2, B^2. Continue in this manner. The same pattern

$$
\begin{array}{ccccc}
 & A & & B & \\
 & A^2 & AB & B^2 & \\
A^3 & A^2B & AB^2 & B^3 &
\end{array}
$$

is also found in Nicomachus with an application to the theory of music.

This construction of geometric progressions must therefore certainly belong to the arithmetic of the Pythagoreans. The same is true of the rest of Book VIII. Indeed, it is exactly those Platonic dialogues, in which the largest number of Pythagorean ideas have been used (Timaeus and Epinomis), that contain numerous allusions to propositions and concepts of Book VIII. For example, in the Timaeus it is stated that between two squares there is one mean proportional, between two cubes, two mean proportionals; and the Epinomis speaks of similar plane and space numbers. (Two "plane numbers" $A \cdot B$ and $C \cdot D$ are called similar if their "sides" are proportional; an analogous definition is given for "space numbers" $A \cdot B \cdot C$ and $D \cdot E \cdot F$). The contents of book VIII must therefore be credited to the Pythagoreans, in particular to Archytas and his circle.

But Book VIII is based on Book VII. We have already seen that Archytas quotes VII 22 almost word for word. The logical genealogy of Archytas' propositions A and B is the following [3]

---

[1] i.e. if $a : b = b : c = c : d \ldots$
[2] i.e. that there are no smaller numbers a', b', c', ... which have the same ratios as a, b, c, ...
[3] For a more detailed elaboration, see my article "*Arithmetik der Pythagoreer*", Math. Ann. *130* (1948), p. 127.

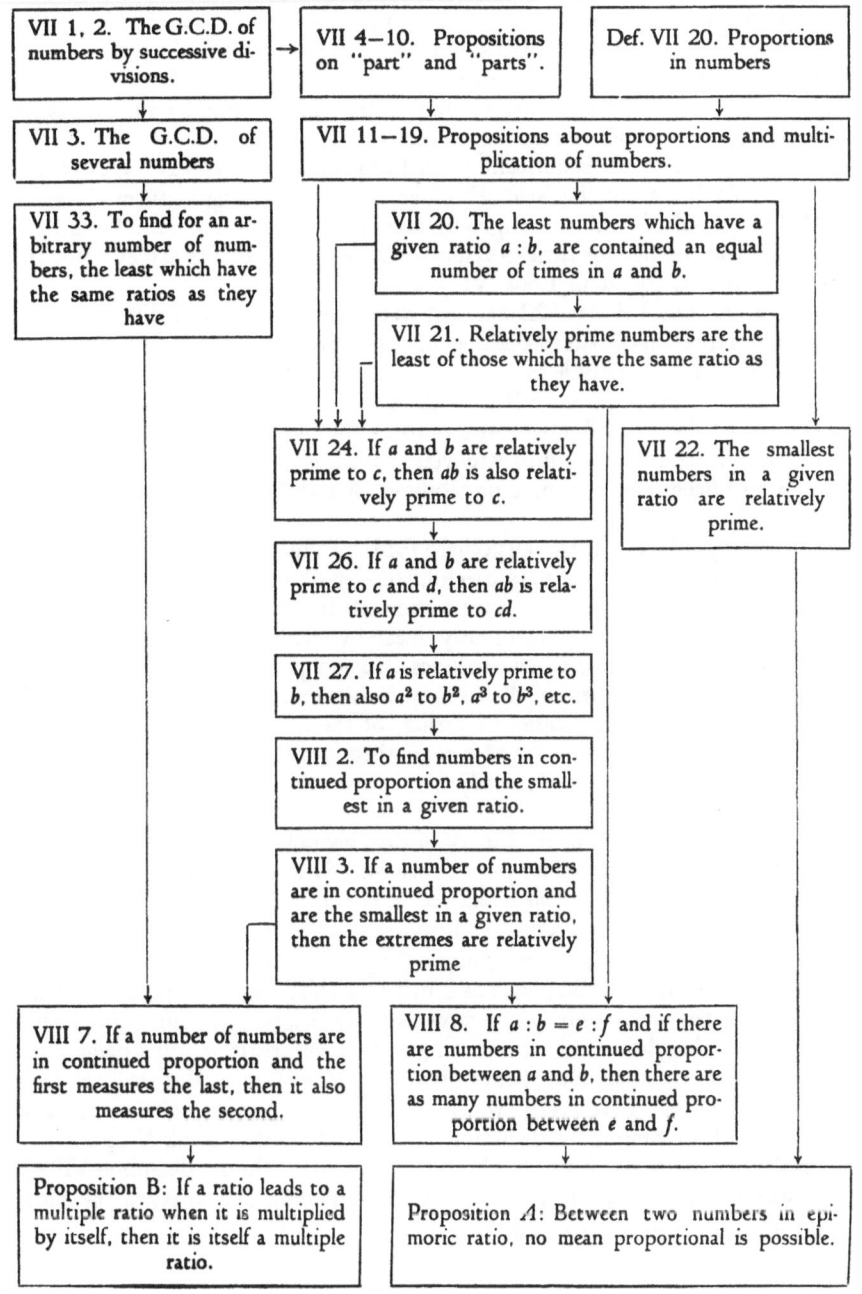

VII 1, 2. The G.C.D. of numbers by successive divisions. →

VII 4—10. Propositions on "part" and "parts".

Def. VII 20. Proportions in numbers

VII 3. The G.C.D. of several numbers

VII 11—19. Propositions about proportions and multiplication of numbers.

VII 33. To find for an arbitrary number of numbers, the least which have the same ratios as they have

VII 20. The least numbers which have a given ratio $a : b$, are contained an equal number of times in a and b.

VII 21. Relatively prime numbers are the least of those which have the same ratio as they have.

VII 24. If a and b are relatively prime to c, then ab is also relatively prime to c.

VII 22. The smallest numbers in a given ratio are relatively prime.

VII 26. If a and b are relatively prime to c and d, then ab is relatively prime to cd.

VII 27. If a is relatively prime to b, then also a^2 to b^2, a^3 to b^3, etc.

VIII 2. To find numbers in continued proportion and the smallest in a given ratio.

VIII 3. If a number of numbers are in continued proportion and are the smallest in a given ratio, then the extremes are relatively prime

VIII 7. If a number of numbers are in continued proportion and the first measures the last, then it also measures the second.

VIII 8. If $a : b = e : f$ and if there are numbers in continued proportion between a and b, then there are as many numbers in continued proportion between e and f.

Proposition B: If a ratio leads to a multiple ratio when it is multiplied by itself, then it is itself a multiple ratio.

Proposition A: Between two numbers in epimoric ratio, no mean proportional is possible.

This analysis shows that in the proofs of VIII 2—3, 7—8 and of Propositions *A* and *B*, all of Book VII, to VII 27 and 33, is involved. Book VII is constructed in such a strictly logical fashion that no pebble can be removed from it, in sharp contrast to the muddy structure of Book VIII with its roundabout proofs, its unnecessary repetitions and its logical errors.

But, how is Book VII historically related to propositions VIII 2—3, 7—8 and *A—B* of Archytas and his school? Three possibilities suggest themselves:

1) The school of Archytas originally accepted propositions VII 21—22 and 27, which they needed for their developments, without proof, as self-evident, and the strict proofs found in Book VII are of later date;

2) Propositions VII 21—22 and 27 were originally proved differently and Book VII represents a later recasting of these proofs;

3) Book VII was in apple-pie order at the time of Archytas, so that he could simply refer to propositions VII 21—22 and 27.

The last of these possibilities appears to be the most probable in view of the fact that, in Boethius' account of Archytas, VII 22 is quoted word for word. But we can go further and eliminate possibilities 2) and 1) entirely, thus leaving 3) only!

1) is impossible. For it was customary for the Pythagoreans to prove number-theoretic propositions in a very careful, step-by-step manner. This can be seen in Archytas' elaborate proof of Proposition *A*, in which even the most trivial syllogisms are worked out punctiliously; it can be seen in the Pythagorean theory of the even and the odd, which includes detailed proofs of the most obvious things, such as the evenness of the sum of an even number of even numbers (IX 21); again in the proof, mentioned earlier, of the irrationality of the diagonal, in which, e.g., the conclusion "if m^2 is even, then m is also even" is very properly obtained by an indirect proof; finally it is seen in the elaborate proofs of Book VIII itself.

2) is also impossible. For the theory of proportions in Book VII is developed from Definition 20: *Numbers are proportional when the first is the same multiple, or the same part, or the same parts,* [1] *of the second that the third is of the fourth.* This definition is an ancient one. It already occurs in Hippocrates of Chios, applied to geometric magnitudes; he calls two circular segments proportional to their circles, if they are the same part of their circles, such as a half or a third. Strictly speaking, such a definition makes sense only for commensurable magnitudes, since, no matter whether one takes the *m*-fold, or the *n*-th part or *m n*-th parts, *m* and *n* being numbers in the Greek sense, one always obtains a rational ratio. We shall see that, later on, definitions were given, which are applicable also to incommensurable magnitudes; for example, the definition of Eudoxus used in Book V of the Elements (V, Def. 5). In Book VII, Euclid retained however the old definition VII 20, obviously because he would otherwise have been compelled to recast the entire

---

[1] e.g., the double, or the third part, or two thirds.

book so as to adapt it to the new definition. It follows that Book VII is not a later reconstruction, but a piece of ancient mathematics.

Another argument in favor of this conclusion is that Book VII is a well-rounded whole without traces of later revisions, such as occur in other books.

Thus we have acquired an important insight: *Book VII was a textbook on the elements of the Theory of Numbers, in use in the Pythagorean school*. It is not an accident that this book has been preserved; it is due to its strictly logical structure. Euclid did not find anything to correct; neither could he make a change, short of tearing the whole thing apart and building it anew — an indication of its firm structure.

In Book VII occur some propositions on prime numbers, but it does not contain the proposition that every number can be written as the product of primes in one and only one way, which plays such an important role in our modern elementary Theory of Numbers. The theory of the G.C.D. and the L.C.M., as well as that of relatively prime numbers are developed without the use of the factorization into prime numbers. The entire development rests on the "Euclidean algorithm" for determining the G.C.D. by means of successive subtractions. The emphasis is not on the divisibility properties of numbers, but rather on the theory of proportions and on the reduction of a ratio to lowest terms.

It is probable that it was the *calculation with fractions* which led to the setting up of this theory. Fractions do not occur in the official Greek mathematics before Archimedes, but, in practice, commercial calculations had of course to use them. The reason why fractions were eliminated from the theory is the theoretical *indivisibility of unity*. In the Republic (525E), Plato says: "For you are doubtless aware that experts in this study, if any one attempts to cut up the 'one' in argument, laugh at him and refuse to allow it; but if *you* mince it up, *they* multiply, always on guard lest the one should appear to be not one but a multiplicity of parts". Theon of Smyrna explains this further [1]: "When the unit is divided in the domain of visible things, it is certainly reduced as a body and divided into parts which are smaller than the body itself, but it is increased in numbers, because many things take the place of one".

When fractions are thus thrown out of the pure theory of numbers, the question arises whether it is not possible to create a mathematical equivalent of the concept fraction and thus to establish a theoretical foundation for computing with fractions. Indeed, this equivalent is found in the ratio of numbers. In place of the reduction of fractions to lowest terms, comes the simplification of a ratio of numbers to smallest terms which is investigated theoretically in Book VII. Reduction of fractions to a common denominator brings the concept of the L.C.M. on the stage; and this is also discussed in Book VII.

The art of computing, including the calculation with fractions, is called "logistics" by the Greeks. Plato distinguishes practical from theoretical logistics, just as he discriminates between practical and theoretical arithmetic (Gorgias

---

[1] *Theonis Smyrnaei Expositio rerum mathematicarum*, ed. Hiller (1878), p. 18

451 A—C). Theoretical logistics deals especially with the study of numbers in their mutual ratios, exactly the sort of thing treated in Book VII, while theoretical arithmetic is concerned with "the even and the odd, how much each amounts to in every individual case". Apparently therefore, Plato considers the old Pythagorean theory of the even and the odd (Book IX) as belonging to theoretical arithmetic, and the theory of the ratios of numbers (Books VII, VIII) as a part of theoretical logistics. He contrasts these theoretical sciences with practical arithmetic, i.e. counting, and practical logistics, i.e. calculation, in particular with fractions.[1]

Ancient arithmetic also includes something which is thought of at present as a topic in algebra, namely

The solution of systems of equations of the first degree.

The solution of the special system of equations

(1)
$$\begin{aligned} x + x_1 + x_2 + \ldots + x_{n-1} &= s \\ x + x_1 &= a_1 \\ \ldots\ldots\ldots\ldots \\ x + x_{n-1} &= a_{n-1} \end{aligned}$$

is known by the name *"flower of Thymaridas"*. The solution is of course

$$x = \frac{(a_1 + \ldots + a_{n-1}) - s}{n - 2}.$$

The Neo-Pythagorean Iamblichus, who atrributes this solution to the old Pythagorean Thymaridas, shows furthermore, how systems of the form

$$x + y = \alpha(z + u), \quad x + z = \beta(u + y), \quad x + u = \gamma(y + z)$$

can be reduced to the form (1).

We see from this that the Pythagoreans, like the Babylonians, occupied themselves with the solution of systems of equations with more than one unknown.

Geometry.

Tannery has called attention to the following remarkable passage in Iamblichus' Pythagorica Vita:

> In the following manner the Pythagoreans explain how it came about that geometry was made known publicly: Through the fault of one of their number, the Pythagoreans lost their money. After this misfortune, it was decided to allow him to earn money with Geometry — thus Geometry came to be designated as "The Tradition of Pythagoras".

In further explanation, Tannery recalls (La géométrie grecque, p. 81) that the Pythagoreans had common ownership of goods. Tannery does not believe

[1] For a further discussion of the Greek concept of number, see J. Klein, *Die griechische Logistik und die Entstehung der Algebra, Quellen und Studien*, B 3, p. 18.

in the original secrecy of the mathematical sciences (I do), but he thinks (and very rightly) that this legend must nevertheless contain a core of truth. Indeed, after the middle of the 5th century, it became quite feasible to earn money with science; there existed a great thirst for knowledge. The sophists, the "teachers of wisdom" received excellent pay; wealthy people took pride in having their sons taught by the best and the most famous sophists. It is therefore quite possible that, in their distress, the Pythagoreans tried to increase their revenues in this manner.

The last sentence of the quoted fragment indicates that a textbook on Geometry must have existed with the title "The Tradition of Pythagoras", some kind of written course of lectures from which the Pythagoreans made money. This also explains how it happened that later writers attributed to Pythagoras all kinds of geometric discoveries, although some of these (e.g. the irrational) were certainly found only much later. It is only necessary to assume that these things occurred in "The Tradition of Pythagoras".

Tannery supposes that the statements of Eudemus concerning the geometry of the Pythagoreans must also have been taken from this book. According to Eudemus, the Pythagoreans discovered the proposition, that in every triangle the angle

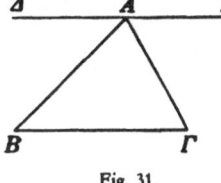

Fig. 31.

sum is equal to two right angles, and proved it as follows:

"Let $AB\Gamma$ be a triangle. Draw through A a line $\Delta E \parallel B\Gamma$. Since $B\Gamma$ and ΔE are parallel, the alternate interior angles are equal, so that $\angle \Delta AB = \angle AB\Gamma$ and $\angle EA\Gamma = \angle A\Gamma B$. Add $\angle BA\Gamma$ to both sides. Then $\angle \Delta AB + \angle BA\Gamma + \angle \Gamma AE$, i.e. two right angles will be equal to the sum of the three

angles of the triangle. Therefore the sum of the three angles of a triangle is equal to two right angles. This is the proof of the Pythagoreans" (Proclus, Commentary on Euclid, I 32).

This shows that the geometry of the Pythagoreans was constructed logically and that they knew not only the proposition concerning the angle sum of a triangle, but also the one on alternate interior angles formed by parallel lines

It is probable that the Pythagoreans had a theory of regular polygons. They certainly knew the star-pentagon, and the dodecahedron; and they also knew that there are only three regular polygons whose angles can fill the space about a point O in the plane, viz. the triangle, the square and the hexagon. It is very likely that this proposition was related to their investigation of the regular polyhedra.

We recall that Plutarch, speaking of the distich "When Pythagoras discovered his famous figure, for which he sacrified a bull", doubted whether the proposition, referred to, is the one about the hypothenuse, or whether it has to do with the adaptation of areas, or perhaps with the problem of constructing a plane figure, equal in area to a second figure and similar to a third. It seems probable to me, that these three subjects were not arbitrarily thrown together, but that

Plutarch, or the source upon which he draws, got these three important propositions from "The Tradition of Pythagoras". As a matter of fact, they are directly connected with one another. Let us start with the last one.

The general problem of constructing a polygon similar to a given polygon and equal in area to another one, can be reduced to the special case of constructing a square equal in area to a given rectangle (Euclid, II 14). This amounts to the construction of the mean proportional between the base and the altitude of the rectangle. In his duplication of the cube, which we shall discuss later on, Archytas uses twice, as an auxiliary, the construction of the mean proportional $x = \sqrt{ab}$, by means of a semi-circle as illustrated in Fig. 32. And, as will be seen presently, the same construction is needed in the "application of areas", which is, according to Eudemus, a find of the Pythagoreans.

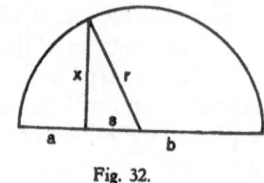

Fig. 32.

Euclid's proof of the proposition $x^2 = ab$ proceeds as follows:

$$x^2 = r^2 - s^2 = (r - s)(r + s) = ab.$$

We see that the "Theorem of Pythagoras" $r^2 = x^2 + s^2$, is applied here. I assume that this proof was taken from "The Tradition of Pythagoras".

But what does Plutarch mean by the "application of areas"? This very important subject deserves separate consideration.

"Geometric Algebra".

When one opens Book II of the Elements, one finds a sequence of propositions, which are nothing but geometric formulations of algebraic rules. So, e.g., II 1:

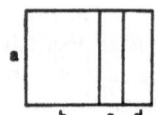

Fig. 33.

"If there be two straight lines, and one of them be cut into any number of segments whatever, the rectangle contained by the two straight lines is equal to the rectangles contained by the uncut straight line and each of the segments,

corresponds to the formula

$$a(b + c + \ldots) = ab + ac + \ldots$$

II 2 and 3 are special cases of this proposition. II 4 corresponds to the formula

$$(a + b)^2 = a^2 + b^2 + 2ab.$$

The proof can be read off immediately from Fig. 34. In II 7, one recognizes the analogous formula for $(a - b)^2$. We have here, so to speak, the start of an algebra textbook, dressed up in geometrical form. The magnitudes under consideration are always line segments; instead of "the product

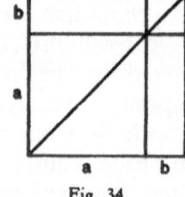

Fig. 34.

ab", one speaks of "the rectangle formed by *a* and *b*", and in place of *a*², of "the square on *a*".

Quite properly, Zeuthen speaks in this connection of a "geometric algebra". Throughout Greek mathematics, one finds numerous applications of this "algebra" The line of thought is always algebraic, the formulation geometric. The greater part of the theory of polygons and polyhedra is based on this method; the entire theory of conic sections depends on it. Theaetetus in the 4th century, Archimedes and Apollonius in the 3rd are perfect virtuosos on this instrument.

Presently we shall make clear that this geometric algebra is the continuation of Babylonian algebra. The Babylonians also used the terms "rectangle" for *xy* and "square" for *x*², but besides these and alternating with them, such arithmetic expressions as multiplication, root extraction, etc. occur as well. The Greeks, on the other hand, consistently avoid such expressions, except in operations on integers and on simple fractions; everything is translated into geometric terminology. But since it is indeed a translation which occurs here and the line of thought is algebraic, there is no danger of misrepresentation, if we reconvert the derivations into algebraic language and use modern notations. From now on we shall therefore quite coolly replace expressions such as "the square on *a*", "the rectangle formed by *a* and *b*" by the modern symbols *a*² and *ab*, whenever they simplify the presentation. Similarly, we shall put proportionalities into a modern dress, e.g.

$$a^2 : ab = a : b.$$

For our purpose it is not necessary to use the abbreviations $O(a, b)$, $T(a)$ and $\Lambda(a, b)$ introduced by Dijksterhuis for *ab*, *a*² and *a* : *b*, provided we take good care, *not to use algebraic transformations, which can not immediately be reformulated in the Greek terminology*. When Cantor (Geschichte der Mathematik I, 3rd or 4th edition, p. 213) derives e.g. from the proportionality

$$a : x = x : y = y : b,$$

the result $$x^2 = ay, \quad y^2 = bx,$$
and then $$x^4 = a^2y^2 = a^2bx,$$
thus obtaining $$x^3 = a^2b,$$

the line of thought is contrary to that of the Greeks; *x*³ can still be interpreted geometrically, viz. as the volume of a cube, but not *x*⁴. It *is* proper to derive

$$x^2 = ay, \quad xy = ab$$

from the continued proportion, and then to write

$$x^3 = axy = a^2b,$$

thus leading to
$$a^3 : x^3 = a^3 : a^2b = a : b.$$

In words: *If x and y are two mean proportionals between a and b, then the cube on a is to the cube on x as a is to b.*

Let us now return from our digression to the geometric algebra of Book II.

As the next formula one might well expect

(1) $$a^2 - b^2 = (a - b)(a + b)$$

or, since Euclid avoids the subtraction of areas,

(2) $$a^2 = (a - b)(a + b) + b^2.$$

It does actually occur, but in a remarkable double form. Propositions II 5, 6 are formulated as follows:

II 5. If a straight line be cut into equal and unequal segments, the rectangle contained by the unequal segments of the whole together with the square on the straight line between points of section is equal to the square on the half.

II 6. If a straight line be bisected and a straight line be added to it in a straight line, the rectangle contained by the whole with the added straight line and the added straight line together with the square on the half is equal to the square on the straight line made up of the half and the added straight line.

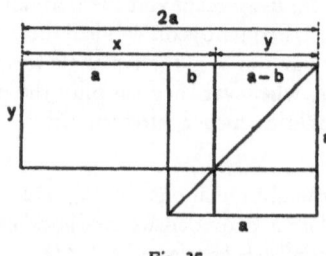
Fig. 35.

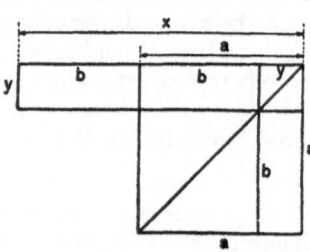
Fig. 36.

Designating the line given in II 5 by $2a$ and the segment between the points of division by b, we may express II 5 by formula (2). But, if the line given in II 6 is denoted by $2b$, and one half of it plus the extension by a, then this proposition is expressed by the same formula (2). But it can not have been the sole purpose of the two propositions to give formula (2) a geometric dress and to prove it in that way; for, why should two propositions be given for *one* formula?

Let us try something else! We call x and y the two unequal parts into which the line $2a$ is divided in II 5. Then II 5 can be expressed by means of the formula

(3) $$\left(\frac{x+y}{2}\right)^2 = xy + \left(\frac{x-y}{2}\right)^2,$$

i.e. by

(4) $$\left(\frac{x+y}{2}\right)^2 - \left(\frac{x-y}{2}\right)^2 = xy.$$

But, if x is used to designate the sum of the line given in II 6 and its extension, and y to denote the extension itself, then this proposition again leads to formula (3).

In both cases the proof is the same: the difference of the squares on *a* and on *b* is a "gnomon", a carpenter's square; this can be transformed into a rectangle *xy* by removing a piece (*ay* in the first case and *by* in the second) on the right side and placing it on the left side.

This gives again two propositions for one formula. Why? What was the line of thought of the man who formulated the propositions in this way?

The answer is found by following out, in the Elements themselves, and in Euclid's other works, the way in which propositions II 5 and II 6 are applied. In the Data (84 and 85), the question is considered how to prove that two segments, *x* and *y*, are known, when their sum *x* + *y* or their difference *x* -- *y*, and also the area of the rectangle formed by them, are given. When the sum *x* + *y* = 2*a* is given, then II 5 is applied to determine half the difference $\frac{1}{2}(x - y) = b$; if the difference is given, then II 6 is used so that half the sum is found from (3).

We see therefore, that, at bottom, II 5 and II 6 are not propositions, but *solutions of problems*; II 5 calls for the construction of two segments *x* and *y* of which the sum and product are given, while in II 6 the difference and the product are given.

The applications in the Elements themselves are consistent with this view. In II 11 a line has to be divided into mean and extreme ratio. In the terminology of area calculations, the problem receives the following form: "To divide a given straight line in such a way, that the rectangle, formed by the entire line and one of the parts is equal to the square on the other part". This leads to the equation

$$y^2 = a(a - y),$$

i.e.

$$y^2 + ay = a^2$$

or

$$y(y + a) = a^2.$$

This last equation calls for the construction of two segments *y* and *y* + *a*, of which the difference *a* and the product *a*² are given. Since the difference is given, the problem is solved by use of II 6.

This interpretation of II 5, 6 as solutions of problems is raised beyond all doubt by the generalizations VI 28, 29, which are quite openly formulated as problems:

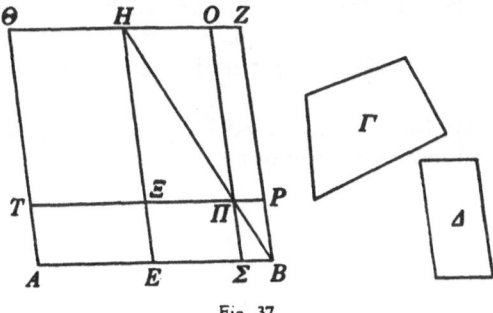

Fig. 37.

VI 28. To apply to a given line *AB* a parallelogram (*AΠ*), equal in area to a given rectilinear figure Γ, lacking a parallelogram $(B\Pi)$ similar to a given (parallelogram) Δ.

VII 29. To apply to a given line AB a parallelogram $(A\Xi)$, equal in area to a rectilinear figure and in excess by a parallelogram, similar to a given (parallelogram) Δ.

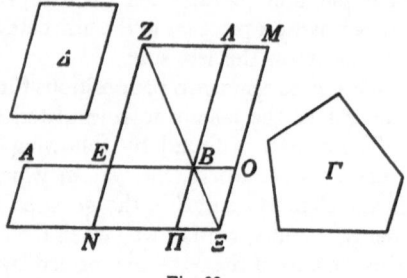

Fig. 38.

Figures 37 and 38 show clearly what is meant. In VI 28 the parallelogram $A\Pi$ (Euclid designates parallelograms by indicating two opposite vertices) must have the same area as the given polygon Γ, while the parallelogram $B\Pi$ on the remaining piece of the given line must be similar to the given parallelogram Δ.

Analogously we explain VI 29.

In the most important applications the given parallelogram is a square. The required parallelogram $B\Pi$, or $B\Xi$, must then be a square. If the base and altitude of the required parallelogram are denoted by x and y, the given area (of Γ) by F and the given line to which the rectangle has to be applied by $2a$, then the conditions in VI 28 are

(5) $$xy = F, \qquad x + y = 2a,$$

and in VI 29

(6) $$xy = F, \qquad x - y = 2a.$$

The solutions of problems (5) and (6), indicated in the generalizations VI 28, 29, is exactly the same as that supplied by propositions II 5, 6. In case (6), we have, by means of (3)

$$\left(\frac{x+y}{2}\right)^2 = xy + \left(\frac{x-y}{2}\right)^2 = F + a^2.$$

When F and $2a$ are given, this area is now known. If it is changed into a square, using II 14, then the side of this square is the half sum $\frac{1}{2}(x + y)$. When $a = \frac{1}{2}(x - y)$ is added to this, x is obtained; when it is subtracted, one finds y. Case (5) is treated analogously.

A comparison with the two normalized Babylonian problems B 1, 2 (see the end of Chapter III) and their solutions

(B1) $$\begin{cases} x + y = s \\ \quad xy = F \end{cases} \qquad \begin{cases} x = \frac{1}{2}s + \sqrt{(\frac{1}{2}s)^2 - F} \\ y = \frac{1}{2}s - \sqrt{(\frac{1}{2}s)^2 - F}, \end{cases}$$

(B2) $$\begin{cases} x - y = d \\ \quad xy = F \end{cases} \qquad \begin{cases} x = \sqrt{F + (\frac{1}{2}d)^2} + \frac{1}{2}d \\ y = \sqrt{F + (\frac{1}{2}d)^2} - \frac{1}{2}d, \end{cases}$$

shows clearly that these are entirely analogous to II 5, 6 and VI 28, 29.

Deficiency and excess, the parallelogram of given shape, which is lacking and which is left over, are called in Greek Elleipsis and Hyperbolè; the application to a line of a parallelogram of given area is called Parabolè. In his commentary on Euclid I 44, Proclus has the following to say about the history of these adaptations:

> These things, so says Eudemus, the peripathetic, are discoveries of the Pythagorean muse, namely the application of areas (parabolè), their deficiency (elleipsis) and their excess (hyperbolè). Later on these names were carried over to the three conic sections. [1]

The solution of problems (5) and (6) plays an extremely important part in Greek mathematics. We have already seen that they are dealt with twice in the Elements, and again in the Data. But, moreover, in Book X, which deals with the theory of irrational segments, an important construction (X 33—35) depends on the elliptic application of areas.

For the solution of quadratic equations, the Greeks reduced them to one of the forms

$$x(x + a) = F, \qquad x(a - x) = F, \qquad x(x - a) = F,$$

which are then solved by means of the application of areas. In the first and third cases, there are two line segments, x and $x \pm a$, of which the difference and the product are given, so that we have an application with a square in excess; in the second case, we have two line segments, x and $a - x$, of which the sum and the product are known, thus leading to an application with deficiency of a square. When the term in x^2 has the coefficient $\gamma = p : q$,

$$\gamma x^2 + xa = F, \qquad ax - \gamma x^2 = F, \qquad \gamma x^2 - ax = F,$$

the more general applications VI 28, 29 are applied, with an excess or a deficiency of given form (e.g. a rectangle).

According to Eudemus therefore, this important part of geometric algebra is a discovery of the Pythagoreans. However, in view of the fact that in both adaptations, with excess and with defect, it is always necessary to change a given area into a square (since a square root has to be extracted), proposition II 14, which solves this problem by use of the Theorem of Pythagoras, must also have been familiar to the Pythagoreans.

It occurs a second time that there are two propositions which express the same algebraic formula. In their formulation these propositions are quite similar to II 5, 6:

II 9. *If a straight line be cut into equal and unequal segments, the squares on the unequal segments of the whole are double of the square on the half and of the square on the straight line between the points of section.*

II 10. *If a straight line be bisected, and a straight line be added to it in a straight line, the square on the whole with the added straight line and the square on the added straight line*

---

[1] It will be made clear later on (in the discussion of Apollonius) what these adaptations have to do with conic sections.

*both together are double of the square on the half and of the square described on the straight
line made up of the half and the added straight line as on one straight line.*

Both propositions lead to the formula

$$x^2 + y^2 = 2\left\{\left(\frac{x+y}{2}\right)^2 + \left(\frac{x-y}{2}\right)^2\right\}.$$

Both can be taken to be solutions of problems. For II 9, the problem is: to
determine x and y when $x + y$ and $x^2 + y^2$ are given. For II 10: given $x - y$ and
$x^2 + y^2$, required x and y. Hence we are concerned with the solution of the systems
of equations

(7) $x + y = s,$ $x^2 + y^2 = F;$
(8) $x - y = d,$ $x^2 + y^2 = F.$

The solutions, indicated in II 9, 10, amount of course to the Babylonian solutions

(9) $x = \frac{1}{2}s + \sqrt{(F/2) - (s/2)^2},$ $y = \frac{1}{2}s - \sqrt{(F/2) - (s/2)^2},$
and
(10) $x = \sqrt{F/2 - (d/2)^2} + d/2,$ $y = \sqrt{F/2 - (d/2)^2} - d/2.$

The proofs make clever use of the "Theorem of Pythagoras".

The fact that all four of the normalized forms of systems of equations, which
we have found in the cuneiform texts, are taken over by Euclid, with their solutions, gives clear evidence of the derivation of the geometric algebra of Book II
from Babylonian algebra. *Apparently the Pythagoreans formulated and proved geometrically the Babylonian rules for the solution of these systems.*

We observe moreover that the simple linear equation

$$ax = F$$

leads, in geometric formulation, to the simple application of an area to a line,
without excess or deficiency. The pure quadratic $x^2 = F$ amounts to the transformation of a given area into a square (II 14). The pure cubic $x^3 = V$, in geometric form, poses the problem of constructing a cube of given volume. The
ancients were concerned with this problem as well; a special case is the famous
"duplication of the cube", to which we shall return later on. The mixed cubics
$x^2(x + 1) = V$ and $x^2(x - 1) = V$ were solved by the Babylonians with the
aid of tables, and the numbers in these tables are exactly the spatial numbers
$n^2(n \pm 1)$, which are called "Arithmoi paramekepipedoi" by Nicomachus (see
Becker, Quellen und Studien B4, p. 181).

Thus we conclude, that *all the Babylonian normalized equations have, without exception, left their trace in the arithmetic and the geometry of the Pythagoreans.* It is out of
the question to attribute this to mere chance. What could only be surmised before,
has now become certainty, namely that the Babylonian tradition supplied the
material which the Greeks, the Pythagoreans in particular, used in constructing
their mathematics.

PLATE 15

PL. 15. The Parthenon, seen from the South-East. Not the least reason why the beauty of the Parthe-non (built in 447–438 B.C.) attains such heights is the fact that the architects Íctinus and Callicrates, whose work was, according to Plutarch, supervised by Phidias, applied to this building their quite advanced, and indeed remarkable, theoretical knowledge of optical phenomena and of perspective. The ancients themselves tell us that Phidias was familiar with optics and geometry. That the architects of the Parthenon had a knowledge of optical phenomena is shown, among other things, by the slight arching of the stylobate (the upper one of the three steps on which the columns stand). Such arching, which occurs also in other Doric temples, can be interpreted as a deliberately applied device to correct the optical illusion of a hollow. Less certain is the explanation of the slop-ing of the columns, (backward) and of the wall of the inner sanctuary (backward), the abacuses of the columns (forward), the architrave and the triglyphs (backward) the cornice and the acroteria (forward). Vitruvius, the Roman architect (de Architectura III, 5, 13), says that the inward incline of the temple columns and the leaning forward of the structural elements above the columns serve to counterbalance an optical illusion, which he explains mathematically.

(Photo Alinari)

PLATE 16

PL. 16. The "Poseidon of Cape Artemision". Greek bronze original in severe style (compare Plate 17), representing a deity, probably Poseidon, according to some Zeus (about 465–460 B.C.). Splendid work of one of the foremost bronze-founders of Greek antiquity. The statue was raised from the sea at Euboea, practically intact. Height of the entire statue 2.14 m. (National Museum in Athens).

Why the geometric formulation?

Why did the Greeks not simply adopt Babylonian algebra as it was, why did they put it in geometric form? Was it their delight in the tangible and the visible, which turned them away from numbers, to occupy themselves with figures instead?

Unquestionably, the enjoyment of what can be seen was a strong motive power in the Greeks. The sculpture reproduced on plate 16, as well as numerous other immortal works of art, show this clearly. And there is ample further evidence in the marvelously vivid descriptions of Homer, and in Plato's image of the cave.

But all of this is insufficient to account for the complete elimination of algebra. We must remember that, according to the reliable report of Eudemus, it was particularly the Pythagoreans who laid the foundations of geometric algebra. But for the Pythagoreans, numbers were "the rock-bottom of the entire universe", the world was made "by imitation of numbers"; the heavens were for them "harmony and numbers". And, as Aristotle says (Metaphysics A5), they attained these views exactly because they applied themselves to mathematics. Would these worshippers of numbers have solved quadratic equations, not in terms of numbers, but by means of segments and areas, purely for the delight in the visible? This is hard to believe; there must have been another push towards the geometrisation of algebra.

Indeed this is not difficult to find: it is the discovery of the irrational, which, as Pappus tells us, actually originated in the Pythagorean school. The diagonal of the square is not commensurable with the side. But this means that, when the side is chosen as the unit of length, the diagonal can not be measured; its length can not be expressed, neither by an integer, nor by a fraction.

Nowadays we say that the length of the diagonal is the "irrational number" $\sqrt{2}$, and we feel superior to the poor Greeks who "did not know irrationals". But the Greeks knew irrational ratios very well. As we shall see later on, they had a very clear understanding of the ratio of the diagonal to the side of the square, and they were able to prove rigorously that this ratio can not be expressed in terms of integers. That they did not consider $\sqrt{2}$ as a number was not a result of ignorance, but of strict adherence to the definition of number. Arithmos means quantity, therefore whole number. Their logical rigor did not even allow them to admit fractions; they replaced them by ratios of integers.

For the Babylonians, every segment and every area simply represented a number. They had no scruples in adding the area of a rectangle to its base. When they could not determine a square root exactly, they calmly accepted an approximation. Engineers and natural scientists have always done this. But the Greeks were concerned with exact knowledge, with "the diagonal itself", as Plato expresses it, not with an acceptable approximation.

In the domain of numbers, the equation $x^2 = 2$ can not be solved, not even in that of ratios of numbers. But it is solvable in the domain of segments: indeed the diagonal of the unit square is a solution. Consequently, in order to obtain exact solutions of quadratic equations, we have to pass from the domain of numbers to that of geometric magnitudes. Geometric algebra is valid also for irrational segments and is nevertheless an exact science. It is therefore logical necessity, not the mere delight in the visible, which compelled the Pythagoreans to transmute their algebra into a geometric form.

Side- and diagonal-numbers.

In the Republic Plato speaks of the number 7 as the "rational diagonal", connected with the side 5. In explanation of this passage, Proclus gives the following definition of "side- and diagonal-numbers", which he attributes to the Pythagoreans and which, indeed, we meet with again in Theon of Smyrna and in Iamblichus.

"As the source of all numbers, unity is potentially a side as well as a diagonal. Now let two units be taken, one lateral unit and one diagonal unit; then a new side is formed by adding the diagonal unit to the lateral unit, and a new diagonal by adding twice the lateral unit to the diagonal unit." Thus we obtain the side-number 2 and the diagonal-number 3. From here on we proceed in similar manner with the numbers found thus far. In this way, we get

$$2 + 3 = 5, \qquad 2 \times 2 + 3 = 7,$$

etc., according to the formulas

(1) $$a_{n+1} = a_n + d_n, \qquad d_{n+1} = 2a_n + d_n.$$

The names side- and diagonal-numbers hint at the fact that the ratio $a_n : d_n$ is an approximation for the ratio of the side of a square to its diagonal. This follows from the identity

(2) $$d_n{}^2 = 2a_n{}^2 \pm 1$$

which, according to Proclus[1], was proved by use of II 10. Proposition II 10, which was quoted above, can indeed be expressed by the formula

$$(2a + d)^2 + d^2 = 2a^2 + 2(a + d)^2$$

If now $d^2 = 2a^2 \pm 1$, then subtraction of this relation from the preceding one gives

$$(2a + d)^2 = 2(a + d)^2 \pm 1.$$

Thus, if (2) is valid for a particular value of n, then it also holds for $n + 1$, but with the opposite sign. But, if we set $a_1 = d_1 = 1$, then (2) holds for $n = 1$; hence it holds for $n = 2$, etc.

We see from this that the Pythagoreans knew the principle of mathematical induction, certainly in essence, and that they applied geometric algebra to problems of the theory of numbers.

[1] Commentary on Plato's Republic II, chapters 23 and 27.

But how did they get the recursion formula (1)? I venture the following conjecture:

Greek mathematics knew the method of successive subtractions (antanairesis) for determining the greatest common measure of two commensurable magnitudes a and b: the smaller one, say a, is subtracted from the larger one, thus giving two new magnitudes a and $b - a$; then the smaller of these magnitudes is again subtracted from the larger one, etc. If a common measure exists, the process leads ultimately to two equal magnitudes $c = d$, which equal the greatest common measure. In Book VII of the Elements, this method is applied to numbers for determining the G.C.D. and, at the beginning of Book X, to arbitrary magnitudes to decide whether a common measure exists and to determine it, in case it does exist. If applied to two incommensurable magnitudes, the process continues *ad infinitum*.

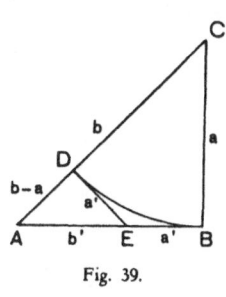

For example, if a is the side and b the diagonal of a square, then one can subtract a from b once (see Fig. 39). The remainder $b - a = AD = DE = EB = a'$ can now be subtracted from $a = AB$, leaving a remainder $b' = AE$. Now, a' and b' are again the side and the diagonal of a smaller square, and from

$$b' = a - a', \qquad a' = b - a$$

follows

$$(3) \qquad a = a' + b', \qquad b = 2a' + b'.$$

Fig. 39.

Formulas (3) already present the same form as the recursion formulas (1). Repetition of this same process of subtraction gives again a smaller side a'' and a smaller diagonal b''. If the process is continued until the difference between, say a''' and b''' has become too small to be observed and if one approximates by setting $a''' = b'''$, then, choosing a''' as the unit of length, a'' and b'', a' and b', and finally a and b are represented by means of (3), in the form of the successive side- and diagonal-numbers.

The problem of approximating to the ratio of the diagonal and the side by means of rational numbers, was proposed and solved by the Babylonians. But the Pythagoreans carried this old problem infinitely farther than the Babylonians. They found a whole set of approximations of indefinitely increasing accuracy; moreover they developed a scientific theory concerning these approximations and they proved the general proposition by complete induction. Again and again it becomes apparent that there were excellent number-theoreticians in the Pythagorean school.

Now we are going to take a look outside this school.

Anaxagoras of Clazomenae

is known chiefly as a natural philosopher. He was held in high esteem in Athens;

Pericles was one of his pupils and remained on intimate terms with him. When the popularity of Pericles was on the wane, shortly before the Peloponnesian war, his political opponents, probably to annoy him, accused Anaxagoras of atheism because he had taught that the sun and the stars were dead glowing rocks. He was condemned and compelled to leave Athens.

Anaxagoras taught that the moon receives its light from the sun and he gave the correct explanation of solar and lunar eclipses. He also estimated the size of the sun; according to him it even exceeded the Peloponnesus in size.

There is no doubt therefore that Anaxagoras had merits as an astronomer; nevertheless his influence on the development of astronomy was not wholly favorable. This was due chiefly to his view that the celestial bodies are inanimate objects which do not move according to mathematical laws, but are dragged along by the vorticial motion of the ether. He reasoned as follows: the planets, and especially the moon, which are nearer to the earth, are pulled along less rapidly than the fixed stars; that is why they fall behind the stars in the diurnal revolution. But what is the reason that, e.g., sun and moon, instead of moving in parallel circles, parallel to the equator, as the fixed stars do, move away from the equator, sometimes to the North, then again to the South, then turn about and again approach the equator? For this Anaxagoras has a mechanistic explanation as well (Hippolytus, Refutatio I 8):

> Sun and moon make a turn about, because they are driven back by the cold air in the North. The moon turns about more frequently because it can not master the cold air

(while apparently the sun, from its own heat, warms and softens the air.)

Astronomically, this point of view is extremely sterile. Instead of resolving the motion of the sun in a diurnal motion, along with the fixed stars, and an annual motion, in opposite direction, along the ecliptic, it is resolved into two components, one parallel and the other perpendicular to the equator; each of these components is accounted for in its own way. One is explained on the basis of the vortex motion, in which the sun lags one lap behind the stars each year, the other by means of the compressed air in the North. It is purely accidental in this theory that these two mutually perpendicular motions have exactly the same period, viz. one year. The fundamental importance of the ecliptic as the sun's orbit, and, approximately, as that of all the planets, does not appear at all. In this respect therefore, Anaximander who discovered the inclination of the ecliptic a century before Anaxagoras, and Alcmaeon, who taught us the proper motion of the planets in a direction opposite to that of the fixed stars, were ahead of Anaxagoras with his modern, enlightened ideas.

Democritus of Abdera

developed these ideas in a still more materialistic direction. He removed from the system the reason (νοῦσ) which had, according to Anaxagoras, set everything in order and had started the vortex motion, and made everything depend upon pure chance, upon pressures and impulses. He also thinks that the planets, and especially the moon, lay behind the stars, because they are nearer to us and are therefore less powerfully dragged along by the vortex motion in the sky. In contrast to this the Pythagoreans held that the planets are divine, animated, rational living beings which are enabled by their reason to describe their eternal circles according to mathematical rules. This point of view is much more fertile for astronomy, because it stimulates man to seek for himself an understanding of these rules and to find these circles.

It is said that, while in prison. Anaxagoras attempted the quadrature of the circle. When we come to a discussion of solid geometry, we shall revert to the treatises on perspective of Anaxagoras and Democritus.

Oenopides of Chios

was somewhat younger than Anaxagoras, according to Proclus; he must therefore have reached the highest point of his life shortly after 450. He is best known as astronomer. Theon of Smyrna records that, according to Eudemus, he invented "the cincture of the zodiac circle and the long year". What is meant by the cincture of the zodiac circle, is not quite clear. The "long year" is used by Plato to designate a long period, after which all the planets return to their initial positions, a common multiple therefore of the periods of all the plantes (including sun and moon). This does not necessarily mean that the concept "long year" has the same meaning for Oenopides that it has for Plato. Later doxographers, such as Aetius, consider the long year of Oenopides as similar to calendar-periods, such as those of 8 and of 19 years. They believe that Oenopides took the long year to be a period of 59 years and the ordinary year as consisting of 365 and 22/59 days. Should this be accepted? The Babylonians had a period of Saturn of 59 years, equal to 2 sidereal periods, but I do not know whether it has any connection with this.

From the statement of Eudemus, we can certainly conclude that Oenopides occupied himself with periods of revolution or with calendar periods and that he made a determination either of the position or of the division of the ecliptic. He did not belong therefore to the followers of Anaxagoras, who assumed that the planets were dragged along without any order and tossed about by wild vortices, but rather to the school of the Pythagoreans, who assigned to the planets fixed orbits in the Zodiac.

In connection with the well-known construction for dropping a perpendicular (Euclid I 12), Proclus informs us that Oenopides was the first to deal with this problem because he considered the construction to be useful for astronomy. Proclus holds (I 23), on the authority of Eudemus, that he also discovered the method of transferring a given angle.

On the basis of this statement it has sometimes been concluded that, at the time of Oenopides, geometry was still in a very primitive state, because even such simple constructions were not yet known. It seems to me that such a conclusion is entirely incorrect. On the contrary, we shall see from the quadrature of the lunules of Hippocrates, who was not much younger than Oenopides, that in his time geometry had reached a very high level, not only as judged by the number of propositions that were known, but also by the demands of rigor in proof.

It would seem to me preferable to assume that, before Oenopides, the need for these constructions had not been felt, that for instance, right angles were constructed in a half circle, using the proposition of Thales, or that they were drawn simply by use of a carpenter's square, and that it was Oenopides who invented the constructions, in view of their application in astronomy.

Squaring the circle.

It has already been mentioned that, while he was in prison, Anaxagoras, occupied himself with the quadrature of the circle. Altogether this problem was very popular towards the end of the 5th century. The comic poet Aristophanes even made a joke about it. In "The Birds", he introduces the astronomer Meton, who says:

> "With the straight ruler I set to work
> To make the circle four-cornered;
> In its center will be the market place,
> Into which all the streets will lead,
> Converging to its center like a star,
> Which, although only orbicular, sends
> Forth its rays to all sides in a straight line."
> "Verily, the man is a Thales!",

scoffs Pisthetaerus, the leader of the birds, and drives Meton away with blows.

It seems clear how the circle is here "made four-cornered" by means of two streets which meet at right angles in the center. Is it not marvelous that a scholary problem was so popular in Athens at that time, that it could be made a source of amusement in the theatre?

Meton's quadrature was only a jest. But, although not accurate, the quadrature of the circle of the sophist

Antiphon

was intended to be serious.

The late compendium Suda, usually called Suidas, contains the following note:

> Antiphon, an Athenian, an interpreter of signs, an epic poet and a sophist. He was called word-cook.

According to Xenophon (Memorabilia 16), Antiphon reproached Socrates for his simple way of living, which could only make him and his imitators unhappy. He also told him that it was not sensible to reveal his theories without charge.

This typical representative of the guild of sophists thought that it is quite easy to square the circle. His method has been preserved due to the fact that Aristotle criticised it casually. In the Physica (185a), Aristotle says that it is not even necessary to refute Antiphon's quadrature of the circle, because it is not based on the recognized principles of geometry. The commentators (Simplicius, Themistius and Joannes Philoponius) then explain how Antiphon proceeded.

He inscribes some polygon in the circle, perhaps a triangle or a square. By bisecting the arcs, he obtains a polygon of double the number of sides, a hexagon or an octagon. He believed that, by continuing in this way, he would ultimately obtain a polygon, whose sides, on account of their small size, would coincide with the circumference. Since for any polygon, one can construct a square of equal area, and since the polygon that has been obtained is to be considered as equal to the circle, he concluded that the circle can also be squared. [1]

Antiphon is of course quite right in thinking that by this method the circle can practically be squared. Equally right are his opponents in saying that the assertion that a polygon can coincide with the circle, is contrary to the principles of geometry.

The most famous among the geometers of the 5th century,

Hippocrates of Chios

(not to be confused with the still more famous physician Hippocrates of Cos), also occupied himself with the quadrature of the circle.

He started as a merchant, but, according to Aristotle, not a clever merchant, for he allowed himself to be relieved of his money by crooked tax officials in Byzantium (according to others, by pirates). Afterwards (probably around 430) he came to Athens; he became very famous as a geometer.

We can get some idea of the way in which he attacked this problem from a statement which Simplicius, the most learned and most reliable among the commentators of Aristotle, gets from his teacher Alexander of Aphrodisias. According

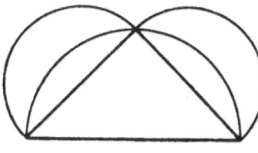
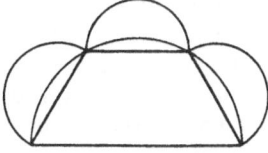

Fig. 40.   Fig. 41.

to this Alexander, Hippocrates began with an isosceles right triangle and proved that the sum of the areas of the two lunules, formed by the semicircles on the

---

[1] See F. Rudio, *Der Bericht des Simplicius über die Quadraturen*, Leipzig 1907. The differences in the statements of the three commentators are insignificant. It is probable that Eudemus is the source for all of them.

right sides and by the semi-circle circumscribed about the triangle, is equal to the area of the triangle. This quadrature of the two lunules agrees with the first of the three quadratures in the fragment of Hippocrates that will be discussed presently. Next Hippocrates took an isosceles trapezoid formed by the diameter of a circle and three consecutive sides of an inscribed regular hexagon. He proved that the sum of the areas of a semicircle on one of the sides of the hexagon and of the three lunules formed by the semicircles on the sides of the hexagon and by the semi-circle circumscribing the trapezoid, is equal to the area of the trapezoid. Therefore, if it were possible to "square" the three lunules, it would also be possible to "square" the semicircle and hence the circle. This led Hippocrates to study the squaring of lunules, bounded by arcs of circles.

The manner in which he squared such lunules can be learned from a famous fragment, copied word for word by Simplicius, according to his own statement, from the History of Mathematics of Eudemus, with the addition of a few clarifying references to Euclid. Allmann, Diels and Usener, Tannery, Rudio, Heiberg and finally Becker [1] have tried to restore the original text of Eudemus, and to eliminate the additions of Simplicius. The purified text begins as follows:

> The squaring of the lunules, considered as remarkable figures on account of their connection with the circle, was first formulated by Hippocratus and his explanation was considered to be in good order. Let us therefore attack the matter and study it.
> He considered as the foundation and as the first of the propositions which serve his purpose, that similar segments of circles are in the same ratio as the squares of their bases. He demonstrated this by showing first that the squares of the diameters have the same ratio as the circles.

It is still an open question whether Hippocrates actually proved this rigorously. We shall see that the proof in Euclid XII 2 comes from Eudoxus. The text continues:

> For the ratio of the circles is the same as that of similar segments, since similar segments are segments which form the same part of the circle.

It has been shown convincingly by Dijksterhuis [2] that this explanation is not due to Simplicius, but to Hippocrates or to Eudemus. Hippocrates uses here the same concept of proportionality, that underlies the Pythagorean theory of numbers: four magnitudes are proportional if the first is the same part or the same multiple of the second that the third is of the fourth. Strictly speaking, this definition is applicable only in the case of rational ratios. Hippocrates had not yet arrived at a rigorous treatment of irrational ratios.

> After having proved this, he raised first of all the question how to square a lunule whose exterior boundary is a semicircle. He accomplished this by circumscribing a semicircle about an isosceles right triangle, and by constructing on the base a circular segment similar to the

---

[1] O. Becker, *Quellen und Studien* B 3, p. 411. See also the charming little book of F. Rudio, *Der Bericht des Simplicius über die Quadraturen des Antiphon und des Hippocrates*, Leipzig 1907.
[2] E. J. Dijksterhuis, *De Elementen van Euclides* I, p. 34, Groningen 1929.

segments cut off by the right sides. Because the segment on the base is equal to (the sum of) the two (segments) on the other (sides), it follows, when the part of the triangle, which lies above the segment on the base, is added to both, that the lunule is equal to the triangle. Since it has now been shown that the lunule is equal to the triangle, it can be squared. Thus, by taking a semicircle as the external boundary of the lunule, he could readily square the lunule.

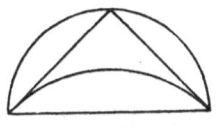

Fig. 42.

Observe that this squaring agrees with the first squaring mentioned by Alexander; in both cases the external boundary of the lunule is a semicircle and the internal boundary a quadrant of a circle.

Next he started from an (external boundary) greater than a semicircle. He constructed a trapezoid, of which three sides were equal to each other, while the fourth side, the longer of the two parallel sides, had a square equal to three times that of any of the others. He circumscribed a circle about the trapezoid, and constructed on the largest side a segment, similar to the segments which each of the other three sides cut from the circle.

Eudemus leaves the squaring of the lunules to the reader. The lunule obviously equals the area of the trapezoid; the proof is similar to that of the previous case.

By drawing the diagonal of the trapezoid one sees that the segment in question is greater than a semicircle. For the line which subtends two sides of the trapezoid must of necessity have a square which is more than twice as great as that on the remaining (side), while the squared longest side of the trapezoid is less than the diagonal and that one of the other sides which, with the diagonal, subtends this longest side. But then the angle which stands on the longest side is acute. Therefore the segment in which it is inscribed exceeds a semicircle; and this is the external boundary of the lunule.

Simplicius draws the accompanying explanatory diagram. In modern symbolism the argument proceeds as follows:

$$B\varDelta^2 = 3AB^2.$$
$$A\varDelta^2 > A\varGamma^2 + \varGamma\varDelta^2 = 2AB^2.$$
$$B\varDelta^2 = 2AB^2 + AB^2 < A\varDelta^2 + AB^2.$$

Apparently, Hippocrates applies the following propositions: The square of a side $(A\varDelta)$ of a triangle $(A\varGamma\varDelta)$, opposite an obtuse angle $(\varGamma)$ is greater than the sum of the squares of the other two sides. And, when

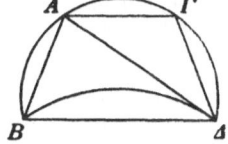

Fig. 43.

the square of a side is less than the sum of the squares of the other two sides, then the opposite angle is acute. And, when the angle inscribed in a segment of circle is acute, then the segment is greater than a semicircle. Obviously he is familiar with the concept of an "angle inscribed in a segment of circle" and he knows that all angles inscribed in the same arc are equal, and that a smaller angle corresponds to a smaller arc.

But supposing that it were less than a semicircle, he proved it on the basis of a construction like the following. Let there be a circle of diameter AB and center K, and let $\varGamma\varDelta$ bisect the (line) BK perpendicularly. And let EZ lie between this (perpendicular bisector) and the circle, directed towards B, while its square is one-and-one half times as great as that of the radius.

Hippocrates applies here one of the "Neusis" constructions, which occur so frequently in Greek mathematics. They require the construction of a line segment (*EZ*) of given length, whose terminal points have to lie on given straight lines or curves (the line *ΓΔ* and the circle) and whose extension is to pass through a given point (*B*). "Neusis" means inclination. We do not know whether he carried out this construction by means of compasses and straight edge — it leads to a quadratic equation in $BZ = x$ — or whether he used a so-called "adjustable straight-edge", i.e. a straight edge on which the distance *EZ* was marked off and which was allowed to slide past and rotate around *B* until the endpoints *E* and *Z* were on the circle and the perpendicular bisector *ΓΔ*. It is certain that Hippocrates must have constructed in advance the length *EZ*, which is the mean proportional between the radius *AK* and $^3/_2 \cdot AK$. He knew therefore how to construct the mean proportional *x* between two given lines *a* and *b*, and he also knew that the square x^2 is equal to the rectangle *ab*.

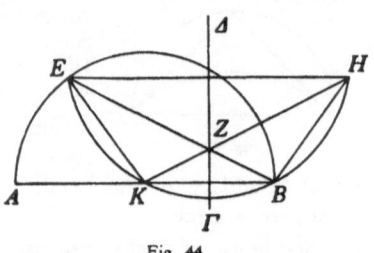

Fig. 44.

Now let *EH* be drawn parallel to *AB* and let *K* be joined to the points *E* and *Z*. And let the line towards *Z* meet the line *EH* in *H*, an let the lines joining *B* to *Z* and to *H* also be drawn. Then it is clear that the extension of *BZ* will pass through *E* and that *BH* will be equal to *EK*. If this is so, then a circle will circumscribe the trapezoid *EKBH*. And the trapezoid will have in its interior a circular segment circumscribed about the triangle *EZH*. And the lunule that results will be equal in area to the rectilinear figure, composed of the three triangles (*EZK*, *HZB* and *KZB*). For, the segments which *EZ* and *ZH* cut off from the rectilinear figure inside the lunule, are equal to the segments outside the rectilinear figure, since each of the two on the inside is $^3/_2$ times as great as the outside ones. If, therefore, the lunule consists of the three segments and the rectilinear figure, except for the two segments, and if the rectilinear figure is itself obtained by adding the two segments but removing the three, then, the two segments being equal to the three, the lunule will be equal to the rectilinear figure.

He proves that the outer boundary of this lunule is less than a semicircle, by showing that the angle inscribed in the outer segment is obtuse. And that this angle is obtuse, he proves as follows: Since the square on the line *EZ* is $^3/_2$ times as great as that on the radii, and the square on the line *KB* is more than twice as great as that on *BZ*, it follows that the square on *KE* is more than twice as great as that on *KZ*. The square on the line *EZ* is $^3/_2$ times as great as that on *EK*. Therefore the square on *EZ* is greater than those on *EK* and *KZ* together. Therefore the angle at *K* is obtuse, so that the segment in which it is inscribed is less than a semicircle.

It is not entirely clear how Hippocrates concludes that KB^2 is greater than $2BZ^2$. The manuscripts add the explanation: "because the angle at *Z* is larger". It seems probable that the line of thought was as follows: from $EZ^2 = ^3/_2 \cdot EK^2$, it follows that $EZ > EK$, so that $\angle EZK$ is acute and $\angle KZB$ obtuse, therefore

$KB^2 > KZ^2 + BZ^2 = 2BZ^2$. Following the text, the argument proceeds now as follows:

$$KE^2 > 2BZ^2,$$
$$EZ^2 = 3/2 \cdot EK^2 = EK^2 + \tfrac{1}{2}EK^2 > EK^2 + KZ^2;$$

therefore the angle at K is obtuse.

> In this manner Hippocrates squared every lunule, whether the outer boundary was a semicircle, or greater or less than a semicircle.
>
> To square a lunule together with a circle, he proceeded as follows:
>
> Let there be two circles with centers at K, and let the square of the diameter of the outer circle be 6 times as great as that of the inner circle; let a hexagon be inscribed in the inner circle and let the lines KA, KB and $K\Gamma$ be the extended from the center until they meet the outer circle in H, Θ and I. And let a segment be described about the line HI similar to the segment cut off by $H\Theta$. Since the square on the line HI is three times as great as that on the side $H\Theta$ of the hexagon, which in turn is six times as great as that on AB, it follows that the segment constructed on HI is equal to the sum of the segments cut off from the outer circle by the lines $H\Theta$ and ΘI, together with (all the segments cut off) from the inner circle by the sides of the hexagon. Therefore the lunule $H\Theta I$ will be as much less than the triangle designated by the same letters as the segments cut off from the inner circle by the sides of the hexagon (are together). The lunule and the segments cut off by the hexagon are together equal to the triangle. And when the hexagon is added to each of these, it follows that the triangle and the hexagon are together equal to the sum of this lunule and the inner circle. By determining the areas of these rectilinear figures, one can therefore also square the circle plus the lunule.

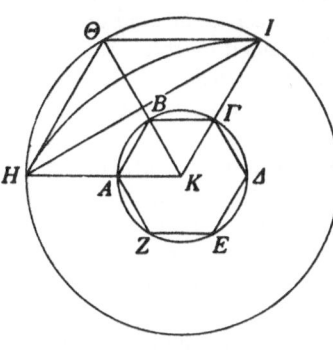

Fig. 45.

Hippocrates obviously knows also the properties of the regular hexagon, such as the fact that the square on the diagonal equal three times the square on the side. He must therefore have known also that the side equals the radius.

Looking at all these developments as a whole, we see in the first place that Hippocrates mastered a considerable number of propositions from elementary geometry. The Elements of Geometry, which he has written according to the Catalogue of Proclus, must have contained a large part of Books III and IV of Euclid, which deal with the circle and the inscribed polygons, as well as the contents of Books I and II, which already were "baby food" for the Pythagoreans. Hippocrates knows the relation between inscribed angles and arcs, and the construction of the regular hexagon; he knows how to circumscribe a circle about a triangle and he knows that a circle can be circumscribed about an isosceles trapezoid. He is familiar with the concept of similarity and he knows that the areas of similar figures are proportional to the squares of homologous sides. He knows

not only the Theorem of Pythagoras for the right triangle (I 47), but also its generalization for obtuse- and acute-angled triangles (II 12—13). Furthermore, he
is able to square an arbitrary rectilinear figure, i.e. to construct a square with the
same area. By means of this he knows how to construct lines whose squares have
the ratio 3 : 2 or 6 : 1 to the square on a given line.

Still more important for our evaluation of the mathematical level, reached during the second half of the fifth century and of Hippocrates in particular, is the
excellent demonstrative technique and the high requirements of rigor demanded
in the proofs. Hippocrates is not satisfied merely to construct the lunules and to
conclude from the drawings that the external boundary is greater than or less than
a semicircle, he wants to prove this rigorously. The operation with inequalities
is a very late achievement of modern mathematics. Euler did not worry much
about the convergence or divergence of his infinite series; he used divergent series
in his calculations, without any scruples and obtained correct results. It was not
until later, during the "critical period" of modern mathematics, that the necessity
of the "epsilontics", the calculations in terms of ε and δ, and the estimation of
remainders were recognized.

During the beginning of the fourth century, Greek mathematics passed through
a similar "critical period". The estimations of Eudoxus, who showed that the
difference between the inscribed and the circumscribed polygons of a circle can
be made arbitrarily small, are on a level with modern estimations of remainder
terms. But, these estimations were possible because Hippocrates had shown how
to give exact proofs of inequalities.

Hippocrates must also have occupied himself with the duplication of the cube,
more exactly with the problem of constructing a cube whose volume has a given
ratio (e.g. 1 : 2) to that of a given cube.

It is an established tradition [1] that Hippocrates was the first to recognize that
the problem would be solved if one could construct two mean proportionals x
and y between two given line segments a and b whose ratio is the given ratio,

(1) $a : x = x : y = y : b.$

For, as we have developed before, if (1) holds then the cube a^3 would have to
x^3 the same ratio as a has to b.

This leads us naturally to

Solid Geometry in the fifth century and Perspective.

Vitruvius reports that Agatharchus was the first to paint perspective wings for
the performances of the tragedies of Aeschylus. We also find the statement in
Vitruvius that he left a treatise on the subject. "This also led Democritus and
Anaxagoras to write on the same subject, i.e. how lines in a natural relation to each
other could be made to correspond to the sharp view and the dispersal of the rays,

[1] See Eutocius in *Archimedis Opera* III, 104. In the next chapter we shall return to the history of the problem and
to the reliability of the tradition.

after a definite point had been selected as center, in order that definite images on the painted wings might create the appearance of buildings, and so that the things pictured on the plane fronts might appear, partly as receding and partly as advancing" (Vitruvius, De Architectura VII, praefatio). [1] See also Plate 19.

Anaxagoras and Democritus were highly respected as men of science. If they have written monographs on perspective, then it is not possible that these would have contained merely practical rules for drawing, such as would probably have been found in the work of the painter Agatharchus, e.g. that parallel lines must be drawn so as to converge to a point; they would have based such rules on the study of solid geometry. The dispersal of rays, mentioned by Vitruvius, must have referred to the rays converging in the eye from the points in space and which cut the image plane in the image points.

From this we conclude that, around 450, solid geometry must have reached a stage of development which made possible the formulation and the solution of problems which arise in perspective.

Democritus

had indeed acquired a reputation as a geometer. There are several mathematical titles among his extant works: On the tangency of circle and sphere, on geometry, on numbers, on irrational line-segments, and on "spreadings" (ekpetasmata), i.e. on mapping a spherical surface on a plane. [2] He takes pride in the fact that in the composition of lines with proofs, no one excelled him, not even the so-called rope-stretchers (probably surveyors) in Egypt. It is reported that he has been in Egypt, Persia, Babylonia, and, according to some, even in India and Ethiopia.

Cone and pyramid.

Archimedes says (in the Introduction to the "Method"), that Democritus was the discoverer of the fact that the volume of a cone, or of a pyramid, is equal to $\frac{1}{3}$ of

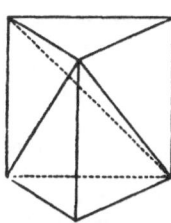

the volume of the cylinder, or the prism, with the same base and the same height, but that he did not prove it rigorously (this was first done by Eudoxus). It is possible that Democritus had learned the formula for the volume of a pyramid in Egypt; but on what argument did he base his conviction that this formula, and the corresponding one for the cone, are indeed correct?

If the proposition is valid for a three-sided pyramid (tetrahedron), then it holds also for an arbitrary pyramid; a limiting

Fig. 46.

process then furnishes the proposition for the cone also. It is a simple matter to divide a three-sided prism into three tetrahedra, which are,

---

[1] In Chapter I, 2, 2, Vitruvius defines scene-painting as a representation of front and sides, in which all lines correspond to the center of a circle.

[2] This explanation of the word is taken from the Geography of Ptolemy (VII 7).

two by two, equal to each other in base and in height. It follows therefore that each of these tetrahedra is equal to ⅓ of the prism, provided one can prove that pyramids of equal bases and equal heights, have equal volumes.

According to Cavalieri, it is possible to convince oneself of the equality in volume of two pyramids (or cones) with equal bases and heights by slicing them by means of planes parallel to the bases, and by considering these slices, approximately, as prims (or cylinders). It is possible that Democritus did something like this. This would at any rate explain the statement of Plutarch that Democritus raised the following question: "If a cone is cut by surfaces parallel to the base, then how are the sections, equal or unequal? If they were unequal" (and, we might add mentally, if the slices are considered as cylinders), "then the cone would have the shape of a staircase; but if they were equal, then all sections will be equal, and the cone will look like a cylinder, made up of equal circles; but this is entirely nonsensical."

Plato on Solid Geometry.

We see therefore that in the 5th century, solid geometry was an active field of study of various mathematicians. If we add the remark that the Pythagoreans knew three of the five regular polyhedra, and that Hippocrates had already reduced the problem of the duplication of the cube to that of constructing two mean proportionals between two given segments, and if we consider furthermore the large amount of solid geometry presupposed in the solution of this problem by Archytas, to be discussed presently, then it becomes really very difficult to account for Plato's complaint, in The Republic, of the poor state of development of solid geometry. He writes (Republic, 528B):

> (Socrates) After plane surfaces, we went on to solids in revolution before studying them in themselves. The right way is next in order after the second dimension to take the third. This, I suppose, is the enlarging of cubes and of everything that has depth.
> (Glaucon) Why, yes, it is; but this subject, Socrates, does not appear to have been investigated yet.
> (Socrates) There are two causes of that; first, inasmuch as no city holds them in honour, these inquiries are languidly pursued owing to their difficulty. And secondly, the investigators need a director, who is indispensable for success and who, to begin with, is not easy to find, and then, if he could be found, as things are now, seekers in this field would be too arrogant to submit to his guidance. But if the state as a whole should join in superintending these studies and honour them, these specialists would accept advice, and continuous and strenuous investigation would bring out the truth. Since even now, lightly esteemed as they are by the multitude and hampered by the ignorance of their students as to the true reasons for pursuing them, they nevertheless in the face of all these obstacles force their way by their inherent charm and it would not surprise us if the truth about them were made apparent.

Taylor [1] looks upon this passage not as a complaint of the condition of solid geometry around 374, when Plato wrote The Republic, but of its state in 422, the year in which the dialogue takes place. To this Dijksterhuis raises the valid

[1] A. E. Taylor, *A commentary on Plato's Timaeus*, Oxford 1928.

objection that in that case, the character of actuality, which marks the passage, would be hard to account for. Is it conceivable that Plato, for the sake of the historical accuracy of the picture, would have spoken so caustically about an undesirable state of affairs which had existed 50 years earlier and which had been improved long since? We prefer to assume, with Eva Sachs [1], that Plato is here, so to speak, concerned with the government of his Academy. He wants to lead the mathematicians in the Academy to a more systematic cultivation of solid geometry under his direction; and it appears from the last sentence of the quotation (perhaps added later) that he succeeded in this.

But what does Plato mean here by solid geometry, or, in his own words, the theory of the "sterea", the spatial bodies? Does it include the entire theory of planes and lines, as developed in Book XI of the Elements, the theory of perspective, about which Anaxagoras and Democritus had written books, the volume of the pyramid, found by Democritus, and so forth? If so, it is incomprehensible that Glaucon can say, with the concurrence of Socrates, that these things do "not appear to have been investigated yet". We are compelled to assume therefore that Plato took solid geometry in a more restricted sense.

This more restricted sense is indicated in Plato's own words: "the enlarging of cubes". One of the famous problems of Greek mathematics is the enlarging of a cube in a prescribed ratio, i.e. the construction of a cube whose ratio to a given cube is the same as that of two given line segments. For the ratio 2 : 1, this gives the "duplication of the cube", the universally known Delian problem.

The duplication of the cube.

In the next chapter I shall discuss the history of this problem in some detail, known to us chiefly from Eutocius' commentary on Archimedes. A few of the main points will be touched upon now.

We have already seen that Hippocrates of Chios reduced the general problem of increasing the volume of a cube in a given ratio to that of constructing two mean proportionals between two given line segments. Archytas of Taras discovered an extremely ingenious three-dimensional solution of this problem. Later, solutions were given by Eudoxus, Menaechmus, Plato(?), Eratosthenes, Nicomedes, Apollonius, Heron, Philon of Byzantium, Diocles, Sporus and Pappus. Obviously, great importance was attributed to this problem.

In the supplement to the Laws of Plato, known as the Epinomis, counted as one of Plato's works but not published until after his death, the problem of the two mean proportionals is indeed considered as *the* problem of solid geometry. [2] First, plane geometry is defined here as the science, which teaches us how to make similar two numbers, which are not themselves similar. Two numbers *ab* and *cd*

---

[1] Eva Sachs, *Die fünf platonischen Körper*, Berlin 1917.
[2] See O. Toeplitz, *Die mathematische Epinomisstelle, Quellen und Studien* B 2, p. 334 and O. Becker, *Quellen und Studien* B 4, p. 191.

(considered as area of rectangles of sides a, b and c, d respectively) are there called similar if their sides are proportional

$$a : b = c : d.$$

According to Euclid VIII 18, two numbers are similar only if a mean proportional exists between them. But plane geometry shows how to construct a mean proportional between any two lines, and to transform any rectangle into a square; hence, even if two numbers, as given, are not similar, they can be made into similar numbers by means of plane geometry.

The Epinomis proceeds now from plane numbers to spatial numbers, consisting of three factors. In this new connection two numbers abc and def are called similar, if their sides for a continued proportion:

$$a : d = b : e = c : f.$$

And now solid geometry is defined as "the new art, which teaches us how to make similar in this sense two numbers which, as given, are not similar". In particular therefore, solid geometry shows how to transform any number into a cube, and hence how to construct two cubes the ratio of whose volumes is equal to that of two arbitrary integers. For the Epinomis, this is evidently *the* problem of solid geometry; no other definition of this subject is given.

Confrontation of this definition with the passage in Plato's Republic, in which the purpose of solid geometry is defined as "the enlargement of cubes and of all things which have depth", shows that for Plato himself, the enlargement of a cube in a given ratio is also the outstanding problem of solid geometry. And now it also becomes clear why he can write that these things do "not appear to have been investigated yet". Apparently the solution of Archytas had not yet been obtained, at least not yet known in Athens, around 375, when Plato was making his plans for The Republic. Perhaps Plato got the news of this solution just before the publication of The Republic, and was then led to the more optimistic tone noticeable at the end of the passage quoted above.

But, how did it come about that Plato and his mathematical friends considered the problem of the enlargement of the cube as of such extreme importance?

Let us first have a look at the state of plane geometry. In the above passage, Plato mentioned as the most important planimetric operations, to change an area into a square (τετραγωνίζειν), the application and the addition of areas. Now, all these operations come from geometric algebra; changing an arbitrary rectilinear area F into a square amounts to solving the pure quadratic $x^2 = F$; the application of an area F to a line a, without excess or deficiency, reduces to the solution of the linear equation $ax = F$; with excess or with deficiency, it leads to the solution of an arbitrary quadratic equation or to the solution of a system of 2 equations with 2 unknowns of the form

$$\begin{cases} x + y = a, \\ xy = F. \end{cases} \quad \text{or} \quad \begin{cases} x - y = a, \\ xy = F. \end{cases}$$

PLATE 17

Pl. 17. Athenian school. Attic bowl with red figures of the painter Duris, about 485–480 B.C., (Staatliche Museen, Berlin.) This represents the interior of a "gymnasion", where the teachers of music and the grammarians gave their lessons. On the left a lesson is taking place in playing the lyre; on the right, in reciting. The other side of the bowl, not shown here, represents lessons in writing and in playing the flute. On the scroll, which the teacher in the center holds up before the pupil, we read the opening words of an epic poem, beginning, as is appropriate, with an invocation of the Muses: "Oh Muses, I begin to sing of the full-flowing Scamander" (one of the two rivers in plain of Troy). Without doubt, the young man in front of the teacher presently has to recite the poem. Is it he who is honored by the inscription visible at the upper edge of the bowl: "Hip(p)odamas is beautiful"? To the right of the pupil, the "paedagogos" sits quietly waiting for the end of the lesson; then he has to take the boy home. The "musike" which the boy was taught (art of the Muses) comprised literature as well as music; the latter played an imporant part in the life of the Greeks.

PLATE 18

PL. 18b. Plato (427–348/7 B.C.) Beautiful Roman copy in marble, from the early years of the Empire, after a Greek original of about 370 B.C., probably by Silanion. Private collection. (See R. Boehringer Platon, Bildnisse und Nachweise, Breslau 1935).

PL. 18a. Statuette of Socrates from the 2nd century A.D., after a bronze original of about 335 B.C. This was perhaps (according to K. Schefold) the famous statue of Lysippus in the Pompeion in Athens. British Museum.

PL. 18d. Euclid (?) Miniature from the manuscript of the Roman surveyors (agrimensores) in Wolfenbüttel. Landesbibliothek Brunswick (M.S. 2403), 6th century A.D. It is possible that this representation derives from a Greek portrait, perhaps of Euclid's own time. The diadem in the hair is antique and seems to have been taken from an ancient original (Schefold).

PL. 18c. Aristotle (384–322 B.C.). Copy in marble (30–50 A.D.) of a famous statue, probably in bronze, made during the master's life (about 325 B.C.). It may be supposed that this was the statue erected by Alexander the Great in honor of his great teacher (according to K. Schefold).

Finally, the "adding" of areas or of lines is after all only the geometric equivalent of addition.

Thus we see, that what Plato calls plane geometry is mainly the geometric algebra of the Pythagoreans. It is not surprising therefore that he looks upon solid geometry as the generalization of geometric algebra to space, i.e. as the geometric interpretation of the calculation with products of three factors each. The first new problem that arises here, is the solution of the pure cubic $x^3 = V$, i.e. the construction of a cube of given volume. It is therefore entirely logical to consider this as the central problem of solid geometry. Obviously it is not the only problem; there are also equations of the form $x^2(x + a) = V$ and other similar ones. That is why Plato adds the words "and everything which has depth" to "the enlargement of cubes", thus opening the way for other problems.

For us, the most important result of this clarification is that we recognize more and more clearly the line of development of geometric algebra from the Babylonians, by way of the Pythagoreans, to the men of Plato's time.

In connection with geometric algebra, at the end of the 5th century, the problem of the *irrational* moves to a position of increasingly central importance. A very important contribution to the clarification of this problem was made by the famous geometer

Theodorus of Cyrene.

Who was this Theodorus? Diogenes Laërtius calls him Plato's teacher. Iamblichus mentions him in his Catalogue of Pythagoreans, but this does not mean very much, because Iamblichus was a freak, who had the tendency to include every famous mathematician among the Pythagoreans. We had therefore better confine ourselves to Plato who introduces the grey Theodorus himself in his dialogue Theaetetus, and who puts highly interesting things about Theodorus' treatment of irrational sides of rational squares into the mouth of the chief protagonist Theaetetus.

Theodorus and Theaetetus.

Theodorus of Cyrene appears as an old man in the dialogue Theaetetus, which is laid in 399; he must therefore have been a contemporary of Hippocrates and Democritus. But he is not too old to give a lecture on incommensurable line segments which arouses the deep interest of the young Athenians Theaetetus and Socrates. As a result of this lecture, Theaetetus started thinking; and he found a general solution of the problem which Theodorus had treated for a few special cases. He classified all line segments, which produce commensurable squares, into those which are commensurable and those which are incommensurable. This can be found in the dialogue Theaetetus.

Theaetetus fell on the battle field in 369. and Plato's dialogue is dedicated to the young hero's memory. In the poem. his character and his merits are spoken of at great length. When, in the dialogue itself, Plato sketches in a brief passage, which gives the impression of having been dragged in, a mathematical discovery of his young friend, and describes in detail what had previously been found by Theodorus and what Theaetetus has added to this, it is reasonable to assume that the separation of the contributions of Theaetetus, from those of Theodorus, is historically correct, even though the detailed circumstances of the dialogue may be fictional. It can not have been Plato's intention to give credit to Theodorus for what is due to Theaetetus, nor vice versa.

The subject matter of Theodorus' lecture is sketched as follows by the person Theaetetus in the dialogue:

"Here our Theodorus drew (or wrote) something about sides of squares (περὶ δυνάμεων τι ἔγραψε) and showed (ἀποφαίνων) that those of three or five feet are not commensurable in length with those of one foot, and in this manner he took up one after another up to the one of 17 feet; here something stopped him (or: here he stopped)."

The expression "commensurable in length" (μήκει σύμμετρος) also occurs in Euclid's Book X and means that a common measure exists. It is applicable only to line segments; hence the word δυνάμεις can not mean "squares", as it does later on in Diophantus, but only "sides of squares". Indeed, a little further on Theaetetus says himself that henceforth he will designate certain sides of squares as δύναμις (in restricted sense). Philologically, δύναμις (impulse, force) can indeed very well mean the side which produces a square; it is not necessary, as Tannery does, to replace the word δύναμις by δυναμένη (creating).

"Those of three or five feet" are the sides of the squares whose areas are three and five square feet. We would represent their lengths by the numbers $\sqrt{3}$ and $\sqrt{5}$, but, in the Greek acceptation of the word, $\sqrt{3}$ and $\sqrt{5}$ are not numbers and the line segments referred to have no lengths that are expressible in numbers; they are incommensurable. It was the purpose of Theodorus' lecture to show this. Let us therefore avoid the modern notation $\sqrt{3}$ and rather denote the side of the square by w_3.

It matters little how Theodorus constructed the sides $w_3, w_5, \ldots w_{17}$ of the squares whose areas are 3, 5, ..., 17 square feet. Perhaps he simply followed Euclid I 14 to transform rectangles of areas 3, 5, ..., 17 into squares; perhaps he used the

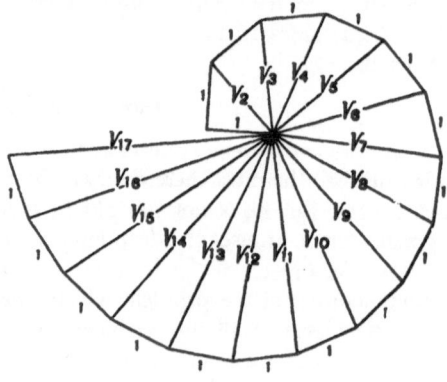

Fig. 47.

Pythagorean Theorem to pass from one to the next, as shown in Fig. 47, taken from the interesting monograph of J. H. Anderhub. [1]

Anderhub uses this figure to explain why Theodorus stopped with w_{17}; but, as we shall presently see, this can be accounted for in another way. More important is the question: how did Theodorus show that w_3, w_5, ... w_{17} and w_1 are incommensurable?

One might try to extend to w_3, w_5 ... w_{17} the "apagoge", i.e. the indirect proof, that we discussed earlier for w_3. This is indeed possible; on the assumption that w_q is commensurable, i.e. that w_1 and w_q have a common measure, one can obtain an equation of the form

$$m^2 = qn^2$$

By means of the theory of numbers of Book VII, which probably was in existence at the time of Theodorus, one can then show that this equation is possible only if q is itself a square, but not for $q = 3, 5, ..., 17$.

But it can not be assumed that Theodorus followed this procedure. For, if he had reasoned in this way, then it is entirely inexplicable why he treated the cases $q = 3$, $q = 5$, etc. each separately, as Plato says explicitly, and then stopped at $q = 17$, because the proof indicated above can very readily be carried through for the general case.

Theodorus was not an arithmetician, but a geometer. Is it not possible that he proved the incommensurability of w_3, w_5, ... w_{17} geometrically? A geometrical criterion for the incommensurability of two line segments is found in Euclid X 2:

"If, when the less of two unequal magnitudes is continually subtracted in turn from the greater, that which is left never measures the one before it, the magnitudes will be incommensurable."

The idea of subtracting in turn comes from the theory of numbers; if applied to two numbers, this process leads to the greatest common divisor, and if applied in general, to two commensurable magnitudes, it produces the largest common measure. Thus, if the process never comes to an end, there can be no common measure.

If Theodorus used this criterium, then it becomes immediately understandable why he had to deal separately with each of the cases w_3, w_5, ..., w_{17}; for the subtraction process runs differently in each of these cases.

How did he show that this process never ends, i.e. that the remainder can never measure the preceding one? Anderhub points out that ἀποφαίνειν does not mean to prove, but to designate, to show, to make clear, to bring into the light. He is therefore of the opinion that Theodorus did not really prove the infinite character of the process, but that he merely showed intuitively, that, at the start of the process, there is always a remainder, and that this continues until the remainder becomes so small, that the limited accuracy of a drawing in sand compels one to stop.

---

[1] Joco-Seria, *Aus den Papieren eines reisenden Kaufmannes*, Ausgabe der Kalle-Werke, Wiesbaden 1941

This does not appear to me as likely. At the time of Theodorus, towards the end of the 5th century, the demands of rigor in proofs had already reached such a level, that, as we saw in the discussion of the fragment of Hippocrates, the simple fact that an arc is greater than or less than a semicircle, was not accepted from a figure, but was considered subject to proof. It does not follow of course that every geometer necessarily was as painfully conscientious, but in the case of Theodorus some further elements have to be considered. From the obvious fact that, in each of the first five or six steps in the subtraction process, there always is a remainder, even though one may be willing to accept the conclusion without proof, nothing whatever follows as to the finitude or infinitude of the process, and hence nothing concerning the commensurability or incommensurability of w_3, w_5, . . ., w_{17}. How then could the mathematician Theaetetus, how could the keen critic Plato assert that Theodorus had "shown" or "made clear" this incommensurability? The most that could have been said is that, by means of his drawing, Theodorus had made the incommensurability of w_3, w_5, . . ., w_{17} probable. But in the dialogue, Theaetetus, trying to make as clear as possible a logical discrimination between the achievements of Theodorus and his own does not point to a difference in probability, but to the fact that there are infinitely many incommensurable sides of squares, which he wanted to bring under one concept, while Theodorus had dealt with a few of them, one by one.

Plato is very fond of appealing to mathematics to show that exact reasoning is possible, not about things which are seen and heard, but about ideal objects which exist in thought only. Speaking about mathematicians, and probably having in mind the proof of the incommensurability of the side and the diagonal of a square, he writes: "And do you not also know that they further make use of the visible forms and talk about them, though they are not thinking of them but of those things of which they are a likeness, pursuing their inquiry for the sake of the square as such and the diagonal as such, and not for the sake of the image of it which they draw? . . . but what they really seek is to get sight of those realities which can be seen only by the mind". (Plato, The Republic, Book VI, 510 DE).

Plato is apparently convinced that the mathematicians will agree with him. And indeed, when we deal with line segments which one sees and which one measures empirically, the question as to the existence of a common measure has no sense; a hair's breadth will measure integrally every line that is drawn. The question of commensurability makes sense only for line segments as objects of thought. It is therefore clear that Theodorus could not appeal to intuition to prove the incommensurability of the sides of his squares.

How then did he discover this fact? By developing further a hypothesis of Zeuthen, I am led to the following conjecture. If one can show that, after a certain number of steps, the subtraction procedure will lead to a pair of lines whose ratio is the same as the ratio of the pair from which one started, then one can conclude that the process will never end. Now this is indeed the case with all sides of squares whose areas are 3, 5, . . ., 17 square feet. For example, let $w = w_3$ be the side of a 3 sq.ft. square, and let $e = w_1$ be the unit of length. Then we have

$$w^2 = 3e^2.$$

Lay off e on w; the remainder is $w - e$. We have now two segments, e and $w - e$. In continuing the procedure we can replace e and $w - e$ by any other two segments that have the same ratio. Now

(1) $$e : (w - e) = (w + e) : 2e,$$

since the product of the means equals the product of the extremes, i.e.

(2) $$(w - e)(w + e) = w^2 - e^2 = 2e^2.$$

The same scheme is now applied to the segments $w + e$ and $2e$; we subtract $2e$ from $w + e$, thus obtaining $w - e$. Now $2e$ and $w - e$ are replaced by two segments in the same ratio:

(3) $$2e : (w - e) = (w + e) : e.$$

The proportionality (3) is proved exactly like (1). If we take now $w + e$ and e, and subtract the latter from the former, we obtain w and e. But these are the segments with which we started. Therefore the ratio is repeated, so that the process is periodic and will never end.

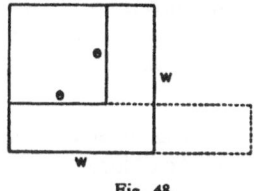

Fig. 48.

The equality (2) can be proved as follows by the use of areas: $w^2 - e^2$, the difference of the squares w^2 and e^2 is a "gnomon", as shown in Fig. 48. If the upper part is removed and then attached on the right, one obtains a rectangle whose sides are $w - e$ and $w + e$. Hence

$$(w + e)(w - e) = w^2 - e^2.$$

This method of proof is entirely analogous to the one used by the Pythagoreans in the application of areas with deficiency or with excess. It follows that, with the tools at his disposal, Theodorus was able to devise this proof or a similar one.

Analogous proofs can be given in all cases $w_3, w_5, w_6, \ldots, w_{17}$. The case w_{17} is especially simple, just as w, because only one proportion is required

$$e : (w - 4e) = (w + 4e) : e.$$

The next case w_{18} is not interesting, because $w_{18} = 3w_2$. But w_{19} is quite complicated, because it requires a sequence of 6 proportions

$$e : (w - 4e) = (w + 4e) : 3e,$$
$$3e : (w - 2e) = (w + 2e) : 5e,$$
$$5e : (w - 3e) = (w + 3e) : 2e,$$
$$2e : (w - 3e) = (w + 3e) : 5e,$$
$$5e : (w - 2e) = (w + 2e) : 3e,$$
$$3e : (w - 4e) = (w + 4e) : e,$$

each of which has to be proved by means of the calculation of areas. This makes it quite understandable why Theodorus closed his explanation with $\sqrt{17}$.

The detailed development in the work of Theaetetus may of course have differed from the one sketched here, but it is highly probable that he used incommensurability criterium X 2 to demonstrate the irrationality of w_3, w_5, . . ., w_{17}.

Theodorus on higher curves and on mixtures.

It seems that Theodorus also occupied himself with curves. For Proclus, who distinguishes the different kinds of "mixtures of the straight and the curved" in lines and surfaces as "synthesis, crasis and synchysis", writes that "Theodorus wrongly speaks of crasis in curves". We do not know exactly what he meant by this.

But we do know that in this period there were others who were concerned with curves.

Hippias and his Quadratix.

The famous sophist Hippias of Elis, who lived around 420, investigated a plane curve that is called the quadratix. Concerning this, Proclus writes in his commentary to Euclid: "Thus Apollonius derived the symptom for each of the conic sections, Nicomedes for the conchoids, Hippias for the quadratrix and Perseus for the spirals" (ed. Friedlein p. 356; German translation p. 414).

Explanation: By the *symptom* of a curve, the ancients meant the condition which a point has to satisfy to lie on the curve. Thus, e.g., the equation

$$x^2 = 2py$$

(or, in the Greek manner: the square on the line x equals the rectangle on the lines $2p$ and y) is the symptom for the parabola, when x and y are the oblique coordinates of a point on the parabola. Here it is supposed that the X-axis is tangent to the parabola at the origin, and that the Y-axis is parallel to the axis of the parabola.

The quadratix of Hippias is the same curve as the one described by Pappus on p. 252 of his great compendium and of which he says: "For the quadrature of the circle, Dinostratus, Nicomedes, and later geometers, used a curve which derived its name from this use, for they called it the quadratix". Hippias is not mentioned in this passage, probably because Hippias did not yet use the curve for the quadrature of the circle but for another purpose, e.g. for the trisection of the angle. In the next chapter, we shall return to Dinostratus' quadrature of the circle; we shall also reproduce there Pappus' description of the quadratrix.

The main lines of development.

Surveying the mathematics of the golden age, we can recognize six lines of development:

1. First comes the systematic foundation of plane geometry: the theory of parallels, the angle sum of the triangle, areas of polygons, proportions,

theorem of Pythagoras, angles and arcs in circles, regular polygons. This development is provisionally brought to a conclusion by the Elements of Hippocrates. There remains as an unsolved problem, to set up a theory of proportions which remains valid for incommensurable magnitudes.

2. The development of the theory of numbers had its root in the number mysticism of Pythagoras, attained a strictly scientific character in the theory of the even and the odd, and led finally to the systematic establishment of the theory of the divisibility and the proportion of numbers, as found in Book VII of the Elements.

3. The third important problem is that of the determination of the area of the circle, which arose naturally from the rules of calculation carried over from the Egyptians and the Babylonians. In the work of Hippocrates we find the careful formulation of the fundamental theorem that the area is proportional to the square of the diameter, and, based on this proposition, an exact quadrature of various lunules.

4. The fourth important line of development is that of algebra. The Pythagoreans carried forward the development of Babylonian algebra ("Flower of Thymaridas") and transformed it into a geometric algebra. In extending this geometric algebra to space, there arose, as the most important problem, the construction of a cube of given volume, the geometric equivalent of the equation $x^3 = V$. This problem was reduced by Hippocrates to the problem of two mean proportionals, but not solved.

5. Solid geometry was developed to a point, at which Anaxagoras and Democritus could outline a theory of perspective. Democritus found the formula for the volumes of the pyramid and the cone, but he did not find a rigorous proof. The Pythagoreans knew only three of the regular solids, viz. tetrahedron, cube and dodecahedron. Hippasus circumscribed a sphere about a dodecahedron. An important problem remained unsolved here, viz. the enumeration and the construction of all the regular solids.

6. Finally the problem of the irrational was formulated. The incommensurability of the diagonal of a square of unit side was proved by means of the Pythagorean theory of the even and the odd; by use of the indefinitely continued antanairesis, Theodorus of Cyrene demonstrated the incommensurability of the sides of squares whose areas are 3, 5, ..., 17. The general problem "which among the sides, that produce commensurable squares, are themselves commensurable" remained unsolved.

Next we shall see how these different lines of development were continued during the fourth century, and how the problems that had remained unsolved became the starting point for further development.

CHAPTER VI

THE CENTURY OF PLATO

The period to be discussed in the present chapter, opens with the death of Socrates by the hemlock cup (399) and closes at the moment when Alexander the Great scatters the seed of Hellenistic culture over the entire world of antiquity.

This period is one of political decay; but for philosophy and for the exact sciences it is an era of unprecedented flourishing. At the center of scientific life stood the personality of Plato. He guided and inspired scientific work in his Academia and outside. The great mathematicians Theaetetus and Eudoxus and all the others enumerated in the Proclus Catalogue, were his friends, his teachers in mathematics and his pupils in philosophy. His great pupil Aristotle, the teacher of Alexander the Great, spent twenty years of his life in the glorious world of the Academia.

Why did Plato attach such tremendous importance to the cultivation of Mathesis? Why did he require all his pupils to obtain a thorough knowledge of mathematics before being initiated into his philosophy? It becomes apparent from the passages concerning the square-itself and the diagonal-itself which were quoted above; it was because in mathematics one can learn that it is possible to reason about things which are neither seen nor heard, but exist in thought only. In The Republic he writes: "The study of mathematics develops and sets into operation a mental organism more valuable than a thousand eyes, because through it alone can truth be apprehended."

Truth, that means the ideas. It is only the ideas which have true Being, not the things which are observed by the senses. Ideas can sometimes be contemplated, in moments of Grace, through reminiscence of the time when the soul lived closer to God, in the realm of Truth; but this can happen only after the errors of the senses have been conquered through concentrated thought. The road which leads to this state is that of dialectics, and the method of proof in dialectics is the indirect proof.

Truth can not be self-contradictory; hence, if a certain preliminary hypothesis leads to a contradiction, then this hypothesis must be rejected. Proceeding in this manner dialectically, from hypothesis to hypothesis, we can conquer the errors in which we are involved and thus free our view for an envisagement of truth.

This is the method applied in the strictly dialectical dialogues, which do not instruct, but which carry on a philosophical conversation. The interlocutor suggests an opinion, Socrates disputes it. Now the point of view is modified, the formulation is made more exact, again Socrates shows that the new formulation leads to a contradiction and is therefore untenable. In this way, matters continue; a positive result is never attained, but the discussion progresses to higher and higher levels, more errors are eliminated, and sometimes, at the end, one reaches a stage in which the truth can be suggested in the form of a myth. But this is no longer dialectics; according to Plato's own words, dialectics is an exact method of proof and no other method of proof ever occurs in the dialogues except that of rejecting assumed hypotheses.

In his Mathematik und Logik bei Plato (Leipzig, 1942), Reidemeister emphasizes that this method of proof by a reductio ad absurdum is derived from mathematics. Plato himself repeatedly cites the proof of the incommensurability of side and diagonal as the typical example of a mathematical argument and points out that it is by means of such an indirect proof that one learns something about the things themselves. Observable things are variable and contradictory, but the true "ideas" which lie behind them can not have two contradictory properties.

Thus, according to Plato, it is in mathematics that one learns to reason concerning things in themselves, that one receives training in dialectics. Mathematical objects lie, in his view, between the visible things and the ideas, and mathematical understanding stands between opinion and philosophical insight. These are the reasons why Plato attributed such eminent importance to mathematical training.

In the older Platonic dialogues, mathematics hardly ever occurs. The man who made Plato acquainted with the exact sciences is the Pythagorean, already mentioned several times, viz.

Archytas of Taras.

The versatility of this remarkable South-Italian Dorian is unparalleled even in that era. By means of an extremely ingenious space construction, he obtained a solution of the famous Delian problem, the duplication of the cube. Not only did he develop some lemmas on proportionalities of numbers and some inequalities concerning three averages, in connection with his theory of music, but the entire Book VIII of the Elements, with its arithmetical theory of continued proportions, similar numbers, etc. is in the main his work. He also made an important contribution to the theory of irrationals. Quite rightly does Ptolemy call him the most important Pythagorean theoretician of music. For, besides computing the numerical ratios for the new musical scales, which came into vogue in his time, by means of systematic applications of the arithmetic and the harmonic means, he also laid the number-theoretical foundation for the theory of music which is found in Euclid's "Sectio Canonis". He also reflected on the connections between the sciences, and on the physical nature of sound (Diels, Fragmente, Archytas B 1). Diogenes Laërtius (VIII 79—83) calls him the first to give a systematic mathematical treatment of mechanics. Vitruvius reports that he wrote about machines; he designed machines himself, a wooden pigeon that could fly (Aulus Gellius, Noctes Atticae X 12) and a children's rattle (Aristotle, Politics E6, 1340b26). He was on terms of intimate friendship with Plato, who received chiefly through

him his initiation in the exact sciences and into Pythagorean philosophy. In his
native town of Tarent he was highly respected as a statesman; during seven
successive years he was elected strategus, although the law allowed only one such
election, and he never lost a battle. After the aristocratic Pythagoreans, his spiritual brethren, had been driven out of Italy, he managed to maintain a semi-auto-
cratic regime under the external forms of a democracy. It is said that, when
Plato was held prisoner by Dionysius, the tyrant of Syracuse, it was his letter
to the tyrant which saved Plato's life.

This variegated picture is concerned only with the external aspects of his life
and work. Greater surprise will be aroused presently, when we shall penetrate more
deeply into his way of thinking and reveal the extraordinary contrast between
his ingenious ideas, his creative imagination and his mastery of geometrical methods
on the one hand, and his lack of logic, his inability to express himself exactly and
clearly, his errors of thought and his prolixity, on the other hand. A critical reading of Euclid's Book VIII and of the Sectio Canonis shows us Archytas struggling
with the demands of rigorous proof and clear exposition, which were required
by the mathematicians of his day, and which are so readily and so naturally met,
e.g., in the older Book VII.

Let us begin with his most beautiful accomplishment, his solution of the Delian
problem, as described by Eudemus in Eutocius (Archimedis Opera III, p. 84,
Heiberg).

The duplication of the cube,

or, more generally, the enlargement of the cube in a given ratio, can be reduced,
as we have already seen, to the construction of two mean proportionals between
two given lines. According to Eudemus, this problem was solved by Archytas in
the following manner:

> Let two line segments, $A\Delta$ and Γ, be given. Then we have to determine two mean pro-
> portionals between $A\Delta$ and Γ. Let a circle $AB\Delta Z$ be described about the larger line $A\Delta$
> and let (a chord) AB be drawn, equal to Γ, and let the extension of this chord meet the
> tangent line to the circle at Δ in Π. Let the line BEZ be drawn parallel to $\Pi\Delta O$, imagine a
> half right cylinder constructed on the semicircle $AB\Delta$, and a vertical semicircle on $A\Delta$,
> lying in the rectangle of the half cylinder. When this semicircle (kept in a vertical position)
> is now rotated, from Δ towards B, while the extremity A of the diameter is kept fixed, then
> it will intersect the surface of the cylinder and mark off a line on it. On the other hand, if
> $A\Delta$ remains in its position and the triangle $A\Pi\Delta$ is rotated in a direction opposite to that
> of the semicircle, [1] then the straight line $A\Pi$ will describe the surface of a cone, and this will,
> during its rotation, meet the line on the cylinder in a certain point. At the same time the
> point B will describe a semicircle on the surface of the cone. Let the position of the moving
> semicircle be $\Delta' K A$ when these lines meet [2] and let the triangle, rotating in the opposite direc-
> tion, occupy the position $\Delta \Delta A$, while the point of intersection is K. And let the semicircle

---

[1] i.e. towards the plane of the rectangle, while the semicircle is moved away from this plane.
[2] The point, which, for greater clarity, is here denoted by Δ', the new position occupied by the moving point,
is designated in the text by Δ, so that two points in the figure are called Δ. This lack of clearness arises from
Archytas' kinematical way of thinking; for him Δ is not a definite point in space, but a moving point.

described by B be BMZ, and let its line of intersection with the circle $B\varDelta ZA$ be BZ.[1] Drop a perpendicular from K to the plane of the semicircle $\varDelta BA$. Its foot will lie on the circum-

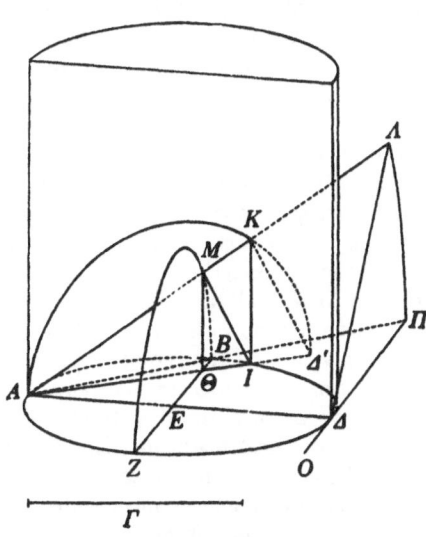

ference of the circle, since the cylinder is a right cylinder. Let it be dropped and let it be called KI, and let the (line) drawn from I to A intersect BZ in Θ, and let the line $A\varDelta$ meet the semicircle BMZ in M. Let the lines $K\varDelta'$, MI and $M\Theta$ also be drawn. Since each of the two semicircles $\varDelta'KA$ and BMZ is perpendicular to the base plane, their common line of intersection $M\Theta$ is also perpendicular to the plane of the circle, so that $M\Theta$ is perpendicular to BZ. Therefore the rectangel on ΘB and ΘZ, and hence that on ΘA and ΘI, is equal to the square on $M\Theta$. Therefore triangle AMI is similar to $MI\Theta$ and to $MA\Theta$, and hence AMI is a right angle. But $\varDelta'KA$ Therefore is a right angle too. the lines $K\varDelta'$ and MI are parallel, and we shall have the proportionality

$$\varDelta'A : AK = KA : AI = IA : AM,$$

because the triangles are similar. Hence the four lines $\varDelta'A$, AK, AI and AM are in continued proportion. And AM is

Fig. 49.

equal to $\varGamma$, since it also equals AB. Thus we have found two mean proportionals AK and AI between two given lines $A\varDelta$ and $\varGamma$.

Is this not admirable? Archytas must have had a truly divine inspiration when he found this construction. The figure which Archytas saw before his mind's eye and which he wanted to construct, is evidently the right triangle $AK\varDelta'$ with the two perpendicular KI and IM, reproduced once again in Fig. 50. Here AM,

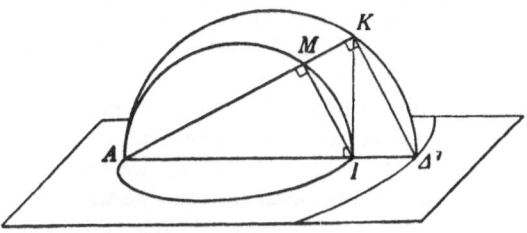

Fig. 50.

AI, AK, and $A\varDelta'$ form a continued proportion, while $A\varDelta'$ is equal to a given line segment $A\varDelta$. Now Archytas observed that, when A is held fixed, the choice of I determines both K and M (since a perpendicular to $A\varDelta'$ erected at I can be made

[1] A superfluous remark. In my opinion, this prolixity is not to be charged to Eudemus, but to Archytas.

to intersect the semicircle on $A\Delta'$). Now he wanted to determine I in such a way that AM shall equal a given line segment Γ. If AI varies from $A\Delta$ to zero, then AM will also vary from $A\Delta$ to zero; hence at some stage AM will be equal to Γ.

To determine this stage, Archytas hit upon the following idea: he varies AI by letting I describe a circle of diameter $A\Delta$ in a horizontal plane and placing the semicircle $AK\Delta'$ in a plane perpendicular to this horizontal plane. The point K will then always lie on a vertical through I, i.e. on a cylinder of which the given circle is the directrix. The point K will thus describe the space curve determined on the cylinder by the rotating semicircle $AK\Delta'$. The semicircle AMI will then describe a hemisphere whose diameter is $A\Delta$ (this is seen most readily by starting with the sphere of diameter $A\Delta$ and determining its intersection with the rotating vertical plane; the curve of intersection will then be exactly the circle of diameter AI). If now AM has a given length Γ, then M will fall on a definite circle on this sphere, so that the line AMK will lie on a definite cone with vertex at A. The point of intersection of the space curve with this cone is the required point K.

In Archytas' diagram, everything is in motion; he thinks kinematically. Already in antiquity it was observed that he introduced mechanical methods into geometry. It is seen furthermore that, as indeed most Greek mathematicians, he did not hesitate to use the principle of continuity. This may be formulated, for instance, as follows: If a continuous variable is first greater than a given magnitude and later less, then, at some time, the variable equals this magnitude. [1]

The style of Archytas.

The quoted section reveals Archytas' wealth of ideas, the vividness of his spatial and kinematic imagination. There is no doubt that his ideas must have had a great effect on others, particularly on Plato. When Aristotle says that Plato derived a great deal from the Pythagoreans, we have to think in the first place of Archytas.

In one respect however, Archytas is far behind Plato; he completely lacked the gift of expressing his ideas briefly and clearly. When Plato occasionally speaks enigmatically, he has his reasons for doing so; when he wants to, he can be clarity itself. On the contrary, in reading in Diels (Fragmente der Vorsokratiker) the longest connected fragment Archytas B 1, one is most struck, not only by the unbearable prolixity, but especially by the lack of clearness, not to say confusion. Archytas does try his best to make clear the ideas he wants to convey, by using many words and by giving elaborate examples, but his reasoning is not logical. He confuses the velocity of the motion which produces a sound with the velocity of propagation of the sound itself, which leads him to conclude from observations, correct in themselves, that higher tones are propagated more rapidly than lower ones. And, the remark that "the same air flows more slowly through an extended

[1] Compare O. Becker, Quellen und Studien B 2 (1933), p. 369 and B 3 (1936), p. 236.

space, more rapidly through a short one" does not excel in clarity; perhaps he means to say that, if the difference in pressure remains unchanged, the air will flow more slowly when a tube is lengthened.

The same prolixity is found in the fragment of Archytas on the theory of numbers, given by Boethius, discussed earlier, which contains the proof that no mean proportional is possible between two numbers in the ratio $(n + 1) : n$. It is of course quite obvious that the least numbers in this ratio differ by 1, but Archytas considers it necessary to give all the minutiae of a proof based on the epimoric ratio. It looks as if he is afraid of stumbling on the slippery paths of logic. And he has good reason for such fear, because he has stumbled more than once in other places. An essential error occurs, for example, in his foundations of the theory of music, at a point where he confuses a proposition and its converse (see my article in Hermes 78).

Book VIII of the Elements.

Archytas' logical weaknesses are found again in Book VIII of Euclid's Elements. Indeed it is highly probable that this book should be ascribed to Archytas. We have seen earlier that the beginning of Book VIII is very closely related to his theory of music. He quotes Proposition VIII 8 himself, and VIII 7 is equivalent to the proposition from the Sectio Canonis, which we have designated before as Proposition B. A number of other propositions and of concepts from Book VIII are quoted by Plato, who must have learned the theory of numbers largely from Archytas. But moreover, the style of Archytas is recognized unmistakably in Book VIII — this clinches the matter. Let us analyze the construction of this book.

The central problem of Book VIII is the following: under what conditions is it possible to find one or more numbers in continued proportion between a and b? According to VIII 8 this possibility depends only on the ratio $a : b$, so that a and b may be assumed to be relatively prime. For this case, the problem is solved in principle by VIII 9, 10: If a and b are relatively prime, then as many numbers can be found in continued proportion between them, as between each of them and unity; and, conversely. Hence a and b are the last terms of two geometrical progressions which have the same number of terms:

$$1, p, p^2, \ldots, p^n = a,$$
$$1, q, q^2, \ldots, q^n = b.$$

It is not explicitly stated that a and b are powers with the same exponent. This may be due to the fact that the concept power was still absent. But it might have been paraphrased by saying that a is produced by multiplying p a definite number of times by itself. At the very start of the proof of VIII 9, it becomes clear that a is produced in this way. The proof is carried through, guided by the special case $n = 3$. In a terribly roundabout way, it shows that p and p^2 are indeed two mean proportion between 1 and p^3.

Once the necessary and sufficient condition for the existence of a fixed number of mean proportionals has in this way been established, the cases $n = 2$ and $n = 3$ (one and two mean proportionals), with which the rest of Book VIII is concerned, can of course immediately be formulated as special instances; a and b are then of course squares and cubes respectively. But the writer does not proceed in so logical and simple a manner. He proves in VIII 11, 12, that between two squares, it is possible to determine one mean proportional, and between two cubes, two mean proportionals. These results are not obtained as special cases form VIII 10, but proved entirely afresh. The converse is not explicitly formulated, but it is hidden in a corollary to VIII 2 and, at the end of the book, in VIII 26, 27, it is formulated once more in a slightly different form. It becomes clear that the whole book is put together in a confused and disorderly manner.

In VIII 18, 19 we find the following generalizations of VIII 11, 12: Between two similar plane numbers ab and cd (i.e. such that $a : c = b : d$), one mean proportional is possible, between two similar space numbers abc and def (in which $a : d = b : e = c : f$) two are possible. And VIII 19, 20 state that this sufficient condition for the existence of one or two mean proportionals is also necessary. The question as to the conditions under which there are possible one or two mean proportionals between two numbers is thus answered not once, but three times. It looks as if the author were constantly fluttering around the problem without succeeding in finding a simple formula entirely satisfactory to himself.

The point of departure for the systematic development of the entire matter is VIII 2: to determine numbers, as many as may be required, in continued proportion, and the least ones in a given ratio. If the given ratio, reduced to lowest terms, is $a : b$, then a^2, ab, b^2 are 3 successively proportional numbers. Upon multiplying them by a and the last one by b as well, one obtains 4 numbers in continued proportion a^3, a^2b, ab^2, b^3. Proceeding in this way, one finds an arbitrary number of terms, the two extremes always being powers of a and b.

The misery in Book VIII arises from the fact that this procedure is not formulated as a proposition, but as a problem. When applying it, one has in every case to refer back to the way in which the geometric progressions are actually formed. Every time the proof of VIII 2 has to be recalled, so that the quotations become very complicated. In other respects as well, the formulations in Book VIII are unnecessarily involved.

But this is not the worst of the matter — there are extraordinary logical errors. Proposition VIII 22 says: If three numbers a, b and c are in continued proportion, and if a is a square, then c is also a square. Proof: From VIII 20 we know that a and c are similar numbers; if a is a perfect square, then c must be one as well. But this conclusion is not justified, because, besides being factored into two equal factors, a might also be factorable into unequal factors; when c is factored into proportional factors, the definition of similar numbers is satisfied, without yielding the conclusion that c can also be factored into two equal factors. A similar remark applies to VIII 23, the analogue for spatial numbers.

Thus we see that in Book VIII, as well as in his other works, Archytas is constantly at odds with logic, trying unsuccessfully to meet its strict demands.

It would be wrong to think that such faulty logic is a general characteristic of the science of this period. This would be contradicted by Book VII, which is older than Book VIII and which has an excellent logical structure. As we have already seen, Hippocrates adhered to the highest standards for a rigorous mathematical proof.

When Aristotle drew up the rules of logic, he merely codified the regularities which he found in the reasoning of mathematicians and philosophers before his time. He draws most of his examples from contemporary textbooks on mathematics. Now it is clear that such textbooks followed, in their logical structure, the example of the original treatises of the great mathematicians, not inversely. But from this it follows that the thinking of Greek mathematicians must have satisfied very strict demands of rigor long before Aristotle. In Archytas' poor logic we have therefore a very personal trait of this, otherwise so excellent, mathematician.

The Mathemata in the Epinomis.

In a lofty tone and in enigmatic verbiage, the Epinomis (a posthumous work of Plato, according to some completed by Philippus of Mende) gives a survey of the mathematical curriculum for the future leaders of Plato's ideal state.

Philologists have vainly broken their teeth on this hard nut — it was only in modern times that mathematicians have been able to throw some light on this matter. [1] It turned out that it is particularly the mathematics of Archytas which is referred to. We shall here reproduce this remarkable fragment as interpreted by Becker, with only absolutely essential comment:

"The greatest and the first is the theory of numbers, not of the numbers which have bodies, but of the generation of the even and the odd, and of the power which this has over the nature of being."

One is naturally led to think here of Aristotle who says that for the Pythagoreans, the elements of the numbers are the elements of all being, and that the even and the odd are the elements of numbers (Metaphysics A5).

"For one who has learned this, there follows what is very foolishly called land measurement (geometria), but which is concerned with making similar numbers, which in their nature are not similar, which is made transparent through the fate of plane figures."

This sentence has been discussed earlier. Two plane numbers ab and cd are called similar, if their "sides" are proportional $(a : b = c : d)$. But two numbers, which "in their nature" (i.e. in the realm of numbers) are not similar, can still be "made

[1] O. Toeplitz, *Die mathematische Epinomisstelle, Quellen und Studien* B 2 (1933), p. 334; O. Becker, ibid. B 4, p. 191; B. L. v. d. Waerden, *Hermes 78*, p. 185.

similar" by means of plane geometry, which means that they can be represented as areas of similar rectangles (e.g. squares). For, between a and b (c and d respectively) one can always construct the mean proportional l (m resp.) and thus transform the areas ab and cd into squares l^2 and m^2.

"But truly, not a human marvel, but a divine miracle will be revealed to him who considers, after this (plane geometry), the triply-extended (numbers) similar in their spatial nature. And again he will be able, by means of another art, called solid geometry by the informed ones, to transform into similar (numbers) those which have been produced as non-similar."

Two spatial numbers abc and def are called similar if $a : d = b : e = c : f$. To transform two non-similar numbers into similar numbers is the generalization of the problem of the duplication of the cube which is, for Plato, the central problem of solid geometry. That these problems constantly hover about the duplication problem, is stated emphatically once more in the following sentence:

"However what is divine and marvelous for those who understand it and reflect upon it, is this that through the power, which is constantly whirling about the duplication and through its opposite according to the different proportions, the pattern and type of all nature receives its mark."

The power, which is constantly whirling about the duplication, or, as it is expressed more briefly in the next sentence, the power of duplication, is the operation of doubling numbers, areas, volumes and ratios. The repeated duplication of a ratio produces a continued proportion such as

$$a : x = x : y = y : b,$$

to which the duplication of the cube can be reduced, and such as the proportion

$$1 : 2 = 2 : 4 = 4 : 8,$$

to which the next two sentences refer. The opposite power is the generation of mean proportionals for the various types of proportions, such as geometric, arithmetic and harmonic mean proportionals, worked out more fully in the sentence after that. By means of these powers nature produces the pattern or idea ($\varepsilon\tilde{\iota}\delta o\varsigma$) and the type or genus ($\gamma\acute{\varepsilon}\nu o\varsigma$), which are likened to hall-marks.

"The first power of duplication progresses as numbers in the ratio $1 : 2$, but the progression by means of second powers ($\varkappa\alpha\tau\grave{\alpha}$ $\delta\acute{\upsilon}\nu\alpha\mu\iota\nu$) is also a duplication."

"By means of second powers" is here the power which produces the ratio $1 : 4$. This becomes clear by analogy with the next sentence:

"The one by means of the spatial and the tangible is once more a duplication and progresses from 1 to 8."

PLATE 19

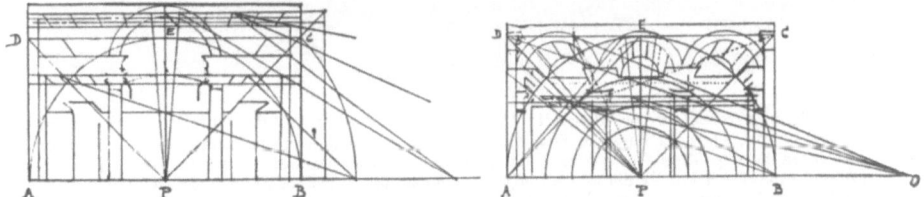

Pl. 19. Wall decorations from Pompeii, belonging to the "second style" (about 60 B.C.). These decorations show in the upper part (the double square *ABCD*) a perspective construction. The most important receding lines, if prolongated, meet in one point *P*. Similarly, the diagonals of the panels on the ceiling converge on one point *O*. If we suppose that the panels are constructed as squares, we may conclude that *OP* is the theoretical distance between the eye and the wall. Since the publication of these perspective wall decorations by H. G. Beyen (Jahrbuch deutsch. archäol. Inst. 1939) there can be no more doubt that the passages from Vitruvius quoted on p. 137 are really pertinent to central perspective. For Vitruvius' relations to the wall painters of his time see De Architectura VII 3–14, where he gives instructions to the decorators. The semi-circles in the figures have been added, because Vitruvius mentions the center of a circle to which the receding lines must converge.

PLATE 20

Pl. 20. Persian astrolabe, constructed in 1221/22 by Muhammed ben Abi Bakr al-Farisi. Museum of the History of Science, Oxford. The outer ring is divided into 360 degrees. A disc, pierced in various places, the "spider", revolves in this ring. The sharp pointers of the spider indicate stars, the eccentric circle the zodiac. The spider can rotate and thus imitate the diurnal rotation of the celestial sphere. The angle of rotation, and hence the time, can be read off on the rim. Behind the spider one sees a plate on which circles are engraved; on the upper half there are 30 circles, representing the horizon and its parallel circles at various altitudes. This plate does not turn when the spider rotates. When the altitude of a star (or of the sun) is observed, one turns the spider until the point representing the star lies on the proper altitude line. The observation of the altitude may be made by means of the two pin-hole sights right and left on the circumference of the spider. When in use, the instrument is suspended vertically from the ring. The instrument is a Greek invention; it is based on stereographic projection. See p. 182.

The last two sentences may be interpreted as follows. Taking a line segment and doubling it, one gets a geometric image of the ratio 1 : 2. But if a square is doubled in length and in width, one gets the ratio 1 : 4. If, finally, a cube is doubled in every direction, then the ratio 1 : 8 appears.

"Finally the (opposite power) of duplication, which tends towards the middle, in which the (arithmetic) mean is as much larger than the smaller one as it is less than the greater, the other (harmonic mean) however by the same part of the extreme terms exceeding them and being exceeded by them; in the middle of 6 to 12, the ratios 2 : 3 and 3 : 4 come together."

The arithmetic and harmonic means of 6 and 12 are 9 and 8. The 8 exceeds 6 by one third of 6 and is exceeded by 12 by one third of 12, in accordance with the definition of the harmonic mean. The ratio $6 : 9 = 8 : 12 = 2 : 3$ and $6 : 8 = 9 : 12 = 3 : 4$ are called in Greek Hemiolion and Epitriton; they correspond to the fifth and the fourth in the theory of music.

"By starting from these (ratios) and moving from the center to both sides, it (the opposite power) presented to mankind melodious consonance and measured charm of play, rhythm and harmony, abandoned to the blessed dance of the Muses."

I have clarified the meaning of this mysterious sentence in my article in Hermes. If, starting from the ratio 2 : 3, one forms again an arithmetical or a harmonic mean, one obtains the ratios 4 : 5 and 5 : 6 which means, in the language of the theory of music, that the interval of the fifth has been divided into a major and a minor third. If the ratio 3 : 4 is similarly divided by means of an arithmetic or harmonic mean, the ratios 6 : 7 and 7 : 8 appear, so that the interval of the fourth is divided into a diminished minor third and an augmented whole tone. Now, it is on these four intervals that Archytas built his three scales; they actually produce therefore the melodious consonance of music in play, rhythm and harmony of which the text speaks.

The lofty conception of nature, formed on the example of numbers, in accordance with the laws of harmony, is thoroughly Pythagorean. There is in this entire world of ideas an unmistakably mystical element. It is not a sober natural science, which is speaking here, but a mystical surrender to the divine Creator of the disciple who has been initiated into the secrets of numbers and of harmony. Because the human spirit, which has the divine spark within itself, is mystically united with the divine spirit, it penetrates into the marvelous plan on which nature has been constructed. In this mystical region, everything flows together; one does no longer distinguish between musical intervals and the corresponding numerical ratios, nor between geometric and material bodies. The human soul communicates with the divine soul; thinking is no longer human, it has become divine. All of this is characteristically Pythagorean. The Aristotelian criticism of this world picture is based on this that the Pythagoreans do not discriminate between mathematical and material things.

This mystical identification is already found in the oldest Pythagorean sayings, which have come down to us, such as:

"What is God? Unity!

or:

"What is the oracle of Delphi? The tetractys, which is indeed the scale of the sirens."

According to Aristotle, the Pythagoreans identify justice, the soul, the spirit with definite numbers. For them the heavens "are harmony and numbers", etc.

This mystical way of thinking, which does not discriminate between related things, is also directly connected with the lack of logic which we have repeatedly found in Archytas. It has left its traces not only in his own writings, but also in the mathematical textbooks which develop his ideas. E.g., at the beginning of the Sectio Canonis, three purely number theoretical propositions are proved; no mention is made of numerical ratios however, but of intervals, thus frequently obscuring the formulation for the reader.

Surveying now once more the mathematics of Archytas by way of the Epinomis, it becomes clear that his theory of numbers, his geometry and his theory of music are all connected; indeed, he himself calls them "related" sciences. In his theory of numbers, he considers continued proportions and he raises the question as to the existence of one or more mean proportionals between given numbers. This leads him to distinguish between similar and non-similar plane and spatial numbers, and he proves, i.a., that no mean proportionals can exist between two numbers in the ratio $(n + 1) : n$. Translated into the language of the theory of music, this means that an epimoric interval (such as the octave, the fifth, fourth or whole tone) can not be split into two or more equal intervals. It is this proposition which forms, for Archytas, the basis of the theory of consonant intervals. It is possible to subdivide such intervals by use of the arithmetic or the harmonic mean, and he uses this method in constructing his scales. Finally, the problem of the mean proportionals can be put into geometric language; this leads to the problem of constructing one or more mean proportionals between two given line segments. And now it turns out that, in contrast with the theory of numbers and the theory of music, this problem does have a solution in geometry. The construction of *one* mean proportional had been known for a long time; that of two mean proportionals was discovered by Archytas in his duplication of the cube.

These results must have been very baffling for a Pythagorean, because the Pythagoreans always assumed that everything was ordered according to numbers. It was exactly the assumption that all tones could be represented by numbers and all intervals by numerical ratios, which enabled Archytas to prove that certain intervals could not be halved. If one were to assume — a natural assumption for a Pythagorean — that all line segments can also be expressed by numbers, one would conclude that, for line segments in certain ratios, no mean proportional could exist either. But the mean proportional of two line segments *can* always be constructed. This forces the conclusion that line segments can not always be represented by numbers, or, in a more exact formulation: ratios of line segments can not always be expressed as ratios of integers. In other words: *there exist incommensurable line segments.*

Is it likely that Archytas himself clearly understood this connection between his theory of continued proportions and the existence of irrationals? I think it is. A study of the Epinomis, which points out so emphatically that numbers which are not similar in their nature (and for which therefore, in accordance with proposition 20 of Book VIII, no mean proportional number is possible) can be made similar by use of the geometric construction of mean proportionals, proves that, certainly for Plato and his disciples, this connection was clear.

Something has to be added. In Book X of the Elements, two proofs occur, so completely different in style from the rest of the book, that they have to be considered as later additions, viz. the proofs of X 9, 10. For, whereas Book X as a whole excels in its strictly logical structure and its extremely brief and elegant proofs, these two proofs not only contain superfluous parts, but also gaps.

In the proof of X 9, it becomes necessary to conclude from $a^2 : b^2 = c^2 : d^2$, in which a and b are line segments, c and d numbers, that $a : b = c : d$. For this purpose it is observed that the ratio of the squares $a^2 : b^2$ is the duplicate ratio of the ratio $a : b$, and that the ratio $c^2 : d^2$ is similarly the duplicate of the ratio $c : d$. Without further motivation, the equality of the ratios is now concluded from the equality of the duplicate ratios; the proof is therefore incomplete. In the proof of X 10, use is made of X 11 — also a defect of course.

The proof of X 10, which Heiberg also considers as "spurious", contains the expression ἐμάθομεν γὰρ ("for we have learned this") which, as observed by Heath, does not occur anywhere else except in the Sectio Canonis. Both proofs use propositions from Book VIII, concerning mean proportionals and similar numbers, which are entirely alien to the methods of proof of Book X. Whoever established the text of these proofs — probably the same uncritical and unoriginal author who is responsible for the Sectio Canonis — knew therefore that these propositions can be applied to the theory of irrationals. It is quite probable that Archytas and his disciples, from whom the theories of Book VIII and of the Sectio Canonis are derived, have already made such applications themselves. [1] Did not Pappus say that the theory of the irrational had its origin in the school of the Pythagoreans?

Connecting with the work of Archytas, we shall now discuss the further development of the famous Delian problem, that stirred the best minds of Plato's time.

The duplication of the cube.

What we know about this problem is to be found chiefly in Eutocius, the commentator of Archimedes. But the tradition is very confused. Cantor, Heath and other historians of mathematics, have observed various contradictions of

[1] See Reidemeister, *Arithmetik der Griechen*, Leipzig 1943. I owe a great deal to this pamphlet and to subsequent correspondence with Reidemeister.

which they have not found a satisfactory resolution. To resolve them, we shall
have to consult the sources from which the story of Eutocius is derived, and, above
all, we shall have to separate the actual history of the problem from the dramatic
story found in Eratosthenes' "Platonicus" of which some fragments have been
preserved.

Eutocius himself does not say much about his sources. He begins with the
statement that he has found the writings of many famous men, in which the
construction of two mean proportionals between two lines is announced. "Of
these we have left aside the writings of Eudoxus of Cnidos", says Eutocius,
"because he says in his preface that he found the solution by the use of curves,
while he does not introduce curves at all into his proof; also, because, after having
established a discrete proportion, he treats it as if it were a continued proportion. It would be foolish to suppose that something like this were possible
even in an author who was but a mediocre geometer, let alone therefore in Eudoxus. In order to present clearly the points of view of the writers, which have
been handed down to us, we shall describe here the method of solution of each
of them."

It is clear that Eutocius had not Eudoxus' original work before him, but
only a poor extract or a compilation which did not clearly reproduce the reasoning,
since in Eudoxus' own treatise curved lines would appear, but no logical errors.

After having discussed a number of solutions by various authors, Eutocius re-
produces what he calls a letter from Eratosthenes to king Ptolemy.

Wilamowitz [1] has shown that this letter can not possibly be genuine; but it
contains very important material.

It starts as follows:

> It is said that one of the ancient tragic poets brought Minos on the scene, who had a tomb
> built for Glaucus. When he heard that the tomb was a hundred feet long in every direction,
> he said:
> "You have made the royal residence too small, it should be twice as great. Quickly double
> each side of the tomb, without spoiling the beautiful shape."
> He seems to have made a mistake. For when the sides are doubled, the area is enlarged
> fourfold and the volume eightfold. The geometers then started to investigate how to double
> a given body, without changing its shape; and this problem was called the duplication of
> the cube, since they started with a cube and tried to double it. After they had looked for
> a solution in vain for a long time, Hippocrates of Chios observed that, if only one could find
> two mean proportionals between two line segments, of which the larger one is double the
> smaller, then the cube would be duplicated. This transformed the difficulty into another
> one, not less great.
> It is further reported that, after some time, certain Delians, whom an oracle had given the
> task of doubling an altar, met the same difficulty. They sent emissaries to the geometers in
> Plato's academy to ask them for a solution. These took hold with great diligence of the pro-
> blem of constructing two mean proportionals between two given lines. It is said that Archy-
> tas solved it with half cylinders, Eudoxus with so-called curved lines.

[1] U. v. Wilamowitz-Moellendorff, *Ein Weihgeschenk des Eratosthenes*, Nachr. Ges. Wiss., Göttingen, Phil. hist,
1894, p. 15.

We notice a certain contradiction between the two parts of the tale; apparently two different versions of the story are reported, one after the other. According to the first version, the duplication of the cube is an old problem, connected with a legend about Minos; Hippocrates of Chios, and others before him, had occupied themselves with it. In the second version, the problem arose from a declaration made to the Delians by an oracle at the time of Plato, a half century after Hippocrates. How can this contradiction be accounted for?

Fortunately, a quotation in Theon of Smyrna (ed. Hiller, p. 2) gives the source of the second legend:

> In his work entitled Platonicus, Eratosthenes relates that, when God announced to the Delians through an oracle that, in order to be liberated from the pest, they would have to make an altar, twice as great as the existing one, the architects were much embarassed in trying to find out how a solid could be made twice as great as another one. They went to consult Plato, who told them that the god had not given the oracle because he needed a doubled altar, but that it had been declared to censure the Greeks for their indifference to mathematics and their lack of respect for geometry.

The same story, in almost the same words, is found in Plutarch (de Ei apud Delphos, 386 E). Elsewhere (de genio Socratis 579 CD), he adds that Plato referred the Delphians to Eudoxus and to Helicon of Cyzicus for the solution of the problem.

It is likely that the Platonicus was a dialogue in which the Delians, Plato, Archytas, Eudoxus and Menaechmus appeared. In this dramatic story, Eratosthenes condensed the entire development of the problem into a short period of time. Of course in this setting, he could not make use of Hippocrates of Chios. In fact the problem is a much older one, for it arose from the translation of the Babylonian cubic equation $x^3 = V$ into spatial geometric algebra. Thus the contradiction between the first and the second account in the "letter" solves itself: the first tale probably derives from historical sources, the second from the Platonicus.

Furthermore, the "letter" contains an extemely important document, an epigram of Eratosthenes, partly in prose, partly in artfully made verses. The epigram was engraved on a marble tablet in the temple of Ptolemy in Alexandria.[1] In the prose section, a solution of the Delian problem is given by the aid of a diagram and a model, to be discussed at a later point, in connection with Eratosthenes. I shall reproduce the verses in Heath's excellent translation:

> If, good friend, thou mindest to obtain from a small (cube) a cube double of it, and duly to change any solid figure into another, this is in thy power; thou canst find the measure of a fold, a pit, or the broad basin of a hollow well, by this method, that is, if thou thus catch between two rulers (two) means with their extreme ends converging. Do not thou seek to do the difficult business of Archytas' cylinders, or to cut the cone in the triads of Menaechmus, or to compass such a curved form of lines as is described by the god-fearing Eudoxus.

[1] For further details we refer to the article by von Wilamowitz, cited above. Von Wilamowitz has convincingly demonstrated the genuineness of the epigram.

Nay thou couldst, on these tablets, easily find a myriad of means, beginning from a small base. Happy art thou, Ptolemy, in that, as a father the equal of his son in youthful vigor, thou hast given him all that is dear to Muses and Kings, and may he in the future, O Zeus, god of heaven, also receive the sceptre at thy hands. Thus may it be, and let any one who sees this offering say "This is the gift of Eratosthenes of Cyrene."

From this we see that Archytas' solution with the cylinders, that of Eudoxus with the curves and that of Menaechmus with the three conic sections, are historical. We already know the solution of Archytas, the one of Eudoxus has been lost. The solution of Menaechmus is described by Eutocius as follows:

According to Menaechmus.

It is required to construct two mean proportionals x and y between two given lines a and b:

(1) $$a : x = x : y = y : b.$$

Suppose that the problem has been solved, and lay off $\Delta Z = x$ and $Z\Theta = y$. From (1) follows in the first place

(2) $$x^2 = ay,$$

so that Θ lies on a parabola whose vertex is Δ.
From (1) follows also

(3) $$xy = ab,$$

so that Θ lies on a hyperbola with asymptotes ΔZ and ΔK. Hence Θ can be constructed as the point of intersection of the parabola and the hyperbola. Next Eutocius proves in detail that, conversely, (1) follows from (2) and (3), so that the point of intersection Θ furnishes indeed the solution of the problem.

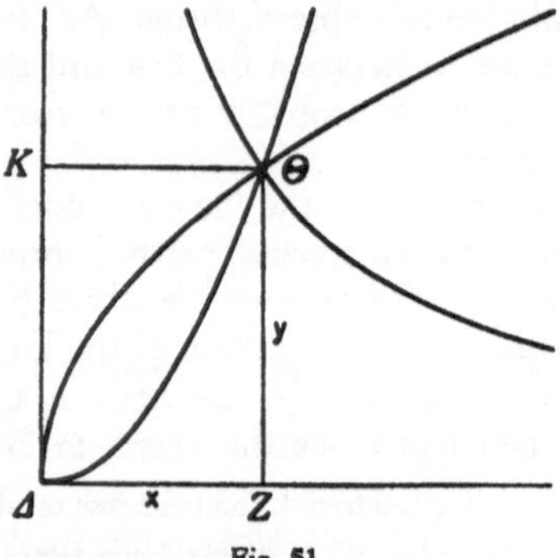

Fig. 51.

Alternately:

From (1) follows

(4) $$y^2 = xb,$$

so that Θ lies also on a second parabola with the same vertex. Hence Θ can also be found as the point of intersection of two parabolas.
Now, after having followed the *real* history of the duplication of the cube for a short distance, let us return to the dramatized story in the Platonicus.

In neither of the two places where he cites Plato's words to the Delians, does Plutarch say, that the Platonicus is his source; he merely speaks of Plato. This justifies the conjecture that two other places in Plutarch, which also refer to the

Delian problem and in which Plato is again the speaker, also come from the Platonicus, The first of these passages, from the eighth book of the Quaestiones conviviales, is the following:

> Plato himself censured those in the circles of Eudoxus, Archytas and Menaechmus, who wanted to reduce the duplication of the cube to mechanical constructions, because in this way they undertook to produce two mean proportionals by a non-theoretical method; for in this manner the good in geometry is destroyed and brought to nought, because geometry reverts to observation instead of raising itself above this and adhering to the eternal, immaterial images, in which the immanent God is the eternal God.

In the life of Marcellus, "Plato" expresses himself similarly, giving vent even more strongly to his horror of taking recourse to "material things, which require extended operations with unworthy handicrafts."

Such words do not appear anywhere in Plato's writings, but they fit very well in the Platonicus.

The construction of Menaechmus, which is known to us, is purely theoretical; it does not involve any instrument, not even compasses and straight edge, but only conic sections. What was then the basis for Plato's crushing criticism? This dilemma is resolved satisfactorily as well, if we separate the true history of the Delian problem from the dramatic account in the Platonicus. In the Platonicus, Plato refers the Delians to Eudoxus and his disciples. These come with mechanical constructions; Plato rejects them. Then the mathematicians, humiliated, reflect on the problem some more, and finally they produce their purely theoretical solution based on the intersection of curves. It is possible that things happened in the reverse order, that they had theoretical solutions, but that, for the benefit of the Delians, they invented mechanical solutions, which were then rejected by Plato. It is this order which is suggested by the words with which Plutarch starts his story in the Life of Marcellus:

> The highly-praised mechanics was first set into activity by the men around Eudoxus and Archytas. They introduced an elegant variation into geometry and supported with intuitive models the problems which were not fully provided with theoretical solutions. For example, the problem of the two mean proportionals, on which so many constructions depend, was solved by both with the help of mechanical methods, by designing certain instruments which could produce mean proportionals, *starting from curves and sections.*

We can now also understand why Eutocius could write that, in the proof of Eudoxus, he could not detect the "curved lines" announced in his introduction. It appears that Eutocius' source quoted the introduction of the genuine Eudoxus text, but then proceeded with a mechanical solution, taken from the Platonicus, which does not involve curves.

Most astounding for all modern commentators is the meachanical solution which Eutocius ascribes to Plato himself. This solution makes use of carpenter's squares with grooves and of adjustable rulers, all of them mechanical aids which Plato condemns so roundly!

"Plato's solution" depends upon the outline diagram shown in Fig. 52. If one succeeds in constructing points N and M on the extensions of the sides OA and

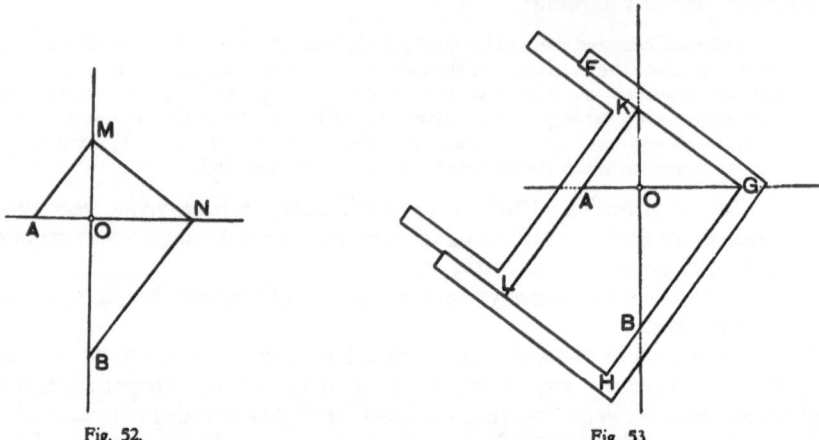

Fig. 52.   Fig. 53.

OB of a right angle, on which given line segments OA and OB have been laid off, in such a way that AM and BN are both perpendicular to MN, then OM and ON are two mean proportionals between OA and OB. To accomplish this, "Plato" imagines a carpenter's square FGH with a movable ruler KL which is kept in a position perpendicular to FG. If one now moves the ruler in such a way that KL passes through A and GH through B, while G remains on the line AO and K on the line OB, then the problem is solved. The text strongly emphasizes the craftsmanship needed to construct the grooves and the crosspieces, so that the angles remain right angles, and the ruler KL exactly parallel to GH, etc.

The consensus of modern opinion is that this solution can not possibly be ascribed to Plato. Heath adds the further argument that the Eratosthenes' epigram does not breathe a word of a solution by Plato. With all this, I am in complete agreement; the real Plato can not have invented this solution. But the Plato of the Platonicus? Can one not hear the irony in the elaborate emphasis on the craftsmanship required for the grooves, etc.? Does one not notice that these words connect closely with the contemptuous remark of "Plato" in Plutarch, about those "who take recourse to material things, which require extended operations with unworthy handicrafts?"

It looks as if, in the Platonicus, Plato said to Archytas, Eudoxus and Menaechmus: You have found mechanical solutions? That is nothing; even I, who is not a geometer, can do this. It only requires an outline diagram, not even a preliminary geometric solution of the problem. Just see, etc. But in this way everything that is good in geometry, is completely spoiled, our attention reverts from pure geometry to observable things, etc., etc.

This reconstruction of Plato's speech not only resolves all apparent contradictions, but it gives us moreover a good idea of the structure of this part of the Platonicus. Various phrases, such as "reverting attention" to material and observable things, become clear. For Eudoxus and Menaechmus had first found purely theoretical, geometric solutions and then, starting from these, worked out mechanical tools for the benefit of the Delians. Thus indeed, they turned their thoughts "back" from pure theory to material things. Eratosthenes chose his words very carefully, and he hit the Platonic style very well. And, in the main, he remained true to history, in so far as he brought Archytas, Eudoxus and Menaechmus on the stage, since they had indeed found solutions. On the other hand, he supplied various playful details and he made the dramatis personae say various things, which they might have said, but never did say.

We come now to the two greatest mathematicians of the time of Plato: Theaetetus and Eudoxus.

Theaetetus

must have been, according to Plato, an extremely attractive young man. In the introductory conversation of the dialogue "Theaetethus", Euclid [1] tells his friend Terpsion that he had seen Theaetetus, being brought back to Athens in dying condition from the camp at Corinth, and that dysentery was added to his wounds. "What a man he is who you say is in danger!" exclaims Terpsion. "A noble man, Terpsion", Euclid agrees, "and indeed just now I heard some people praising him highly for his conduct in the battle". "That is not at all strange", Terpsion continues, and then Euclid recalls a prophecy of Socrates, that Theaetetus would certainly become famous if he reached the age of manhood.

This prologue takes place in 369, the year of our hero's death. In the dialogue, which plays 30 years earlier, on the day on which Socrates is summoned before his judges, Theaetetus is still very young. The old Theodorus of Cyrene, who comes to see Socrates in Athens, relates that he has there become acquainted with a young man so marvelously gifted as he had never seen in all his long life. He is not beautiful ("now don't be angry with me" Socrates, "but he is like you in his snub nose and protruding eyes"), but unbelievably quick of understanding, gentle and yet brave as none else. Theodorus had never considered such a combination of excellent qualities possible. "But this boy advances toward learning and investigation smoothly and surely and successfully, with perfect gentleness, like a stream of oil that flows without a sound, so that one marvels how he accomplishes all this at his age."

Now Theaetetus appears himself. The friendly manner, the modesty with which he conducts himself, the logical acumen of his answers, completely confirm the impression created by the praise of Theodorus. "Most excellent, my boys!", says

[1] Not the author of the Elements, but the philosopher Euclid of Megara.

Socrates, "I think Theodorus will not be found liable to an action for false witness."

In the dialogue Plato gives an example of a mathematical investigation made by Theaetetus. The example is rather dragged in; it has to serve as an introduction to a philosophical discussion, but it does not fit very well. I can find but one possible explanation for this: Plato wanted to take a subject with which the mathematician Theaetetus had actually occupied himself. Therefore, we can safely take the dialogue as a historical source for the reconstruction of the mathematics of Theaetetus.

After having explained what Theodorus had revealed as to the irrationality of the sides of squares whose areas are 3, 5, . . . ,17 square feet, Theaetetus continues:

> "Now it occurred to us (Theaetetus and another young man), since the number of roots appeared to be infinite, to try to collect them under one name, by which we could henceforth call all the roots."
> "And did you find such a name?"
> "I think we did. But see if you agree... We divided all numbers into two classes. The one, the numbers which can be formed by multiplying equal factors, we represented by the shape of a square and called square or equilateral numbers ... The numbers which lie between these, such as three and five and all numbers which cannot be formed by multiplying equal factors, but only by multiplying a greater by a less or a less by a greater, and are therefore always contained in unequal sides, we represented by the shape of the oblong rectangle and called them oblong numbers."
> "Very good; and what next?"
> "All the lines which produce a square whose area is a square number we called lengths (μῆχος), and those which form the oblong numbers we called (in a restricted sense) sides of squares (δυνάμεις), as being incommensurable in length with the first, but only (commensurable) in the areas which they produce. And in the same way for spatial bodies."
> "Most excellent, my boys!"

Here ends the mathematical part. Following Socrates, Plato attaches the greatest value to exact definitions; that is why the definitions of oblong numbers and of equilateral or square numbers are given in such detail. What is for us more important than these trivial definitions, is the proposition given at the end, very briefly but nevertheless clearly:

Line segments, which produce a square whose area is an integer, but not a square number, are incommensurable with the unit of length.

The analogous proposition for spatial numbers, briefly indicated by Theaetetus in his last short sentence, would have to be stated as follows:

Line segments which produce cubes whose volume is an integer but not a cubic number, are incommensurable with the unit of length.

How would the proof of such a proposition be made with classical methods? Indirectly of course. Supposing that two lines, which produce squares of areas n and 1, had commensurable lengths, one would try to show that n is a square number.

The proof necessarily breaks up into a geometrical and a number-theoretical part. From the assumed commensurability of the sides, one first deduces geometrically that the ratio of the sides is equal to that of two numbes p and q, and that the ratio of the areas of the squares which they produce is equal to that of p^2 and q^2, so that

(1) $$n : 1 = p^2 : q^2.$$

or, equivalently,

(2) $$p^2 = nq^2.$$

Then it has to be shown, number-theoretically, that it follows from (1) and (2) that n is a square number. Analogously for third powers.

The geometric part of the proof is found in Book X of the Elements. Propositions 5, 6 and 9 of this book are:

X 5. *Commensurable magnitudes have to one another the ratio which a number has to a number.*

X 6. *If two magnitudes have to one another the ratio which a number has to a number, the magnitudes will be commensurable.*

X 9. *The squares on straight lines commensurable in length have to one another the ratio which a square number has to a square number; and squares which have to one another the ratio which a square number has to a square number will also have their sides commensurable in length.*

The corresponding proposition for cubes is not found in Euclid, but it can be formulated by analogy with X 9.

There is moreover a striking similarity between the terminology of the dialogue Theaetetus and that of the beginning of Book X; the same phrase "commensurable in length" (μήκει σύμμετρος) occurs in both, and to the expression „potentially commensurable" (δυνάμει σύμμετρος) corresponds in Plato the circumlocution "commensurable relative to the areas which they produce". In sound and in meaning, the words μῆκος and δύναμις in Plato are closely related to the phrases μήκει σύμμετρος and δυνάμει σύμμετρος in Euclid. And while in Plato, Theaetetus begins by recalling the accomplishments of Theodorus, Book X opens with three propositions concerning the "successive subtractions" of unequal magnitudes, among which occurs as X 2 exactly the criterium for incommensurability which Theodorus had most probably used in his investigation.

It becomes clear from all these things that the dialogue Theaetetus and the beginning of Book X belong together and supplement each other. Book X contains the more detailed mathematical development of matters briefly indicated in the dialogue. Theaetetus needed propositions X 5, 6 and X 9 to prove the proposition he enunciated in the dialogue; they enabled him to translate the *geometric* assumption of the incommensurability of the sides into an *arithmetic* property of the numbers which represent the areas of the squares.

This view receives ample confirmation from a scholium, i.e. a marginal note, on X 9, which says that this theorem had been discovered by Theaetetus.

It is Zeuthen's judgment [1] that the greatest merit of Theaetetus is however not to be found in the geometric analysis of the irrationality problem, but in the arithmetic part, in the proof therefore that (2) can hold only if n is a square. According to this judgment, Theaetetus had discovered and proved several propositions from the arithmetical Books VII and VIII which were needed for this proof.

I do not share Zeuthen's view on this. For, as we have seen before, Book VII is of older date and forms the foundation of the Pythagorean theory of numbers. But, once one has at his disposal the propositions of Book VII, the number-theoretical part of Theaetetus' proof has no further difficulties. One can, for instance, reason as follows: q and p in (1) can of course be taken to be relatively prime. Then it follows from VII 27 that p^2 and q^2 (and p^3 and q^3 as well) are relatively prime, so that, by VII 21, they are the least of the numbers in that ratio. But in the left member of (1) occur n and 1 which are also relatively prime and hence also the least in this ratio. It follows that $n = p^2$ and $1 = q^2$, so that n is a square. The case of cubes proceeds analogously.

In my opinion the merit of Theaetetus lies therefore not in his contribution to the theory of numbers, but in his study of incommensurable line segments which produce commensurable squares.

He introduced the exact concepts "commensurable in length" and "potentially commensurable", which led to the classification of line segments into commensurable "lengths" and incommensurables "sides of squares"; and in proposition X 9 he formulated the necessary and sufficient conditions on two squares under which their sides are commensurable in length. This enabled him furthermore to answer the question as to which sides of squares are commensurable with the unit of length, and thus to solve in complete generality a problem which Theodorus had only been able to handle for squares of areas of 3 to 17 square feet. Finally, he was able to extend the entire theory without difficulty ("like a stream of oil that flows without a sound") to the sides of commensurable cubes.

In order to get a picture of the other accomplishments of Theaetetus, we shall now proceed first to a discussion of the further contents of Book X.

Analysis of Book X of the Elements.

At the very start of Book X, in Definition 3, a fixed line segment is introduced which we shall denote by the letter e, and which plays the same role as the unit of length ("that of 1 foot") in the dialogue Theaetetus.

We shall call a line segment (or an area) *measurable*, if it is commensurable with the fixed line segment e (or with the square e^2). In Euclid measurable areas are called *expressible* (ῥητός). *Expressible lines* on the other hand are all those lines which produce measurable squares, not only measurable lines therefore, but also non-measurable lines, such as the sides of squares of areas 3, 5, . . . investigated by

[1] H. G. Zeuthen, *Oversigt K. Danske Vidensk. Selsk.* 1910, p. 395.

Theodorus. This terminology exhibits a first consequence of the principle which classifies line segments according to the squares which they produce. All other line segments are called *unreasonable* [1] (ἄλογος). Apparently, these "unnameable" magnitudes had not been either recognized or named in the earlier stages of the theory.

This did take place later on. In X 21, 36 and 73, Euclid defines three fundamental irrational lines, the *medial*, the *binomial* and the *apotome* [2].

A *medial area* is the area of a rectangle whose sides a and b are expressible, but incommensurable. We would say, it is an area $\sqrt{r}$, where r is a rational number. A straight line whose square is equal to such an area is called a *medial line*. In our notation $\sqrt[4]{r}$. This line satisfies the equation $x^2 = ab$ and is therefore a mean proportional between a and b; hence the word medial.

The sum $a + b$ of two expressible but incommensurable lines is called *binomial* ("one with two names"), their difference $a - b$, *apotome* ("the one cut off").

It is proved in Book X that all these new types of lines are "unreasonable", and are mutually exclusive, also that a binomial can be represented as a sum $a + b$ in only one way, and an apotome in only one way as a difference $a - b$ of expressible lines. All these proofs are based on one fundamental idea which runs as a guiding thread through the entire book: *to prove properties of any type of line, one constructs a square on this line and one investigates the properties of this square*. For instance, to prove that a binomial can not be a medial, it is shown that the square on a binomial can not be a medial area.

Properly speaking, this basic idea already turns up in the first part of Book X and in the dialogue Theaetetus, for Theaetetus derived the incommensurability of certain line segments from the ratio of their squares.

The other classes of irrationals, which are introduced immediately after the binomial and the apotome, can best be understood if one starts from the following problem: under what conditions is the square root of a binomial (or an apotome) itself a binomial (or an apotome)?

A binomial $a + b$ is not an area but a line segment. Therefore the Greek ideas do not permit the extraction of a square root. In order to define a magnitude which corresponds to our $\sqrt{a + b}$, it is necessary first to make $a + b$ into an area, by constructing a rectangle of base e and height $a + b$. It is this which happens every time in X 54—59 and 91—96. The question now becomes therefore: Is this area equal to that of the square of a binomial $u + v$, i.e. Is

(3) $(u + v)^2 = e(a + b)?$

[1] In modern translations one frequently finds the words "rational" and "irrational" in place of "expressible" and "unreasonable". I have chosen these other words here, because in modern mathematics, "rational" is synonymous with "measurable", and "irrational" with "non-measurable".

[2] It is known from a statement of Eudemus that Theaetetus had studied these three irrationalities and that he had related them to the three means, viz. the medial to the geometric mean, the binomial to the arithmetic mean, and the apotome to the harmonic mean. See G. Junge and W. Thomson, *The commentary of Pappus on Book X*. Harvard Semitic Series VIII, Cambridge (Mass.), 1930.

Development leads to

(4) $$(u^2 + v^2) + 2uv = ea + eb.$$

Now, $u^2 + v^2$ is a measurable area and $2uv$ a medial area. From the fact that the binomial can be split in one way only into expressible terms, the conclusion follows readily that the two terms on the left of (4) must separately be equal to the terms on the right, i.e. a being the larger of these terms

(5) $$u^2 + v^2 = ea, \text{ and } 2uv = eb.$$

In X 33—35, the solution of these equations is indicated, for various cases, by the use of geometric algebra. For example, in X 33, ea is a measurable area and eb a medial area; in X 34, the situation is reversed, in X 35 they are both medial. The method remains the same in all these cases: setting $ea = c^2$ and $eb = cd$, we have in place of (5)

(6) $$u^2 + v^2 = c^2, \text{ and } uv = \tfrac{1}{2}cd.$$

Now, new unknowns are introduced by setting

(7) $$u^2 = xc, \quad v^2 = yc.$$

Since $u^2 + v^2 = c^2$, one can interpret u and v geometrically as the sides ZA and ZB of a right triangle ZAB, whose hypothenuse is $AB = c$; the new unknowns x and y are then the projections of the sides on the hypothenuse (see Fig. 54).

From (6) and (7) we obtain for x and y the equations

(8) $$x + y = c, \quad xy = (\tfrac{1}{2}d)^2,$$

so that x and y are the roots of a quadratic equation, or, in Greek terminology, x and y are the sides of a rectangle equal to the square on $\tfrac{1}{2}d$, which "applied" to $AB = c$ leaves a square as deficiency". In X 33—35, this application is made every time according to the rules of the game. In the construction, the square root

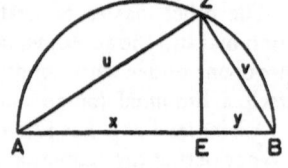

$$w = \sqrt{(\tfrac{1}{2}c)^2 - (\tfrac{1}{2}d)^2} = \tfrac{1}{2}\sqrt{c^2 - d^2}$$

occurs, which yields the required line segments x and y when added to and subtracted from $\tfrac{1}{2}c$.

Fig. 54.

What is now the condition which the given areas $ea = c^2$ and $eb = cd$ must satisfy in order that u and v may be expressible and thus u^2 and v^2 measurable? For u^2 and v^2 we have the equations

$$u^2 - v^2 = (x - y)c = 2wc, \quad u^2 + v^2 = (x + y)c = c^2,$$

which may be put in the form

(9) $$u^2 + v^2 = ea, \quad u^2 - v^2 = e\sqrt{a^2 - b^2}.$$

Therefore these two areas must be measurable. This gives the following conditions:

(I) a and e commensurable,

(II) $\sqrt{a^2 - b^2}$ and a commensurable.

If these conditions are satisfied, the root of $e(a + b)$ is again a binomial, viz. $u + v$.

To solve equations (5), the Babylonians would simply square the second equation; they would then know both sum and product of u^2 and v^2. The Greeks could not apply this method, because, for them, the square of an area does not have any meaning. This is why they introduced the auxiliary line c, and replaced the areas u^2 and v^2 by the lines x and y whose sum and product are then known. From the final conditions (I) and (II), the auxiliary line c has again been eliminated. It is to this roundabout procedure that all the complications in the proofs are due. The line of thought is simple and at bottom purely algebraic.

Conditions (I) and (II) give rise to a division of binomials into 6 subclasses. In Euclid these are defined as follows:

If (II) holds, and

a is commensurable with e, so that b is not commensurable with e: *first binomial*;

b is commensurable with e, so that a is not commensurable with e: *second binomial*;

a and b both incommensurable with e: *third binomial*;

If (II) is not satisfied, and

a is commensurable with e, but not b: *fourth binomial*;

b is commensurable with e, but not a: *fifth binomial*;

a and b both incommensurable with e: *sixth binomial*.

It is possible to express conditions (I) and (II) in terms of u and v, instead of by means of a and b; hence the discrimination of the 6 cases can also be so expressed. In accordance with (9), condition (II) means that $u^2 - v^2$ and $u^2 + v^2$ are commensurable or, what amounts to the same, that u^2 and v^2 are commensurable. The six cases can therefore be formulated as follows:

Let u^2 and v^2 be commensurable, but not u and v and

case 1) $u^2 + v^2$ and hence both u^2 and v^2 measurable, but not uv;

case 2) $u^2 + v^2$ and hence both u^2 and v^2 non-measurable, but uv measurable;

case 3) $u^2 + v^2$ and hence both u^2 and v^2 non-measurable, and uv non-measurable.

Let u^2 and v^2 be incommensurable, and

case 4) $u^2 + v^2$ measurable, but uv only medial;

case 5) $u^2 + v^2$ medial, non-measurable, but uv measurable;

case 6) $u^2 + v^2$ and uv both medial, not measurable and not commensurable.

In X 29—35 it is shown that in each of these six cases, line segments exist which satisfy these conditions. Moreover X 29—32 also serve to show that line segments a and b exist, which do or do not satisfy condition (II).

There follows a long set of propositions (X 36—72) concerning the properties of $u + v$, and an equally long one (X 73—110) concerning the properties of $u - v$, in each of the 6 cases. In case 1), $u + v$ is a binomial and $u - v$ an apotome, because in this case conditions (I) and (II) are valid. In case 4), $u + v$ is called a *major* and $u - v$ a *minor*. In all other cases the line segments $u + v$ and $u - v$ are given names to indicate the type of square which they produce; e.g. in case 6), $u + v$ is called "producing two medial areas", because the square

$$(u + v)^2 = (u^2 + v^2) + 2uv$$

is the sum of two medial areas, $u^2 + v^2$ and $2uv$.

Thus the 6 cases lead to 12 types of irrational segments $u + v$ and $u - v$, and to 13 types when the medial is counted in. All these segments are "unreasonable" because their squares are unreasonable; and all 13 types are mutually exclusive, because in all these cases the squares have mutually exclusive properties. For example, the square on a medial segment is a medial area, etc.

Book X does not make easy reading. As early as 1585, Simon Stevin wrote: "La difficulté du dixième livre . . . est à plusieurs devenue en horreur, voir jusqu'à l'appeler *le croix des mathématiciens*, matière trop dure à digérer et en laquelle n' aperçoivent aucune utilité". [1] Up to X 28 it goes fairly well, but when the existence proofs start with X 29 ("To find two potentially commensurable expressible straight lines, such that the difference of the squares described on them is equal to the square on a straight line commensurable with the first", etc.) one does not see very well what purpose all of this is to serve. The author succeeded admirably in hiding his line of thought by starting with his constructions, even before having introduced the concept of binomial which does throw some light on the purpose of these constructions, and by placing at a still later point the division into 6 types of binomials.

But who is this author? Has the same Theaetetus who studied the medial, the binomial and the apotome, also defined and investigated the ten other irrationalities, or were these introduced later on?

It seems to me that all of this is the work of one mathematician. For, the study of the 13 irrationalities is a unit. The same fundamental idea prevails throughout the book, the same methods of proof are applied in all cases. Propositions X 17 and 18 concerning the measurability of the roots of a quadratic equation precede the introduction of binomial and apotome, but these are not used until the higher irrationalities appear on the scene. The theory of the binomial and the apotome is almost inextricably interwoven with that of the 10 higher irrationals.

Hence — the entire book is the work of Theaetetus.

This conclusion finds further confirmation in the close connection between the tenth and the thirteenth books. The latter contains

---

[1] *Oeuvres Mathématiques de Simon Stevin de Bruges, où sont insérées les Mémoires Mathématiques desquelles s'est exercé le Très-haut et Très-illustre Prince Maurice de Nassau*, etc. Leiden 1634, p. 10a.

The theory of the regular polyhedra.

A scholium in Book XIII states:

"In this book, the thirteenth, are treated the five so-called Platonic figures, which however do not belong to Plato, three of the aforesaid five figures being due to the Pythagoreans, namely the cube, the pyramid and the dodecahedron, while the octahedron and the icosahedron are due to Theaetetus. They are named after Plato, because he mentions them in the Timaeus. This book also carries Euclid's name because he embodied it in the Elements."

The evident intention of this last sentence is to say that Euclid did not revise this book, but took it in its entirety from an older work. Indeed, as Tannery has observed, Book XIII strongly creates the impression of being an entirely independent treatise, complete in itself. It opens with 12 propositions on the golden section and on the regular pentagon and triangle inscribed in a circle. These topics have been fully dealt with in Books II and IV, but now everything is taken up anew and in a different manner.

If Euclid had written Book XIII himself, he would certainly have used the diagram of II 11 (division of a line into mean and extreme ratio), which would have shortened the arguments considerably.

"Time and again, results are obtained implicitly", Dijksterhuis[1] observes rightly, "which had already been obtained explicitly in Book II." The author of Book XIII evidently knew the methods of "geometric algebra", but he was not acquainted with its systematic development in Book II.

On the other hand, he did know the contents of Book X very thoroughly, for he makes constant use of the classification of irrationals! Thus we read in XIII 11: *If in a circle which has its diameter expressible, an equilateral pentagon be inscribed, the side of the pentagon is the irrational line called minor.* Analogously, we find in XIII 16, 17 that the edge of the regular icosahedron inscribed in a sphere is a minor and the edge of the dodecahedron an apotome.

Thus Book XIII refers to Book X; but the reverse also takes place. Indeed it is very probable that the occasion for the careful study and classification of irrationalities in Book X is found in the fact that these irrationalities occur as sides of regular polyhedra. In modern notation, the side of the pentagon is $\frac{1}{4}\sqrt{10 - 2\sqrt{5}}$. This can be formulated in classical form as follows: it is the side of a square whose area is equal to the difference between a measurable and a medial area. The question was raised whether such magnitudes can be expressible, or if not, whether they can be represented as a binomial or an apotome; this gave rise to the theory of Book X.

Thus we conclude that the author of Book XIII knew the results of Book X, but that moreover, the theory of Book X was developed with a view to its applications in Book XIII. This makes inevitable the conclusion that the two books are due to

[1] See E. J. Dijksterhuis, *De Elementen van Euclides* II, Groningen, 1930, p. 250.

the same author. We already know his name: Theaetetus. The 10th century compendium "Suda", usually called "Suidas", mentions under Theaetetus that he was the first to write on the so-called 5 solids (or that he was the first to construct them).

Our conclusion is confirmed by an undeniable similarity in style and in method of proof. As Book X, so does Book XIII employ chiefly the methods of geometric algebra; proportions appear very rarely in either book. In both books geometric constructions are cleverly interwoven with algebraic calculations. The author never reveals the line of thought which leads him to his constructions by the way of algebraic analysis. He starts at the other end; first he constructs (rapidly and simply, but obscurely) the magnitudes to which he is led by algebraic analysis, then he constructs from them the required figure (such as one of the regular polyhedra) and he ends by showing that this figure has the desired properties, e.g. that it has a circumscribed sphere of unit diameter.

Let us take as an example the construction of the regular tetrahedron inscribed in a sphere of given diameter AB (XIII 13). Right at the start, the statement is made that the square on the diameter will be one-and-one-half times as great as the square on the side of the "pyramid", i.e. of the tetrahedron. To construct this side and the other magnitudes, required for the construction, in the shortest and most elegant manner, he divides the given diameter AB in the ratio $2:1$ by means of a point Γ, draws a semicircle on AB, erects a perpendicular to AB in Γ, meeting the semicircle in Δ and draws $A\Delta$. Then

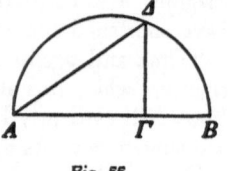

Fig. 55.

$A\Delta^2 = \frac{2}{3}AB^2$, so that $A\Delta$ is the required side; moreover $A\Gamma$ is the height of the pyramid, and $\Gamma\Delta$ the radius of the circumcircle of the base. Apparently, the author knew all of this in advance, but he does not betray the source of his knowledge. Now he constructs an equilateral triangle in a circle of radius $\Gamma\Delta$, erects a perpendicular to the plane of the circle at its midpoint, makes it equal to $A\Gamma$ and completes the pyramid. Then he proves that the faces are equilateral triangles and that the entire solid can be inscribed in a sphere of diameter AB.

Very beautiful is the construction of the icosahedron. It starts again by the construction, with the aid of a semicircle on the given diameter d, of a line segment $B\Delta = r$, such that

$$r^2 = d^2/5.$$

Next, a regular pentagon $EZH\Theta K$ is inscribed in a circle of radius r. The midpoints of the arcs EZ, etc. form a second regular pentagon $\Lambda MN\Xi O$, so that OE is the side z_{10} of the decagon. Now perpendiculars to the plane of the pentagon are erected at the points $\Lambda MN\Xi O$, and on each of them the length r is laid off. Thus another such pentagon $\Pi P\Sigma TY$ is obtained in a plane parallel to that of the first drawing. Each of the vertices $\Pi P\Sigma TY$ is connected with the two adjacent vertices of the first pentagon. Thus a wreath is obtained, consisting of 10 triangles, which are proved to be equilateral. Then the solid is completed, above and below,

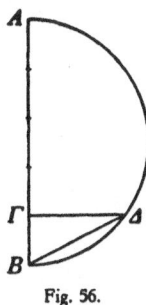

Fig. 56.

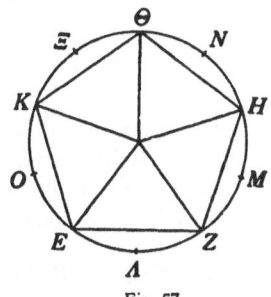

Fig. 57.

by a regular five-sided pyramid of height z_{10}. The lateral edges of these pyramids are

$$\sqrt{z_{10}^2 + r^2} = z_5.$$

Thus all the faces of the icosahedron are equilateral triangles, and the distance between two opposite vertices always turns out to be

$$r + 2z_{10} = \sqrt{5r^2} = d,$$

so that the solid can be inscribed in a sphere of diameter d.

The interested reader is referred to Dijksterhuis, De Elementen van Euclides II, or to Euclid himself, for the construction of the dodecahedron.

The fragment of Theaetetus concludes with a proposition concerning the relative size of the edges of the 5 solids. Euclid adds the statement that no other solid can be constructed, except the five that have been mentioned, which is bounded by equal, equilateral and equiangular polygons. As Dijksterhuis remarks, the proposition is incorrect in this formulation; indeed the proof is decidedly inferior in rigor to the other proofs. I see no reason to ascribe the statement to Theaetetus; it has indeed a very loose connection with the rest.

The theory of proportions in Theaetetus.

The formulation of propositions X 5, 6 and 9 of Theaetetus we could take from the Elements; this is not possible for the proofs. Indeed, Euclid's proofs contain logical errors for which we can not hold Theaetetus responsible. Moreover these proofs depend on the theory of proportions developed in Book V, following Eudoxus, which was not yet known to Theaetetus.

In Definition 5 of Book V we read: "Magnitudes are said to be in the same ratio, the first to the second and the third to the fourth, when, if any equimultiples whatever be taken of the first and the third, and any equimultiples whatever of the second and the fourth, the former equimultiples alike exceed, are alike equal to, or alike fall short of, the latter equimultiples respectively taken in corresponding order." This says that the proportionality $a : b \ c : d$ means that

$$na > mb \text{ implies } nc > md,$$
$$\text{and } na = mb \text{ implies } nc = md,$$
$$\text{and } na < mb \text{ implies } nc < md,$$

no matter how the integers m and n are chosen.

As we shall see presently, this ingenious definition is due to Eudoxus. When Theaetetus, in his early youth, found propositions X 5, 6 and 9, he could not yet make use of these definitions. He must have started from another definition of proportionality.

The older definition of Book VII: "Numbers are proportional when the first is the same multiple, or the same part, or the same parts, of the second that the third is of the fourth", is applicable to numbers, but not to incommensurable line segments, because a is neither a multiple of b, nor a part, nor parts, if the line segments a and b have no common measure. The definitions are therefore of little value to Theaetetus for his non-measurable line segments. Neither can I imagine that he, strict logician as he was, might use the ratio concept naively, without a sharp definition. But what definition of proportionality did he start from then?

A remarkable passage from the Topica of Aristotle (158b) throws some light on this question:

> It appears also in mathematics that the difficulty in using a figure is sometimes due to a defect in definition, e.g. in proving that the line which cuts the area parallel to one side (of a parallelogram) divides similarly both the line which it cuts and the area; whereas if the definition be given, the fact asserted becomes immediately clear; for the areas have the same antanairesis from them as have the sides: and this is the definition of "the same ratio".

The area here referred to, is a rectangle, or a parallelogram and the assertion is that the areas, into which a line parallel to the base divides the rectangle, are proportional to the parts into which the height is divided. If these parts are called b and c, then it is a question of showing that

$$ab : ac = b : c.$$

Now Aristotle calls this "immediately clear", if the "definition" is stated. Evidently he is speaking of a definition of the concept of proportionality, because the definitions of the concepts rectangle, parallel, etc. are of no importance in this connection. According to Aristotle this definition would therefore have to be the following: two areas and two lines are proportional, if the areas and the lines have the same antanairesis.

But what is the antanairesis? The lexicon derives ἀνταναίρεσις, deduction, from the verb ἀντ-αν-αιρεῖν, subtract, literally "balancing against each other", which is used especially for sums of money, for instance in drawing up the balance sheet. The commentator Alexander of Aphrodisias adds here that by antanairesis,

Aristotle means the same thing as by anthyphairesis. This actually brings us some help, because in Euclid VII 2 and X 2, 3, the verb ἀνθυφαινεῖν means "to take away in turn" the smaller of two numbers or line segments from the larger one, for the determination of the greatest common divisor.

On the basis of this etymology, Zeuthen, Dijksterhuis and Becker have given, independently of one another, the following explanation of the words of Aristotle. The definition of proportion which Aristotle has in mind, is the following: two magnitudes a and b are proportional to c and d, when the antanairesis, the subtracting in turn of the smaller from the larger, proceeds with a and b in the same way as with c and d, i.e. if a can be subtracted from b (or b from a) as often as c from d (or d from c), the remainders again equally often, etc. On the basis of this definition, it does indeed become "immediately clear" that the rectangles ab and ac are proportional to b and c; for, if it is possible to lay off the height b a certain number of times on the height c, then the rectangle ab can be taken away from the rectangle ac equally often, etc.

This definition of proportionality makes it easy to prove a number of properties of proportions, e.g. that from

$$a : b = c : d,$$

we can conclude $\quad\quad c : d = a : b,$

and $\quad\quad\quad\quad\quad b : a = d : c,$

and $\quad\quad\quad\quad\quad (a + b) : b = (c + d) : d,$

and, if $a > b$, $\quad\quad (a - b) : b = (c - d) : d.$

But one property, viz. the interchange of the means, causes difficulty. And now it is curious that, according to Aristotle, it was indeed exactly the proof of this proposition, which at first led to difficulties. "Formerly", says Aristotle in Anal. Post. 15, "this proposition was proved separately for numbers, for line segments, for solids and for periods of time. But after the introduction of the general concept which includes numbers as well as lines, solids and periods of time" (viz. the concept of magnitude), "the proposition could be proved in general".

The proof for numbers can be found in Book VII (VII 13). What was the old proof for line segments?

O. Becker has advanced an ingenious hypothesis for this.[1] From the proportionality

(1) $\quad\quad\quad\quad\quad a : b = c : d,$

one deduces first the equality of the areas

(2) $\quad\quad\quad\quad\quad ad = bc,$

then interchanges b and c, and finally returns to the proportionality

(3) $\quad\quad\quad\quad\quad a : c = b : d.$

[1] O. Becker, Eudoxus-Studien, Quellen und Studien B 2, p. 311.

It is this last step which involves the proposition on rectangles, that Aristotle talked about:

(4) $a : c = ad : cd,$

since

$$a : c = ad : cd = cb : cd = b : d$$

follows from (4) and (2).

The fragments from Aristotle, which have been cited, make it very plausible, that (3) was indeed derived from (2) in this manner.

The deduction of (2) from (1) involves another proposition. For, from (1) follows

$$ad : bd = a : b = c : d = bc : bd;$$

now it remains to prove: *P. If in a proportion the consequents are equal then the antecedents (ad and bc) are equal as well.*

The proof of P. requires use of the so-called "lemma of Archimedes", that is formulated as follows: *Q. If A and B are comparable magnitudes (e.g. both line segments), and if A is less than B, then a certain multiple nA exceeds B.* In Euclid the lemma of Archimedes is usually applied in the following form

R.[1] *Let A and Γ be comparable magnitudes, and A larger than Γ. If a piece larger than one half of A is taken from A, from the remainder a piece larger than one half of it, and so forth, then at some time, the remainder will be a magnitude less than Γ.*

This is exactly the first proposition of Book X, the work of Theaetetus. This indicates that we are on the right track. For, why does this proposition X 1 appear here? In Book X, it is used exclusively in the proof of proposition X 2: *If, when the less of two unequal magnitudes is continually subtracted in turn from the greater, that which is left never measures the one before it, the magnitudes will be incommensurable.* The proof of X 2 by means of X 1 is very elegant, but there is not the slightest difficulty in proving X 2 without X 1, e.g. as follows: if the magnitudes A and B had a common measure E, then the remainders, obtained in the alternate subtractions, would always be multiples of E, indeed constantly diminishing multiples, so that the sequence of remainders would have to end after a finite number of steps.

Consequently, X 1 is not necessary as a preliminary for X 2. But what purpose does X 1 then serve? It serves for the proof of proposition P, which in turn is needed for setting up the theory of proportions!

All becomes clear now. Evidently, Theaetetus began his book with an exposition of the theory of proportions, based on the antanairesis-definition. Following his usual procedure, he started with lemmas which would be needed later on; among these is R = X 1. In propositions X 2, 3 he established the theory of the infinite or finite antanairesis, thus obtaining at the same time a criterium for the commensurability of two line segments or two areas. It is probable that the next thing was the theory of proportions, based on the antanairesis-definition to which Aristotle refers. Euclid omitted this part, because he had already given another

---

[1] O. Becker has shown (loc. cit.) how P can be derived from R.

theory of proportions in Book V. The Aristotle fragment from the Anal. post., cited above, makes clear why this new theory (due to Eudoxus) was given preference. Theaetetus then proceeded to develop his theory of expressible magnitudes and of their ratios, on the basis of his theory of proportions (X 4, 5 and 9—13). In this part, Euclid replaced the proofs by others of which the method was borrowed from Archytas and his school. The next main division of Book X, which is concerned with the 13 kinds of irrational lines, was left practically unchanged by Euclid, except that he and his followers added a number of less important propositions and remarks, intended to clarify the very difficult subject.

Of still greater stature and more famous than Theaetetus is his younger contemporary, one of the most brilliant figures of his time:

Eudoxus of Cnidos.

Eudoxus was famous not only as a mathematician, but also as a medical man and especially as an astronomer; he was moreover an eminent orator, a philosopher and a geographer. In jest his friends called him Endoxus, the renowned.

The chronicle of Apollodorus, a chronological classic of later date, mentions that Eudoxus "flourished" about 368, which would mean that he was born around 400. He died at the age of 53 years, in his native town of Cnidos on the Black Sea, highly honored as a lawgiver.

He studied mathematics with Archytas in Tarentum and medicine with Philistium on the island of Sicily. When he was 23 years old, he went to Athens to learn philosophy and rhetoric. He was so poor, that he had to live in the harbor-town Piraeus, a walk of two hours each way from Plato's Academy. Some years later his friends enabled him to undertake a journey to Egypt. From Agesilaus, king of Sparta, he received a letter of recommendation to the Pharaoh Nectanebus. It is reported that in Egypt he learned astronomy from the priests of Heliopolis and that he made observations himself in an observatory, situated between Heliopolis and Cercesura, still in existence in the days of the Emperor Augustus. After that he established a school at Cyzicus on the sea of Marmora, which attracted a large number of pupils. Around 365 he came once more to Athens with his pupils. At that time he stood in high repute; he held discussions on philosophical questions with Plato, on the ideas and on the supreme good. He held the opinion that the ideas are present in, or "blended with" observable things, thus causing the being-thus of things, just as pure white is present in visible white and thus produces being-white. He also taught that pleasure, joy, is the highest good, "since all living beings strive after it". This does not by any means justify the conclusion that he advocated a dissolute life; on the contrary, on the testimony of Aristotle, he was a model of moderation, goodness and strength of character. But Plato did not agree with Eudoxus' views on the ideas, nor with his doctrine of pleasure as the highest good; in the Philebus he combats these tenets with a variety of arguments.

This shows that, although his philosophic views were closely related to those of Plato, Eudoxus was opposed to Plato on a number of points. The story of Diogenes Laërtius, that Eudoxus and Plato were enemies, is almost certainly exaggerated, just as the opposite statement of Strabo, that Eudoxus was beloved by Plato. It is best in accord with the character of the two men, as they become known to us through the best sources, to say that they respected each other, that they debated and that they collaborated to the good of the sciences to which they were both devoted.

This appears also from the following statement, which the reliable Simplicius takes from the astronomer Sosigenes. Eudemus relates that Plato proposed to the astronomers the question as to the uniform circular motions, which could serve to "save", i.e. to explain, the planetary phenomena. Eudoxus was the first to give an answer to this question.

Eudoxus as astronomer.

From communications of Simplicius and Aristotle, Schiaparelli was able to reconstruct almost completely the extremely ingenious planetary system designed by Eudoxus. The following bird's-eye view of the system will reveal its extraordinarily ingenious construction; further particulars can be found in the treatise of Schiaparelli. [1]

The spherical earth is at rest at the center. Around this center, 27 concentric spheres rotate. Of these the exterior one carries the fixed stars, the others serve to account for the motion of sun, moon and the 5 planets. Each planet requires four spheres, the sun and moon three each.

It is not difficult to describe the motion of the 3 concentric spheres which govern the motion of the moon. The exterior one of these three rotates in one day about the poles of the equator, just as the sphere of the fixed stars; in its motion it carries the two others along. This explains why the moon shares the diurnal motion of the stars. An oblique circle lies on this first sphere, the ecliptic. The second sphere is carried along by the motion of the first; it has moreover a slow rotation in the same sense about the poles of the ecliptic. This motion serves to account for the "recession of the nodes" of the lunar orbit. On the second sphere is a great circle, slightly inclined to the ecliptic; this is the orbit of the moon. The third sphere rotates about the poles of this circle; this sphere also shares the motions of the first two spheres, and it carries the moon. The period of revolution of this third motion is the draconitic period of the moon. The lack of uniformity in the motion of the moon is not accounted for in this system.

The model for the sun is similar to that for the moon. But the model for planetary motion is incredibly clever. Let us consider, as an example the four concentric spheres for Jupiter.

[1] L. Schiaparelli, *Die homozentrischen Sphären des Eudoxus*, Abh. Gesch. Math. I, Leipzig 1877. See also Th. Heath, *Aristarchus of Samos*, Oxford 1913.

The exterior sphere again has only one motion, that of the fixed stars. The second shares this motion and has an additional one, the third has three and the fourth four motions. The second sphere rotates about the poles of the ecliptic in the sense opposite to that of the diurnal motion; this serves to explain Jupiter's sidereal period of 11—12 years. The rotations of the third and the fourth spheres are used to account for the alternate direct and the retrograde motions of Jupiter. Their rotational velocities are equal but opposite in sense; the period is the synodic period of Jupiter (13 months). The poles about which the third sphere rotates with respect to the second, lie on the ecliptic; but they do not coincide with the poles about which the fourth sphere rotates. Consequently the axis of rotation of the fourth sphere is inclined to that of the third. This accounts for the fact that these two rotations in opposite senses do not completely neutralize each other. If there were nothing but these last two motions, the planet, which is attached to the fourth sphere, would describe a horizontal figure-eight curve (see Fig. 58), whose transverse axis lies in the plane of the ecliptic. This curve can also be considered as the intersection of the sphere with a thin cylinder tangent to the sphere; it was called the Hippopede. As a result of the slowly advancing motion of the second sphere, the center of this figure-eight curve describes the entire ecliptic in 11 to 12 years, thus producing a to-and-fro motion in a loop, exhibiting

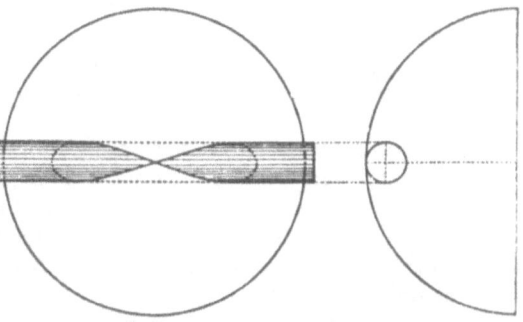

Fig. 58. Front and lateral views of the Hippopede as the intersection of a sphere and a cylinder.

great similarity with the apparent motion of Jupiter with respect to the fixed stars.

Callippus improved the system by adding two each to the number of spheres for sun and moon, and one to each of the others. Aristotle adopted all these improvements and supplemented them with "retrograde" spheres. But all these emendations failed to meet the fundamental objection to the entire system, viz. that it did not supply an explanation for the variable luminosity of the planets, because in this model their distances from the earth remained constant.

Still greater fame than he reaped with his theoretical astronomy, which aroused admiration in the profession, but had to make way a half century later for other, better systems, Eudoxus garnered from his description of the constellations and of the rising and setting of the fixed stars. An important part of this is occupied by data concerning the 12 signs of the zodiac, and the stars which rise and set at the same time as the initial points of these signs, of importance for the determination of time during the night. There is little doubt that Eudoxus gathered this knowledge from a rotating celestial globe, a σφαῖρα. One of his two works on the celestial sphere, the Phaenomena[1], was later put into verses by the poet Aratus, which were read and admired throughout the ancient world, and have also been translated into modern languages.

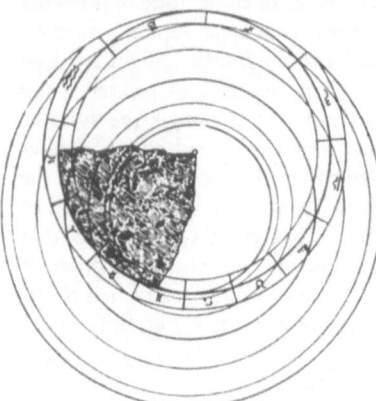

Furthermore Eudoxus designed a permanent calendar, giving data on the first and last appearances of the most important fixed stars, on the equinoxes and the solstitia. A papyrus has been preserved which contains an excerpt of this work. He also wrote on the construction of celestial globes.

It is said that Eudoxus also invented an astronomical instrument, namely the "Arachne", the spider. But, Vitruvius, the Roman architect, from whom this statement is obtained, adds: "Some say however that Apollonius (invented the spider)" What is this spider?

Fig. 59. Reconstruction of the bronze disc of Salzburg (see page 183 and plate 21)

There is an astronomical instrument, called the astrolabe, much in use among the Arabians and the Persians. A fine Persian specimen, from the year 1223, is shown on Plate 20. This instrument contains a revolving disc, which looks somewhat like a spider and which is indeed called Arachne in the Greek treatises on this instrument. On the Arachne one sees an excentric circle, representing the ecliptic, and also some pointers, which carry the names of bright fixed stars. The entire spider is a "stereographic projection" of a part of the sky, i.e. a central projection on the equatorial plane from the south pole. The instrument can be made to imitate the daily rotation of the fixed stars by turning the spider in the ring which surrounds it.

Behind the spider one can see a removable plate, on which circles have been scratched. One of these circles represents the horizon, others circles parallel to the horizon at different elevations. All these circles are represented in stereographic projection; indeed the stereographic projection of a circle is again a circle. To use the instrument at different geographical latitudes, this plate has to be changed.

Several problems can be solved by means of the astrolabe. For example, after having observed the altitude of a star or of the sun, one can rotate the spider until the pointer which carries the name of this star, or the point on the ecliptic which corresponds to the position of the sun, lies on the circle of this altitude. Thus one can determine the time of day, or of the night. For further particulars, the reader may consult H. Michel, Traité de l'astrolabe, Paris 1947.

Ptolemy and Theon of Alexandria have written treatises on the astrolabe. A dis-

[1] Its contents are known chiefly from the critique of Hipparchus (in *Arati et Eudoxi phaenomena commentarii*, ed. Manitius) and from the poem of Aratus.

cussion of these treatises and further references to the literature are found in an article by O. Neugebauer, Isis 40 (1949), p. 240.

There is another instrument, related to the astrolabe, in which a "spider" is found. It is some kind of "bad-weather-clock", driven by hydraulic means. The Roman architect Vitruvius has described it, and a piece of a large bronze disc was found in Salzburg, which agrees exactly with his description [1] (Plate 21). In this instrument, the "spider" was a network of wires, which represented the horizon and the hour circles, while the rotating disc behind it represented the sky with the zodiacal signs (see fig. 59), exactly the reverse of the astrolabe. Both instruments are based on the same principle, viz. stereographic projection.

The inventor of this principle must have been a mathematician of stature. Our sources mention Abraham, Eudoxus, Apollonius, Hipparchus and Ptolemy as the inventors. I believe that it was Apollonius, but it might equally well have been Eudoxus or Hipparchus. Ptolemy is out of the question, because Vitruvius lived about 130 years earlier than he.

Eudoxos was also famous as a medical man, but we know nothing further about this. His work in geography includes a "journey around the earth', which contains i.a. the history of the foundation of several cities. But it is high time now to turn to his most important accomplishments, to his mathematics.

Sources for the mathematical work of Eudoxus are rather meagre and uncertain; at present however there is quite general agreement as to what has to be ascribed to Eudoxus (Plate 22).

Exceedingly vague is the statement in the Proclus catalogue, that Eudoxus "increased the number of so-called general theorems". Does Proclus perhaps refer to the propositions in Book V of the Elements, concerning general magnitudes and their ratios, which hold equally well for line segments, areas, angles, etc? The word "general" might also lead one to think of "general understandings" (or axioms), which are enumerated as follows at the beginning of Book I of the Elements:

"Things which are equal to the same thing are also equal to one another. If equals be added to equals, the wholes are equals. If equals be subtracted from equals, the remainders are equal. Things which coincide with one another are equal to one another. The whole is greater than the part."

Aristotle already knows "the so-called general axioms" which form the foundations for all demonstrative sciences and must necessarily be accepted by any one who wants to gain knowledge. As an example he always quotes the third of the Euclidean axioms which have just been cited. It is therefore quite possible that Euclid took these axioms, used constantly in the theory of ratios of Book V, from Eudoxus.

Proclus says moreover: "Eudoxus added three proportions to the three others and he continued the investigation of the section, begun by Plato, making use of analyses."

[1] See A Rehm, *Zur Salzburger Bronzescheibe*, Jahreshefte österr. archäol. Inst. Wien 6 (1903), p. 41.

The "section" referred to by Proclus is probably the "golden section". It is treated three times in Euclid, first in Book II (proposition 11) in connection with the Pythagorean geometric algebra, then in Book VI (proposition 30), along with the theory of proportions, and finally in Book XIII (propositions 1—6). As we have seen, this last set of propositions is due to Theaetetus; it is therefore possible that VI 30 comes from Eudoxus.

We are brought a bit further along by Proclus' statement that, in his Elements, Euclid "collected many of the discoveries of Eudoxus and completed many of those of Theaetetus." We have already excised the work of Theaetetus from the Elements; it consists of Books X and XIII. There is no doubt that the material in Books I—IV (foundations of plane geometry without proportions) and in Book XI (foundations of solid geometry) is of older date. The arithmetical books VII—IX can here be left out of consideration. The "discoveries of Eudoxus" will therefore have to be traced to Books V, VI (theory of proportions) and to Book XII.

Can we specify them more closely? Let us first have a look at Book XII.

The exhaustion method.

Book XII starts with the proof that the ratio of the areas of two circles is equal to that of the squares on their diameters (XII 1, 2). It will be recalled that Hippocrates' quadrature of lunules was based on this proposition, but that Hippocrates was not able to prove it rigorously. The proof in Book XII rests on two pillars: the theory of proportions of Book V, and the concept of the indirect proof, by means of enclosing the circle between inscribed and circumscribed polygons, whose areas differ by less than an arbitrarily given area. For the method used in this kind of proof, the unhappily chosen name "exhaustion method" is used, based on the idea that the circle would finally be exhausted by inscribed polygons of a constantly increasing number of sides. In reality however the circle is never exhausted; the attempt at "wearing out", to use the terminology of Dijksterhuis, is indeed abandoned after a finite number of steps in the proof of XII 2, and it is shown that what remains of the circle is less than an arbitrarily given area.

The proof depends on X 1: "Two unequal magnitudes being set out, if from the greater there be subtracted a magnitude greater than its half, and from that which is left a magnitude greater than its half, and if this process is repeated continually, there will be left some magnitude which will be less than the lesser magnitude set out."

The proof of the proposition on the ratio of the areas of two circles starts as follows. Suppose that the two circles were not proportional to the squares on the diameters $B\Delta$ and $Z\Theta$, then the ratio of $B\Delta^2$ to $Z\Theta^2$ would be equal to that of the first circle to an area Σ, which would have to be either greater or smaller than the second circle.

Suppose first that Σ is smaller. Inscribe then in the second circle a square $EZH\Theta$.

This is larger than one half of the circle, because a circumscribed square is exactly twice as large as the inscribed square, while the circle is less than the circumscribed square. Now bisect the arcs *EZ*, etc. in *K*, *Λ*, *M* and *N* and construct the octagon *EKZΛHMΘN*. Then each of the triangles *EKZ*, etc. is larger than one half of the

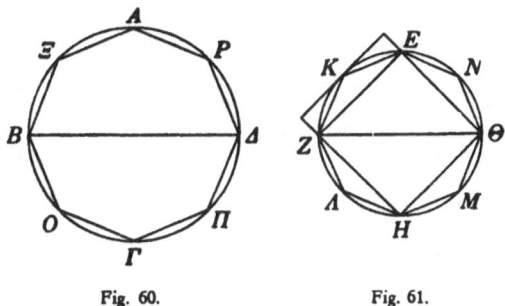

Fig. 60.   Fig. 61.

corresponding circular segment, because double each triangle is a rectangle, which is larger than the circular segment. If this process is repeated, by bisecting the arcs each time, then (by X 1) there must ultimately remain circular·segments, whose sum is less than the difference between the circle *EZHΘ* and the area *Σ*. The inscribed polygon that has been obtained is then larger than the area *Σ*. Now construct a similar polygon *AΞBOΓΠΔP* in the first circle. Then the ratio of $BΔ^2$ and $ZΘ^2$ is equal to that of the polygon in the first circle and the similar polygon in the second circle. The ratio of the first circle to the area *Σ* is then the same as that of the polygon in the first circle to the polygon in the second. Interchanging the means one finds the proportion:

First circle : First polygon = *Σ* : Second polygon.

But the first circle is larger than the first polygon, while *Σ* is less than the second polygon; this is a contradiction.

The case in which *Σ* is larger than the second circle is reduced to the first case by an interchange of the two circles; hence it also leads to a contradiction.

Therefore the areas of the two circles have the same ratio as $BΔ^2$ and $ZΘ^2$, quod erat demonstrandum.

At the beginning of this proof, the existence of a fourth proportional is tacitly assumed; the same thing takes place in all the analogous proofs in Book XII. This assumption is not necessary; a somewhat subtler formulation of the proof could have avoided it. In spite of this imperfection, the proof remains a scientific accomplishment which compels admiration. It contains, with full exactness, the modern limit concept, for the inscribed polygons approach the circle in the strict sense, that their difference can be made less than an arbitrary area.

Proposition X 1 which is used in this proof, in turn rests on a tacitly assumed

postulate, which was formulated in the following way by Archimedes in De
Sphaera et Cylindro:

> "The larger of unequal lines, areas or solids exceeds the smaller in such a way that the
> difference, added to itself, can exceed any given individual of the type to which the two
> mutually compared magnitudes belong."[1]

In the Quadratura parabolae, Archimedes formulates the same postulate for
areas; he adds that the older geometers had also used this "lemma."

> "for[2], by use of this lemma, they proved:
> 1° that circles have double the ratio of their radii,
> 2° that spheres have triple the ratio of their radii,
> 3° that every pyramid is one third of a prism on the same base and with the same height;
> and by means of a similar lemma they proved
> 4° that every cone is one third of a cylinder on the same base and with the same height.

What Archimedes attributes here to the "older geometers" is exactly the brief
contents of Book XII. For XII 3—7 produce exactly the result mentioned under
3°, XII 10 says the same thing as 4° and XII 18 the same as 2°. The remaining propositions are conclusions from or preparations for these four main theorems, and
the proofs do indeed depend in each case on the lemma, mentioned by Archimedes.

But, who invented the method used in all these proofs? No one but Eudoxus.
For, in the preface to De Sphaera et Cylindro, Archimedes ascribes the proofs of
3° and 4° to Eudoxus.

As an example of Eudoxus' ingenuity, I shall briefly reproduce the proof for the
pyramid (3°).

A preparatory step is XII 3: Every triangular pyramid $AB\Gamma\Delta$ can be divided
into two equal and similar triangular pyramids
$AEH\Theta$ and $\Theta K\Lambda\Delta$ and two equal prisms
$BZK\text{-}EH\Theta$ and $HZ\Gamma\text{-}\Theta K\Lambda$, the sum of the
prisms being greater than the sum of the
pyramids.

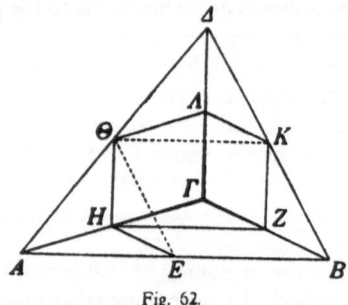

Fig. 62.

Next comes XII 4: If two triangular pyramids with equal height are given, and if each
of them is divided into two prisms and two
pyramids, as indicated above, then the sum
of the prisms in the first pyramid is to the
sum of the prisms in the second pyramid as
the bases of the given pyramids.

The proof depends on the fact that prisms with equal heights have the ratio of
their bases; this follows readily from previously proved propositions.

[1] Compare Dijksterhuis, *Archimedes* I, p. 139.

[2] This word "for" creates the impression that the "ancients" had used the lemma, without having formulated
it explicitly. Neither does Euclid formulate it explicitly. It is however hidden in V, Def. 4: "Magnitudes are said
to have a ratio to one another, which are capable, when multiplied, of exceeding one another". Euclid assumes
everywhere tacitly that two line segments, two areas, etc. always have a ratio.

Now comes the key proposition:

XII 5. *Pyramids which are of the same height and have triangular bases are to each other as the bases.*

The proof, by means of the exhaustion method, is entirely analogous to that of

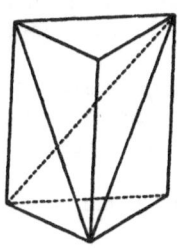

XII 2. The place of the inscribed polygons is now taken by the prisms which arise from the continued division of the pyramids. At every step, the prisms absorb more than half of the pyramids, so that in the end there is a remainder which is less than any prescribed solid — this is the basis of the proof.

It is shown in XII 7, that a triangular prism can be divided into three triangular pyramids, which have, two by two, equal bases and equal heights, and are therefore equal. Each of these pyramids is therefore one third of the prism; this proves the principal result for triangular pyramids. The same result follows for pyramids with an arbitrary number of sides, by division into triangular pyramids.

Fig. 63.

The theory of proportions.

Up to this point we could appeal to Archimedes; but for Book V this indisputable authority leaves us in the lurch. There *is* in Book V a scholium, whose author is unknown, according to which "it is said that the general theory of proportions, explained in this book, was found by Eudoxus." Nevertheless, according to the same scholium, "it is right that this book carries the name of Euclid, because the exposition in the form of the Elements and its adaptation to the system of geometry, is generally considered to be due to Euclid."

The scholium sounds plausible, particularly since, as we have seen, Proclus also attributes a number of "general theorems" to Eudoxus and because there is moreover a certain unmistakable relatedness between the methods of Book V and those of Book XII. Both operate constantly with proportions, both are based on the postulate of Eudoxus, cited above, and both use, in a rigorous manner, approximations and inequalities which are used ultimately to prove equalities by a reductio ad absurdum. Furthermore, the scholia in Euclid usually square with what is known from other sources. We are therefore prepared to join the scholiast and to attribute Book V to Eudoxus.

The definitions on which this book is based, are the following:

Def. 3. A ratio is a sort of relation in respect of size between two magnitudes of the same kind.

Def. 4. Magnitudes are said to have a ratio to one another which are capable, when multiplied, of exceeding one another.

Def. 5. Magnitudes are said to be in the same ratio, the first to the second and the third to the fourth, when, if any equimultiples whatever be taken of the first and third, and any equimultiples whatever of the second and fourth, the former equimultiples alike exceed, are alike equal to, or alike fall short of, the latter equimultiples respectively taken in corresponding order.

This admirable definition, which we have reproduced and interpreted before (see p. 175–6), makes it possible for Eudoxus to develop the theory of proportions in masterly fashion, in complete generality for arbitrary magnitudes. As a supplement there appears

Def. 7. When, of the equimultiples, the multiple of the first magnitude exceeds the multiple of the second, but the multiple of the third does not exceed the multiple of the fourth, then the first is said to have a greater ratio to the second than the third has to the fourth.

That is: if $ma > nb$, but $mc > nd$, where m and n are natural numbers, then

$$a : b > c : d.$$

On this definition is based the following auxiliary proposition:

V 8. *Of unequal magnitudes, the greater has to the same a greater ratio than the less has; and the same has to the less a greater ratio than it has to the greater.*

Following Dijksterhuis, the proof of the first part can be formulated as follows in modern algebraic notation: let $a > b$ and let c be a third magnitude comparable to a and b. A multiple of $a - b$ will exceed c

$$k(a - b) > c,$$

and a multiple of c will exceed kb. Suppose that this is accomplished by $(m + 1)c$ but not by mc, so that

$$(m + 1)c > kb \geqq mc.$$

By addition we find

$$ka > c + mc = (m + 1)c.$$

But

$$kb < (m + 1)c.$$

The proposition now follows from def. 7, viz.

$$a : c > b : c.$$

The second part is proved similarly.

By a reductio ad absurdum, we obtain from V 8:

V 9. *Magnitudes which have the same ratio to the same are equal to one another; and magnitudes to which the same has the same ratio are equal.*

V 10. *Of magnitudes which have a ratio to the same, that which has a greater ratio is greater; and that to which the same has the greater ratio is less.*

After these preliminary propositions, we find the principal properties of proportions. From the definition of proportions, we conclude almost at once:

V 12. *In a continued proportion*

$$a : b = c : d = e : f \ldots,$$

PLATE 21

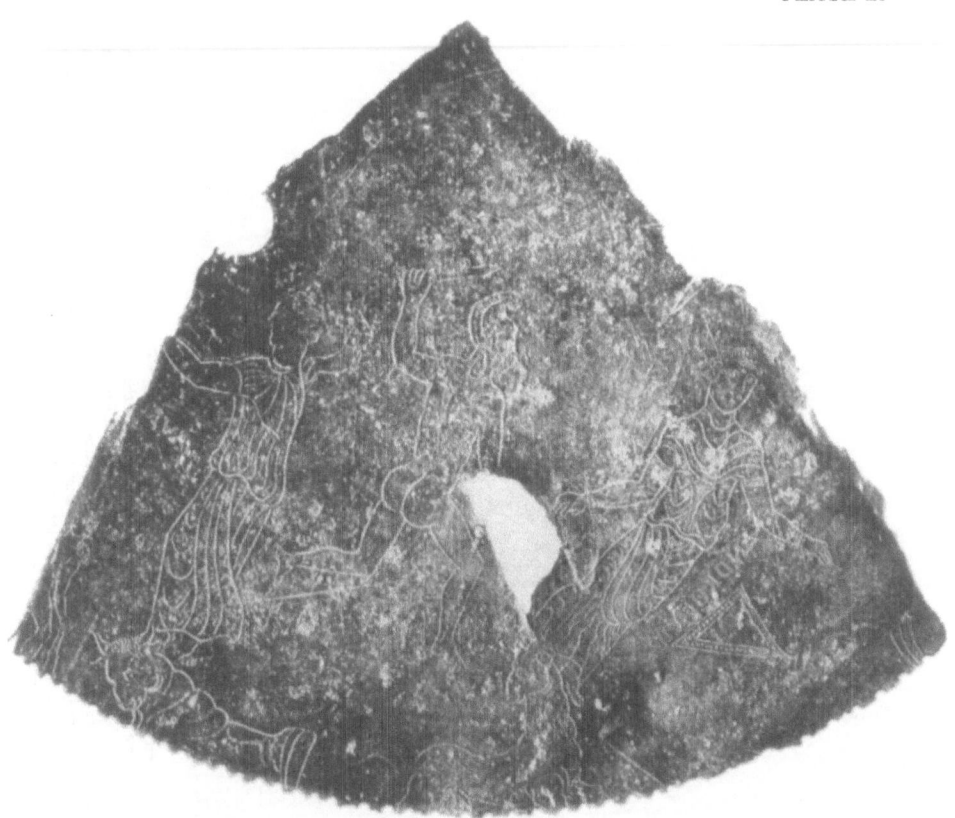

PL. 21. Bronze disk, found in Salzburg. Part of the dial of a Roman water clock, reconstructed by A. Rehm (see Fig. 59 and p. 183 with footnote). The whole disc had a diameter of 1.22 m. It was turned by water force so as to make one rotation a day. The circle, at which the disc is broken off, represents the ecliptic. Along the circle were 182 or 183 holes, one for every 2 days of the year. A knob, representing the sun, had to be moved from one hole to the next every two days. The hour of the day was indicated by the position of the knob with respect to the hour lines on the "spider", which was placed before the rotating disc.

PLATE 22

PL. 22. "Art of Eudoxus". Part of a Greek papyrus, now in the Louvre, written in Egypt between 331 and 111 B.C., called "Teaching of Leptinus". A popular astronomical text. The name Leptinus is in the last line but one. In the last column, we see a circle representing the sky, in which the names of the 12 zodiacal signs are written. The 3 preceding columns (col. 21 – 23 of the papyrus) contain an excerpt from the calendar of Eudoxus (see p. 183), with additions taken from calendars of Democritus, Euctemon and Callippus. Thus we read in col. 23, line 3 – 8: From fall equinox to winter solstice according to Eudoxus days 92, to Democritus days 91, to Euctemon days 90, to Callippus days 89. In coll. 22, line 7 – 8 we read: From Orion's to Sirius' setting days two. The columns 21 – 23 form the most valuable part of the teaching; days 90, to Callippus days 89. In coll. 22, line 7 – 8 we read: From Orion's to Sirius' setting days two. The columns 21 – 23 form the most valuable part of the teaching; the rest of the text and the drawings are rather primitive. On the back, 12 verse lines are written, the initial letters forming the words „Art of Eudoxus".

See P. Tannery, Mém. scientif. II p. 407, znd K. Weitzmann, Illustrations in Roll and Codex p. 49.

the sum of any number of antecedents has the same ratio to the sum of their consequents as any antecedent has to its consequent:

$$(a + c + e) : (b + d + f) = a : b.$$

As a special case (equal terms), one finds for every natural number *m*

V 15. $ma : mb = a : b.$

This is preceded by an auxiliary proposition derived from V 8—10:

V 14. *If* $a : b = c : d$ *and* $a > c$, *then* $b > d$: *and if* $a = c$, *then* $b = d$, *and if* $a < c$, *then* $b < d$.

Proof: From $a > c$ follows by means of V 8

$$a : b > c : b,$$

and hence $c : d > c : b;$

but from this we obtain, by use of V 10:

$$d < b.$$

The other cases are treated similarly.

And now follows, in one step for arbitrary magnitudes, and in an amazingly simple manner, the extremely important proposition on the interchange of the means, mentioned by Aristotle:

V 16. *If four magnitudes be proportional, they will also be proportional alternately.*
From $a : b = c : d$ follows $a : c = b : d$.

Proof. From $a : b = c : d$ follows $ma : mb = nc : nd$. Hence, we conclude from V 14 that, if $ma > nc$, then $mb > nd$; and if $ma < nc$, then $mb < nd$; and if $ma = nc$, then $mb = nd$. Therefore $a : c = b : d$.

It this not a masterpiece of logic?

The remaining usual properties of proportions follow now without any difficulty.

This concludes our discussion of the most important mathematical accomplishment of the great Eudoxus. We do not know to what extent he is responsible for the creation of Book VI (geometrical applications of the theory of ratios). He has also given a solution of the Delian problem by means of the intersection of "curves", but this solution is not known to us.

Theaetetus and Eudoxus.

I can not resist the temptation of comparing the modes of thinking of Theaetetus and Eudoxus, the two great geometers of this period of florescence. They are both extremely keen, logical thinkers, they are both very ingenious in finding geometrical constructions. But there is a characteristic difference: Theaetetus thinks in a way which we moderns would call "algebraical", Eudoxus on the other hand is a typical "analyst". For Theaetetus, every line segment, rational or irrational, is a

separate entity with definite algebraic properties, constructible in a definite manner, medial or apotome, or whatever you like; sharply distinguished from other
line segments, no matter how closely they approach it in size. For Eudoxus on
the other hand, line segments are continuously variable magnitudes, which can
approach limits and which can be approximated arbitrarily closely by other line
segments. Theaetetus determines a ratio by means of a sequence of integers, which
arise in the process of alternate subtractions, derived from arithmetic. But Eudoxus determines a ratio by means of its place among the rational ratios which
enclose it on both sides. Theaetetus determines the edges of the regular polyhedra
from the algebraic properties which follow from their construction. Eudoxus
encloses the circle between inscribed and circumscribed polygons, and from this
procedure he obtains indirectly the properties he wants to prove; he proceeds
similarly with the pyramid, the cone and the sphere.

After these stars of the first magnitude, we come to the lesser lights of Plato's
circle. The Proclus catalogue mentions in the first place Amyclas, of whom we
know nothing further, and then Menaechmus and his brother Dinostratus. The
greatest of these three is undoubtedly

Menaechmus,

a pupil of Eudoxus, who acquired fame as an astronomer and as a geometer.
According to Proclus he increased the number of spheres in the planetary theories
of Eudoxus and Callipus. He wrote on the foundations of geometry, on the two
meanings of the word element, and on the difference between theorems and problems. When Alexander the Great asked him, whether there did not exist for him
a shortcut to geometry, he answered: "O King, for travellers through the country,
there are royal roads and roads for ordinary citizens, but in geometry there is but
one road for all." The same story is told about Euclid and king Ptolemy, but most
probably the original version related to Menaechmus. As always with such anecdotes, one does not know whether they are based on actual occurences.

The most important accomplishment attributed to Menaechmus is the discovery of the conic sections, which he used to solve the Delian problem. His two
solutions have already been discussed.

Menaechmus did not yet use the words parabola and hyperbola, which were
introduced only later on by Apollonius. But he did construct his curves as plane
sections of cones, as becomes clear from the epigram of Erastothenes (see p. 161–2).
The ancient names of the conic sections are:

> section of a rectangular cone (parabola),
> section of an obtuse-angled cone (hyperbola),
> section of an acute-angled cone (ellipse).

The only cones considered by the "ancients" were cones of revolution; they distinguished them according to the right, obtuse or acute vertical angle, and used each type to generate *one* kind of conic section, cutting the cone by a plane perpendicular to a generating line. [1]

Thus they obtained the parabola from the rectangular cone, the hyperbola from the obtuse-angled cone and the ellipse from the acute-angled one.

It is not known, how Menaechmus obtained from this method for generating the conic sections, their "symptoms", i.e. their equations

$$xy = ab \text{ (hyperbola)}, \qquad y^2 = xb \text{ (parabola)},$$

which he needed for his duplication of the cube. Later, in the section on Apollonius, we shall return to the history of the conic sections.

Dinostratus.

In the large compendium of Pappus, which must have been written in the time of the emperor Diocletian (284—305), it is mentioned that Dinostratus and Nicomedes used for the quadrature of the circle a curve, which for that reason was called the quadratrix (τετραγωνίζουσα). It appears from other sources that Hippias of Elis had already found this curve; but it seems that Dinostratus, the brother of Menaechmus, was the first who used it for squaring the circle.

Pappus describes as follows the construction of this curve:

Describe a circular arc *BEΔ* about *Γ* in a square *ABΓΔ*. Let the straight line *ΓB* rotate uniformly about *Γ* so that *B* describes the arc *BEΔ*, and let the line *BA* move uniformly towards *ΓΔ*, remaining parallel to *ΓΔ*. Let both uniform motions take place in the same time, so that both *ΓB* and *BA* will coincide with *ΓΔ* at the same moment. These two moving lines intersect in a point which moves along with them and which describes a curve *BZΘ*. If *ΓZE* is one definite position of the rotating line and *Z* the point of intersection with the line which moves parallel to itself, then, according to the definition, *BΓ* will be to the perpendicular *ZΛ* as the entire arc *BΔ* is to the arc *EΔ*.

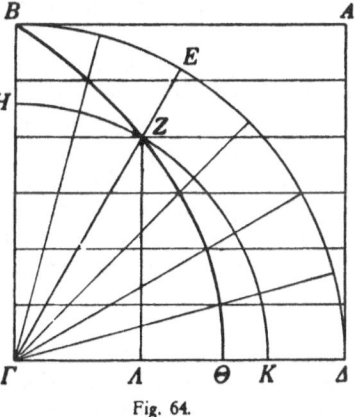

Fig. 64.

It is clear that, once this curve has been drawn, it can be used for the division of an angle in an arbitrary number of equal parts. But it also gives the length of the quadrant *ΔB* and hence the quadrature of the circle. For Pappus proves that the arc *ΔEB* is to the line segment *BΓ* as *BΓ* is to *ΓΘ*, where *Θ* is the terminal point of the quadratrix.

The proof is made by the same indirect method, which is characteristic of Eu-

[1] See Pappus, *Collectio*, VII 30, and Eutocius, *Commentary on the Conica of Apollonius* II (ed. Heiberg), p. 168.

doxus' exhaustion proofs. If the asserted proportion does not hold, then the fourth proportional ΓK is either greater than or less than $\Gamma\Theta$. If $\Gamma K > \Gamma\Theta$, describe an arc of circle KZH about Γ, which intersects the quadratrix in Z, extend ΓZ to E and draw the perpendicular $Z\Lambda$. Then, since ΓK was defined as the fourth proportional, we have

$$\text{arc } \Delta EB : B\Gamma = \Gamma\Delta : \Gamma K = \text{arc } \Delta EB : \text{arc } KZH,$$

so that $B\Gamma = \text{arc } KZH.$

But, from the definition of the quadratrix, it follows that

$$B\Gamma : Z\Lambda = \text{arc } HZK : \text{arc } ZK;$$

and, if the antecedents in this proportion are equal, then the consequents are also equal, so that

$$\text{arc } ZK = Z\Lambda,$$

which is impossible.

In case the fourth proportional is less than $\Gamma\Theta$, a contradiction is reached in similar manner, Λ and K being interchanged.

The critique on the definition of the quadratrix made, according to Pappus, by Sporus, appears to me as only partially justified. Sporus says: how can the uniform motions be defined as long as the ratio of the line AB to the arc $BE\Delta$ is unknown? Indeed, the velocities must be in this ratio. To this one can reply that the two motions, and hence the curve, are *theoretically* determined, because the ratio of the velocities has been defined theoretically as the ratio of AB to the arc $BE\Delta$, and that the curve can readily be drawn *practically* by constantly bisecting line AB and also the arc $BE\Delta$, thus obtaining an arbitrary number of points on the curve. Therefore, neither theoretically, nor from the point of view of the draughtsman, is there anything to object to in the definition of the curve. But there is something valid in the remark of Sporus, that the terminal point Θ of the curve is not defined, since, at the end of the motion, both moving lines coincide with $\Gamma\Delta$, so that they have no definite point of intersection. On the other hand one can say, that the point Θ is completely defined theoretically by the relation

$$\text{arc } \Delta EB : B\Gamma = B\Gamma : \Gamma\Theta,$$

and that practically the point Θ can also be determined very accurately by drawing a smooth curve through the other points. That this smooth curve passes exactly through the point Θ, defined as above, follows rigorously from the proof given by Pappus.

A careful drawing, which I made, gave the result

$$\Gamma\Delta : \Gamma\Theta = 3.14 : 2.$$

Having thus determined the length of the circumference, how does one find the area, i.e. the quadrature of the circle? Apparently, Dinostratus knew the proposition, later proved rigorously by Archimedes, that the area of a circle is equal to that of a triangle, whose base is equal to the circumference and whose height equals the radius, or some equivalent proposition.

A genuine academician, an astronomer after Plato's heart was

Autolycus of Pitane.

We shall see presently that Autolycus was older than Euclid. This agrees with the tradition, according to which he was the teacher of Archesilaus, the founder of the "middle Academy", which succeeded Plato's old Academia. We can therefore place Autolycus at 320 or 310.

Autolycus is especially important, because he is the most ancient mathematician-astronomer of whom two works have been completely preserved. To appreciate the significance of these works, it will be well to recall what Plato says in The Republic (Book VII 529 A) about astronomy as a science.

Socrates relates how Glauco had first praised astronomy for its practical value, how he himself had taunted him and how Glauco then wanted to sing the praises of astronomy after the manner of Socrates:

— For it is obvious to everybody, I think, that this study compels the soul to look upward and leads it away from things here to those higher things.

— It may be obvious to everybody except me, for I do not think so. . . . You seem to me in your thought to put a most liberal interpretation on the "study of higher things", for apparently if anyone with back-thrown head should learn something by staring at decorations on a ceiling, you would regard him as contemplating them with the higher reason and not with the eyes. Perhaps you are right and I am a simpleton. For I, for my part, am unable to suppose that any other study turns the soul's gaze upward than that which deals with being and the invisible. But if anyone tries to learn about the high things of sense, whether gaping up or blinking down, I would never say that he really learns — for nothing of the kind admits of true knowledge — nor would I say that his soul looks up, but down, even though he study floating on his back on sea or land.

— A fair retort, he said; your rebuke is deserved. But how, then, did you mean that astronomy ought to be taught contrary to the present fashion if it is to be learned in a way to conduce to our purpose?

— Thus, said I: these sparks that paint the sky, since they are decorations on a visible surface, we must regard, to be sure, as the fairest and most exact of material things; but we must recognize that they fall far short of the truth, the movements, namely of real speed and real slowness in true number and in all true figures both in relation to one another and as vehicles of the things they carry and contain. These can be apprehended only by reason and thought, but not by sight . . . Then, we must use the blazonry of the heavens as patterns to aid in the study of those realities, just as one would do who chanced upon diagrams drawn with special care and elaboration by Daedalus or some other craftsman or painter. For anyone acquainted with geometry who saw such designs would admit the beauty of the workmanship, but would think it absurd to examine them seriously in the expectation of finding in them the absolute truth with regard to equals or doubles or any other ratio.

— How could it be otherwise than absurd?, he said.

— Do you not think, said I, that one who was an astronomer in very truth would feel in the same way when he turned his eyes upon the movements of the stars? He will be willing to concede that the artisan of heaven fashioned it and all that it contains in the best possible manner for such a fabric; but when it comes to the proportions of day and night, and of their relations to the month, and that of the month to the year, an of the other stars to these and one another, do you not suppose that he will regard as a very strange fellow the man who believes that these things go on for ever without change or the least deviation — though they possess bodies and are visible objects — and that his unremitting quest is the realities of these things?

— I at least do think so, he said, now that I hear it from you.

— It is by means of problems, then, said I, as in the study of geometry, that we will pursue astronomy too, and we will let be the things in the heavens, if we are to have a part in the true science of astronomy . . .

To us, accustomed as we are to modern empirical natural science, all of this sounds very strange; but, at bottom, Plato is entirely right. Theoretical astronomy does not consider actual celestial bodies, but, like geometry and mechanics, it is concerned with theoretical, idealized objects, such as material points and perfect spheres, which move in space in accordance with mathematical laws and to which observable celestial bodies correspond only approximately.

What does not agree with our present views is this that Plato calls the idealized motions the "true" ones. Nevertheless this is easily understood. Plato starts from the position that exact judgments are possible only concerning idealized objects. One can not know anything with certainty about observable objects, because through their variability they constantly escape us. We can discover some of the causes of change, but never the totality of all causes. Hence, where no judgment, based on exact foundations, is possible, no "truth" can exist.

Truth does occur in mathematics because it is exact. And truth is divine. "God always geometrizes" says Plato. Following the pattern of the true, divine, mathematically pure motions, the Creator has, according to Plato, ordered the visible universe.

Where shall we find an ideal astronomy, such as Plato demands, a geometry of moving points, circles and spheres? For this science was certainly cultivated in the Academia. The works of Eudoxus and Callipus, of Heraclides of Pontus are argely lost. But we do have Autolycus' book

On the rotating sphere,

in which we do indeed find an astronomy to Plato's taste: pure kinematics of points and circles on a uniformly rotating sphere.

Proposition 1 is:

When a sphere rotates uniformly on its axis, then all points of the sphere which do not lie on the axis, describe parallel-circles about the poles, about which the sphere rotates, in planes perpendicular to the axis.

In proposition 4 a great circle is introduced which does not take part in the rotation, the "bounding" circle (ὁρίζων = bounding, hence our word "horizon"). In proposition 4 this circle is perpendicular to the axis, in proposition 5 it passes through the poles, but beginning with proposition 6, the horizon is an arbitrary oblique circle. On the rotating sphere there is also an oblique circle which does take part in the rotation, apparently the ecliptic. For, in a similar work of Euclid, the *Phaenomena*, that shows many points of contact with Autolycus, the rotating oblique circle is openly called the "circle through the centers of the signs of the zodiac". But Autolycus remains entirely abstract; visible things such as the zodiac in the sky, are entirely left out of consideration, true to Plato's ideal. Thus proposition 11 is:

> When an oblique circle bounds the visible part of the sphere and when another oblique great circle touches greater parallel circles than those which the bounding circle touches, then this second circle will rise and set on that part of the bounding circle that lies between the parallel circles which it touches.

Euclid (Phaenomena VII) quotes this proposition in the following form:

> "The circle of the zodiac rises and sets on that part of the horizon which lies between the tropical circles, because it touches larger circles than those which the horizon touches."

This shows that Autolycus antedates Euclid.

Less abstract is Autolycus' second book:

On the rising and setting of stars.

This book deals with the sun and the 12 signs of the zodiac, with visible and invisible rising and setting of stars, in and below the zodiac. Euclid treats the same topics in his Phaenomena. He proves i.a. that some signs of the zodiac require a longer time than others for rising and setting.

One can not get very far beyond such generalities before the invention of trigonometry. It was Hipparchus who first determined accurately the times of rising and setting of the signs by means of trigonometric computations with the aid of his table of chords.

Frequently, both Autolycus and Euclid quote, without proof, propositions concerning circles on the sphere. From this fact, Hultsch concludes that a spherical geometry, a Sphaerica, must have existed before Autolycus. We are able to get some idea of what this must have been like from the much later Sphaerica of Theodosius[1], which has been preserved and in which these propositions are found again with their proofs.

At the end of the 4th century, all the mathematics, as cultivated in the school of Plato, was collected in the work of

Euclid

in a manner which was destined to remain exemplary for thousands of years. His "Elements" constitute one of the greatest successes of world literature; an entire world has learned geometry from it. Even at the present time, geometry is taught

---

[1] Théodose de Tripoli. *Les sphériques*, traduit par P. ver Eecke, Bruges 1927.

in English schools from an English adaptation of the Elements, and school geometry is known in England as "Euclid". And indeed, his remarkable didactic
qualities fully entitle Euclid to this fame. He is the greatest schoolmaster known in
the history of mathematics.

What do we know about his life? From a statement of Pappus we learn that
he taught mathematics in Alexandria. [1] He was older than Archimedes († 212),
who quotes somewhere a proposition from the Elements. [2] This indicates that the
date "about 300" can not be far from the truth. Proclus, in his Catalogue, calls
him a contemporary of king Ptolemy I (305—285) to whom he had the courage to
say to his face that there exists no royal road to geometry. Pappus (loc. cit.)
praises his scrupulous honesty, his modesty and the kindness of his judgment on
anybody who aided the development of mathematics, no matter in how modest
a way. Stobaeus relates the following interesting story: "some one who had begun
to read geometry with Euclid, when he had learnt the first theorem, asked Euclid,
'But what shall I get by learning these things?' Euclid called his slave and said
'Give him threepence, since he must make gain out of what he learns'." (Compare
Heath, The thirteen books of Euclid's Elements, 2nd ed., I, p. 3).

This is all we know about the man himself. But of much greater importance are
his works, which bear witness even to-day to his unequaled gifts as a teacher. For
example, what excellent judgment he shows in postponing to Book V the difficult
theory of proportions of Eudoxus and in treating the most important topics of
school geometry in Books I—IV of the Elements without the use of proportions!
This makes it possible, even for mediocre pupils, to get hold of the first four books
without being scared off at the start by things which are beyond them. And there
are other such instances.

According to Proclus, Euclid was a disciple of the Platonic school. His works
do indeed fit perfectly into the tradition of that school. As shown in The Republic
and in other works, Plato prescribed for every one, as a preparation for philosophy, the study of the four subjects: arithmetic, geometry, harmony and astronomy. It is exactly these four which are dealt with in the works of Euclid, the first
two in the Elements, the third in the Sectio Canonis, and, in the book entitled
"Phaenomena", astronomy in the sense of Plato, the theory of uniformly rotating
spheres.

The "Elements"

constitute the conclusion of a sequence of similar works, the Elements of Hippocrates, of Leon, of Theudius, all mentioned in the Proclus Catalogue. Both Leon
and Theudius belonged to the circle of Plato's Academia. We may well suppose
therefore that their Elements were used for the teaching of mathematics in the
Academia.

[1] Pappus, Collectio VII 34: "Apollonius lived for a long time in Alexandria with the pupils of Euclid".
[2] Archimedes, On the sphere and the cylinder, I, prop. 2.

Whenever Aristotle presupposes in his hearers a knowledge of a geometrical proposition, this apparently means that it occurs in these works. We see therefore that also in this respect, Euclid continues the tradition of the Academia.

Euclid is by no means a great mathematician. We have already seen that the most important and the most difficult parts of the Elements have been taken from other authors, especially from Theaetetus (Books X and XIII) and from Eudoxus (Books V and XII). Like the arithmetical Books VII and IX, these parts are on a very high mathematical level, while other parts, especially the middle one of the arithmetical books (Book VIII) and the related Sectio Canonis, fall far below this level. These contain logical errors and the formulations in them are sometimes confused. Euclid's level is apparently determined by that of the predecessor whom he follows. When he is guided by a first-rate author, such as Theaetetus or Eudoxus, he is himself excellent; but when he copies from a less eminent author, his standard goes down. Euclid is first of all a pedagogue, not a creative genius.

It is very difficult to say which original discoveries Euclid added to the work of his predecessors.

In connection with the Pythagorean Theorem, Proclus says in his commentary on Euclid: "I admire the writer of the Elements, not only that he gave a very clear proof of this proposition, but that, in the sixth book, he also explained the more general proposition by means of an irrefutable argument." It has been concluded from this statement that the proof of the Pythagorean Theorem, found in the Elements, is due to Euclid himself; but this conclusion goes a bit too far. Did Proclus really know Euclid's predecessors? Could he compare their proofs with those of Euclid? We know nothing about this. It can be said that Proclus attributed to Euclid himself the more general proposition of the sixth book. This proposition says: "In a right triangle, a figure of arbitrary shape described on the hypothenuse, is equal to the sum of similar figures, described in similar manner on the right sides".

I am sorry, but it seems to me that Proclus exaggerates his praise of Euclid. The proof of VI 31 in the Elements is "irrefutable" only for rectilinear similar figures, since, according to VI 20, these have the ratio of the squares of homologous sides. For curvilinear figures, the exhaustion method of Book XII would have to be called upon. Moreover, as Dijksterhuis observes, the proof is not quite complete not even for rectilinear figures.

Proposition VI 32 is also formulated very carelessly. It is rather remarkable that a pedagogue of Euclid's excellent qualities is sometimes so illogical. But this is not an isolated phenomenon.

To become more intimately acquainted with the Elements, the interested reader has available the excellent work of Dijksterhuis, or the translations of Thaer and of Heath. We turn now to a brief discussion of Euclid's other works.

The "Data"

is a book of great importance for the history of algebra. The work consists of propositions of the following form: when certain magnitudes are given or determined, then other magnitudes are also determined[1], for example:

> Proposition 2. If a magnitude A and the ratio $A : B$ are determined, then B is also determined.
>
> Proposition 7. If $A + B$ and $A : B$ are determined, then A and B are also determined.

Def. 11 introduces the concept "a definite amount greater than in ratio". A magnitude X is said to be a definite amount greater than in (definite) ratio to Y, if X, after having been diminished by a definite amount C, has a definite ratio to Y:

$$(X - C) : Y = \text{determined}.$$

In modern terminology, one would say that X and Y are linearly related:

$$X = aY + C.$$

Propositions 10—21 correspond to certain operations to which such linear equations can be subjected, such as substitution. A typical example is the following:

> Proposition 19. If X is a definite amount greater than in ratio to Y, and Y is a definite amount greater than in ratio to Z, then X is a definite amount greater than in ratio to Z.

Next come simple propositions about straight lines given "in position" and about triangles given "in shape", e.g. when two angles are given, or one angle and the ratio of the including sides. Then analogous propositions concerning polygons and their areas, such as:

> Proposition 55. If an area is determined in shape and in magnitude, then each side is determined in magnitude.

This leads quite naturally again to the domain of geometric algebra. Propositions 57, 58 refer to the application of areas; they state that the application of definite areas to definite line segments. without defect or with defect or excess of a definite form, gives a definite result. From this, propositions 84 and 85 deduce that two straight lines are determined, when their difference and their product, or their sum and their product, are given. We have discussed these propositions at an earlier point; it is again a question of the old-Babylonian normal problems

(1) $$x - y = a, \qquad xy = F$$

and

(2) $$x + y = a, \qquad xy = F,$$

in geometric dress.

Proposition 86 treats a more troublesome set of equations, viz.

(3) $$xy = F, \qquad y^2 = ax^2 + C.$$

---

[1] The text speaks everywhere of "given", but Dr. Dijksterhuis has called my attention to the fact that "determined" better expresses the meaning.

It is formulated as follows:

> **Proposition 86.** When two straight lines AB and $A\Delta$ subtend a definite rectangle and if the square on AB is a definite amount greater than in ratio to the square on $A\Delta$, then AB and $A\Delta$ are determined.

The set of equations (3) is strongly reminiscent of Babylonian algebra problems. The Babylonians would simply have squared the first of these equations, thus obtaining the product and the difference of y^2 and ax^2. This is not possible in geometric algebra, but Euclid accomplishes his purpose by introducing a new unknown z.

Namely, in order to subtract the given area C from y^2, it is applied to the width y, i.e. it is transformed into a rectangle

$$C = yz.$$

This leads, in place of (3) to a set of three equations:

(4) $$xy = F, \qquad yz = C, \qquad y(y - z) = ax^2.$$

From the first two of these, the ratio

$$z : x = C : F$$

is determined. Thus $z^2 : x^2$ is known. But the third equation determines the ratio $x^2 : y(y - z)$; hence $z^2 : y(y - z)$ is known, consequently also $z^2 : [4y(y - z) + z^2]$, i.e. $z^2 : (2y - z)^2$. Therefore the ratio $z : (2y - z)$ is known and hence $z : 2y$, from which follow $z : y$ and also $z^2 : yz$. But yz is also given; therefore z^2 and hence z are known, etc.

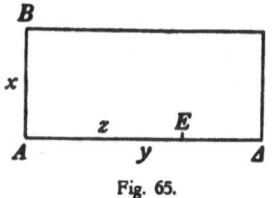

Fig. 65.

All this is of course formulated in geometrical language; instead of y, z and $y - z$, Euclid writes $A\Delta$, AE and $E\Delta$, etc. But I have made no other changes in his reasoning.

Propositions 88—95 refer to circles. E.g., 91 and 92 state: If a line is drawn from a point A outside or inside a given circle, which cuts the circle in B and in Γ, then the rectangle $AB \cdot A\Gamma$ is determined.

I hope that what has been presented gives an idea of the character of the Data.

On the division of figures. [1]

Of this opusculum only a version in Arabic has been preserved. It deals with the problem of dividing a given rectilinear figure into two equal parts, or, more generally, into two parts that have a given ratio, by means of a line in a given direction, or through a given point. If the point lies on the perimeter, the solution is very elementary; if it lies inside or outside, then the solution of a quadratic equation is required, carried out of course by means of the application of areas.

[1] English translation by R. C. Archibald, *Euclid's Book on Division of Figures* (1915).

The familiar Babylonian problem of bisecting a trapezoid by means of a line parallel to the base, appears also. Just as the Babylonians, Euclid finds that the square on the line of division is equal to one half of the sum of the squares on the parallel sides.

Lost geometrical writings.

Pseudaria (on errors of reasoning in mathematics).
Conica (On conic sections).
Τόποι πρὸσ ἐπιφανείᾳ (surfaces as loci, or loci on surfaces).
Porismata. What is this?
According to Pappus [1], Porismata are a kind of proposition which fall between theorems and constructions. Pappus gives some examples. I have no judgment of my own concerning the value of the various attempts at reconstructing the contents of the Porismata, and at clarifying the concept Porisma. [2]

Euclid's works on applied mathematics

have been preserved. [3]
The *Optica* deals with perspective.
The *Catoptrica* is concerned with reflections by mirrors.
The *Sectio Canonis* (*Κατατομή κάνονοσ*) with the theory of music.
Phaenomena with elementary theoretical astronomy (rotation of the celestial sphere, the rising and setting of the sections of the ecliptic).
Euclid has thus actually collected the whole of elementary mathematics as known at the time of Plato in textbooks. All of us, classical and modern mathematicians, are his pupils. He concludes a brilliant period of flowering of mathematics and, at the same time, he and his Alexandrian school initiate the subsequent, if possible still more brilliant, period of florescence, the period of Archimedes, Eratosthenes and Apollonius, the Alexandrian era.

[1] Pappus, *Collectio*, VII 13 (ed. Hultsch II, p. 648).
[2] See Th. Heath, *History of Greek Mathematics* I, p. 431 et seqs., and particularly Cauchy, *Les porismes d'Euclide*.
[3] All Greek writings of Euclid are found in the superb edition of Heiberg and Menge.

CHAPTER VII

THE ALEXANDRIAN ERA
(330—200 B.C.)

With the insight of a man of genius, but also with a thorough knowledge of geographic conditions and of the possibilities of transportation, the young hero Alexander (Plate 26) had selected the location of the future world center Alexandria, and given the start to the construction of the city. With the shortest possible delay, Alexandria became a flourishing commercial metropolis and also a cultural center of the highest rank. The kings Ptolemy Soter, Ptolemy Philadelphus (Plate 26) and Ptolemy Euergetes, who succeeded each other during the years from 305 to 322, established not only a powerful kingdom, but they promoted the arts and the sciences in a truly regal manner. The first Ptolemy founded the Musaeon, which brought together poets and scholars of the very first order, receiving ample salaries from the royal treasury. It included a world-famous library, to which Ptolemy Euergetes added the entire collections of books of Aristotle and of Theophrastus. All those who cultivated letters and science flocked to Alexandria, philologists (going by the name of grammarians), historians, geographers, mathematicians, astronomers, philosophers and poets.

A refined culture reigned at this Hellenistic royal court. The works of Homer were analysed and purged of interpolations, the science of chronology was founded, poetry was developed and refined. In astronomy, careful observations were made and theories were established to explain the observations, such as that of the epicycle and the excenter. The culmination of this development was the great Syntaxis Mathematica, the "Almagest" of Ptolemy (140 A.D.); but the foundations of all these theories were laid in the Alexandrian era. The same men who brought about the great development of astronomy, Aristarchus, Archimedes, Eratosthenes and Apollonius, were also the foremost mathematicians of their day and brought mathematics to a state of unequaled florescence. In astronomy, Hipparchus (130 B.C.) and Ptolemy completed the work of their great Alexandrian predecessors.

The arts and sciences of this period have a character totally different from that of the classical era. The classical art of a Homer, a Pindar, a Sophocles was written for all the people and was enjoyed by all the people. The exceptionally popular Histories of Herodotus were recited publicly in Athens and, so tradition has it, at the Olympic games as well. The indictments of Anaxagoras and of Socrates for atheism show to what extent the people of Athens were concerned with and excited about philosophy. It goes without saying that, in general, the citizens of Athens did not read mathematical treatises, but in his lectures on logic, Aristotle could presuppose a knowledge of elementary geometry; and, in "The Birds", Aristophanes could make an allusion to the quadrature of the circle. In the leading

circles of the principal Greek cities, there existed an unprecedented urge for knowledge, a hunger for culture. The sophists received royal honoraria; Plato's Academy and the school of Eudoxus in Cyzicus drew pupils from near and far.
Every one knew: knowledge is power! Indeed, it may well be doubted whether
Alexander could have conquered the world without his technically perfect engines
of siege, without a staff of geographers and of financial-economic experts, without
the Greek naval architecture, bridge construction and army organization, without
the training in politics given by men like Thucydides, Plato and Aristotle.
Wherever the Greeks lived, in Asia Minor, on the Ionian isles, in Italy, in North
Africa, in Hellas and on the shores of the Black Sea, the arts and sciences flourished,
free and untrammeled, in times of prosperity and in times of distress.

But during the era of Hellenism, the arts and sciences were no longer a matter
for the people, but for the scholars and the lovers of art, for the highly cultivated
circles, at the princely courts of Alexandria, Syracuse, Seleucia. Archimedes resi-
ded at the court of Hiero and of his son Gelon in Syracuse; Eratosthenes was the
librarian in Alexandria, and Apollonius lived there as well. Science was largely
tied to libraries; the study of books was imperative for the mastery of the necessary foundations and methods. When royal favor and royal support ceased,
the arts and sciences languished. The period of flowering of mathematics in the
3rd century B.C. was followed by centuries of decay, relieved only now and then
by a temporary revival. We shall have to return to the inner causes of this decay,
but the external circumstances are evident. The kings who came after Ptolemy III
did not wish to devote money to science.

At the end of the preceding chapter we discussed Euclid, the oldest of the
Alexandrian mathematicians, because he represents less the initiation of a new
florescence than the conclusion of the previous one. We proceed now to the
period of highest flowering of ancient mathematics, to the era of the keen-witted
mathematician-astronomer Aristarchus, of the unsurpassed geniuses Archimedes
and Apollonius, of the versatile scholar Eratosthenes and of the ingenious geometer Nicomedes.

Aristarchus of Samos

lived at the beginning of the 3rd century, for in 280 B.C. he observed the summer
solstice. As related by Archimedes in the "sand-counter", Aristarchus advanced
the bold hypothesis that the earth rotates in a circle about the sun. Most astronomers
rejected this hypothesis, as Archimedes tells us also. Indeed, in view of the status
of mechanics at that time, there are weighty arguments against the motion of the
earth. Such arguments are already found in Aristotle and, developed more fully,
in Ptolemy. If the earth had such an enormously rapid motion, says Ptolemy, then
everything that was not clinched and riveted to the earth, would fall behind and
would therefore appear to fly off in the opposite direction. Clouds, for instance,

would be overtaken by the rotation of the earth and would hence lag behind. From the point of view of Greek dynamics, there is nothing to be said against this, since the Greeks did not know the law of inertia and required a force to account for every motion. If the earth does not drag the clouds along, they have to lag behind. We do not know how Aristarchus met these arguments.

Another argument against the rotation of the earth around the sun, is that it would require uninterrupted changes in the apparent distances between the fixed stars. To meet this difficulty, Aristarchus assumed that the sphere of the fixed stars has a radius so large that the earth's orbit could be considered as a single point in comparison with it. We know this also from the "sand-counter".

The only one of the works of Aristarchus which has been preserved, is the very interesting short treatise "On the distances of sun and moon". It is a great merit of Sir Thomas Heath that he called attention to the mathematical value of this treatise and that he published a translation with an excellent historical-astronomical commentary. [1]

From certain "hypotheses", derived from observation, Aristarchus deduces in this book the following propositions in a rigorously mathematical manner.

1. The distance earth—sun is more than 18 times but less than 20 times the distance earth—moon.

2. The diameters of sun and moon have the same ratio as their distances.

3. The ratio of the diameter of the sun to the diameter of the earth is greater than $19 : 3$ and less than $43 : 6$.

The proof of 3. uses eclipse observations in an ingenious way. It would carry us too far afield to go into this more deeply; but we do want to discuss the proof of proposition 1. This proof is important for the history of trigonometry, because it gives for the first time a method for approximating the sine of a small angle (viz. sin 3°). It depends on the hypothesis, that *at the instant at which the moon is exactly in the second quarter, i.e. when the plane, which separates the light part of the moon from the dark part, passes through our eye, the angle between the lines directed to sun and to moon is exactly a right angle less 1/30 of a right angle.* In reality this angle is not 87°, but 89°50', but the mathematical derivation does not lose any of its elegance from this discrepancy. In short, it amounts to the following:

At the instant referred to in the hypotheses, the centers A, B and Γ of sun, earth

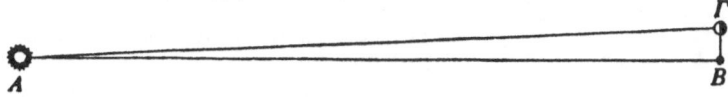

Fig. 66.

and moon form a triangle $A B \Gamma$, with a right angle at Γ, and in which the angle A is equal to 1/30 of a right angle. The question is to prove that the hypothenuse

[1] T. Heath, *Aristarchus of Samos, The ancient Copernicus*, Oxford 1913.

AB of such a triangle is more than 18 times, but less than 20 times the side *BΓ*. In the proof Aristarchus uses inequalities which can be formulated as follows in modern terminology: Let α and β be acute angles and $\alpha > \beta$. Then we have

(1) $$\frac{\tan \alpha}{\tan \beta} > \frac{\alpha}{\beta} > \frac{\sin \alpha}{\sin \beta}.$$

He does not speak of tangents and sines, but of right sides and of chords of circular arcs, and he does not prove the inequalities but supposes them to be known.

Indeed, the tangent inequality occurs already in Euclid's Catoptrica (proposition 8). Aristarchus applies the tangent inequality to two angles of R/4 and R/30, where R is a right angle. If *BΔ* is the diagonal of the square *ABEΔ*, *BZ* is the bisector of angle *EBΔ* and $\angle EBH = R/30$, then the tangent inequality gives

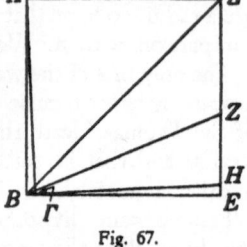
Fig. 67.

(2) $$ZE : HE > 15 : 2.$$

To estimate *EZ*, Aristarchus calculates as follows (his procedure is somewhat abridged here):

$$BΔ^2 : BE^2 = 2 : 1 > 49 : 25,$$
$$ΔZ : ZE = BΔ : BE > 7 : 5,$$
(3) $$ΔE : ZE = (ΔZ + ZE) : ZE > 12 : 5 = 36 : 15.$$

From (2) and (3) ,we obtain

$$ΔE : HE > 36 : 2 = 18 : 1,$$

so that, a fortiori:

(4) $$BH : HE > 18 : 1.$$

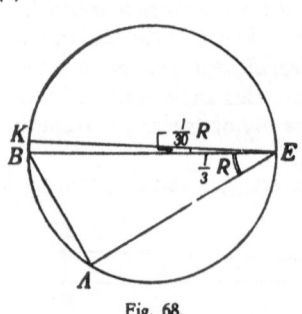
Fig. 68.

He applies the sine inequality to the chords of two arcs, equal to $1/6$ and $1/60$ of a circumference, corresponding to inscribed angles of R/3 and R/30 respectively. If *BK* and *BΔ* are these chords, and *BE* is a diameter, we find therefore

$$BΔ : BK < 10 : 1,$$
(5) $$BE : BK = 2BΔ : BK < 20 : 1.$$

The proof of the proposition is contained in (4) and (5).

Related to this reasoning is the method used in

Archimedes' measurement of the circle

to determine the perimeters of the inscribed and circumscribed regular polygons of 96 sides, to such a degree of accuracy as to enable him to conclude that the perimeter of the circle is less than $3^1/_7 d$, but more than $3^{10}/_{71} d$, where $d = 2r$ is the diameter. Since we shall presently be discussing Archimedes, I shall briefly explain his procedure here.

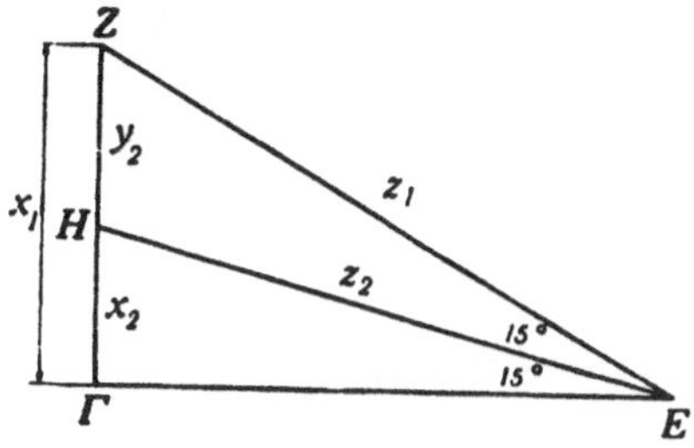

Archimedes starts with two bounds for $\sqrt{3}$, without stating where he got these[1].

$$\frac{265}{153} < \sqrt{3} < \frac{1351}{780}.$$

To calculate the side of the circumscribed polygon, Archimedes starts with a radius $E\Gamma = r$, a tangent line ΓZ and an angle ΓEZ equal to $\frac{1}{3}$ of a right angle. Setting $\Gamma Z = x_1$ and $EZ = z_1$, we have

Fig. 69.

(6) $r : x_1 = \sqrt{3} : 1 > 265 : 153.$

(7) $z_1 : x_1 = 2 : 1 = 306 : 153.$

Now, if $EH = z_2$ is the bisector of angle E, then the segments x_2 and y_2 in which H divides the side ΓZ, are proportional to the sides r and z_1, so that

$$(z_1 + r) : (x_2 + y_2) = r : x_2,$$
(8) $$r : x_2 = (z_1 + r) : x_1 > (306 + 265) : 153 = 571 : 153.$$

Archimedes' method of obtaining (8) closely resembles the derivation of (3) by Aristarchus. Archimedes now continues as follows:

$$z_2{}^2 : x_2{}^2 = (r^2 + x_2{}^2) : x_2{}^2 > (571^2 + 153^2) : 153^2 = 349450 : 23409;$$

hence, since $\sqrt{349450} > 591^1/_8$, we find

(9) $z_2 : x_2 > 591\ ^1/_8 : 153.$

Now the angle $HE\Gamma$ is bisected and $r : x_3$ and $z_3 : x_3$ are calculated just like $r : x_2$ and $z_2 : x_2$; this leads to

$$r : x_3 > 1162\ ^1/_8 : 153,\ \text{and}\ z_3 : x_3 > 1172\ ^1/_8 : 153.$$

Bisecting once again, Archimedes finds that

$$r : x_4 > 2334\ ^1/_4 : 153,\ \text{and}\ z_4 : x_4 > 2339\ ^1/_4 : 153,$$

and finally $r : x_5 > 4673\frac{1}{2} : 153.$

The side of the circumscribed polygon of 96 sides is $2x_5$ the diameter is $2r$; hence the ratio of the diameter to the perimeter of the 96-gon is larger than

$$4673\frac{1}{2} : 96 \times 153 = 4673\frac{1}{2} : 14688.$$

[1] For conjectures on this point see C. Müller and O. Toeplitz in *Quellen und Studien*, 2, p. 281 and p. 286. Compare also M. Cantor, *Gesch. der Math.*, I (3rd ed.), p. 316; Th. Heath *History of Greek Math.*, II, p. 51; K. Vogel, *Jahresber. D. Math. Ver.*, 41, p. 8.

But the last of these numbers is less than $3^1/_7$ times the first, so that the perimeter of the circumscribed 96-gon, and a fortiori that of the circle is less than $3^1/_7 d$.

By consideration of the inscribed 96-gon, Archimedes proves in a similar manner that the perimeter of the circle exceeds $3^{10}/_{71} d$.

The inequalities (1), already used by Aristarchus, occur again in Ptolemy (150 A.D.) who used them in an essential manner in the computation for his famous

Tables for the lengths of chords.

Ptolemy calculates sexagesimally, takes the diameter of the circle as 120^p and calculates to begin with[1], that

the side of the inscribed hexagon which subtends an arc of $60°$, is equal to 60^p;

the side of the inscribed square, subtending an arc of $90°$, is equal to $\sqrt{2r^2} = 84^p51'10''$;

the side of the inscribed triangle, subtending an arc of $120°$, is equal to $\sqrt{3r^2} = 103^p55'23''$;

the side of the inscribed decagon, which subtends an arc of $36°$, is equal to $t = \frac{1}{2}(\sqrt{5r^2} - r) = 37^p4'55''$;

the side of the inscribed pentagon, which subtends an arc of $72°$, is equal to $\sqrt{r^2 + t^2} = 70^p32'3''$.

Starting with these chords, he can now calculate the chords of the supplementary arcs, because the sum of the squares of supplementary chords equals the square of the diameter. Thus, for the chord of an arc of $144°$, the supplement of $36°$, he finds:

$$\text{chord } 144° = 114^p7'37''.$$

Now he deduces the well-known "Theorem of Ptolemy": the rectangle on the diagonals of a chordal quadrilateral is equal to the sum of the rectangles on the two pairs of opposite sides.

When we learn this proposition in school, we do not learn what use Ptolemy made of it. It served to compute the chord $B\Gamma$, which subtends the difference of the two arcs AB and $A\Gamma$, when the chords subtending the arcs AB and $A\Gamma$ and those subtending the supplementary arcs $B\Delta$ and $\Gamma\Delta$ are known. Thus, Ptolemy found, e.g., the chord of an arc of $72° - 60° = 12°$.

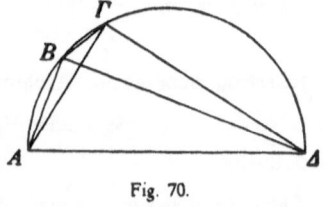

Fig. 70.

By taking two chords AB and $A\Gamma$ in opposite senses, he could calculate in the same manner the chord which corresponds to the sum of two arcs. His method

[1] Following the German translation of Manitius, we use the following notation: $1^p = 1/120$ of the diameter, $1' = p/60$ and $1'' = 1'/60$.

amounts to our formulas

$$\sin(\alpha \pm \beta) = \sin \alpha \cos \beta \pm \cos \alpha \sin \beta.$$

By use of the Pythagorean Theorem, Ptolemy obtains furthermore a formula, which is equivalent to our modern formula

$$2\sin^2 \tfrac{1}{2}\alpha = 1 - \cos \alpha,$$

and which enables him to determine the chord of one half of an arc whose chord is known. Thus he finds successively the chords for arcs of 6°, 3°, 1½° and ¾°:

$$chord\ 1\tfrac{1}{2}° = 1°34'15'', \qquad chord\ \tfrac{3}{4}° = 0°47'8''.$$

In order to calculate the chord of arcs of 1° and ½°, he derives the following proposition:

If two unequal chords are constructed in a circle, the ratio of the longer chord to the shorter is less than that of the corresponding arcs.

We have seen that this theorem was already known to Aristarchus and that he used it in his estimate for sin 3°.

From this proposition follows firstly that

$$chord\ 1° : chord\ 3/4° < 1 : 3/4,$$
$$chord\ 1° < 4/3 \times 0°47'8'', \qquad chord\ 1° < 1°2'50''.$$

and secondly that

$$Chord\ 1\tfrac{1}{2}° : chord\ 1° < 1\tfrac{1}{2} : 1,$$
$$Chord\ 1° > 2/3 \times 1°34'15''$$
$$Chord\ 1° > 1°2'50''.$$

"Since the chord of an arc of 1° is on the one hand less, and on the other hand more than one same amount, we can set the chord equal to 1°2'50'''", thus reasons Ptolemy, "without appreciable error". He takes therefore

$$Chord\ 1° = 1°2'50''.$$

By use of the half-angle formula he finds 0°31'25″ for Chord ½°. Then, by means of the addition formulas, he computes successively the chords of arcs 2°, 2½°, etc. to 180°.

It is in this way that Claudius Ptolemy obtained his table of chords. But he was not the first to calculate such a table.

For Hipparchus, who lived three centuries before him (150 B.C.) has written a work "On the theory of the lines in the circle" which certainly contained a table of chords, probably increasing 1° at a time. Hipparchus used the table for the computation of the instants of rising and setting of fixed stars and of the signs of the zodiac. Tannery conjectures that Hipparchus and Ptolemy derived their method of calculation from Apollonius. But this is a mere conjecture. We only know that Apollonius gave a better approximation of π than Archimedes; and this better approximation may indeed have been the starting point for a table of chords.

During the middle of the 3rd century B.C. there lived in Syracuse, far from the world center Alexandria, the greatest mathematician of antiquity, the brilliant genius

Archimedes.

According to his own report, he was the son[1] of the astronomer Phaedias, who had written a treatise on the diameters of sun and moon. It is not known whether he was of distinguished origin; it is true that he moved in the highest circles at the court of Syracuse and that he was on terms of friendship with king Hiero and with his successor Gelon.

It is probable that he has been in Egypt. Diodorus states that there he invented a hydraulic screw, called Cochlias, for raising water, also in use in the Spanish silver mines (Plate 23). His friendship with the Alexandrian astronomer Conon of Samos also points to a residence in Alexandria. From Syracuse he used to communicate his mathematical discoveries to Conon by letter, at first usually without proofs, because, as he wrote himself in the introduction to the work "On Spirals", "he would not begrudge mathematicians the pleasure of finding out for themselves". But to play a trick on his conceited colleagues in Alexandria, he sometimes slipped in a few false propositions, "so that those who pretend to have discovered every-thing themselves, without supplying proofs, may fall into a trap by asserting to have found something which is impossible".

Was this spiteful remark intended for Eratosthenes? One is inclined to think so, when one reads the ironically eulogistic introduction to the "Method", addressed to Eratosthenes. It is said that Archimedes once gave Eratosthenes a practically insoluble problem, which called for the numbers of white, black, yellow and dappled cows and bulls, which had to satisfy nine conditions[2].

Sometimes years passed before Archimedes, urged by his friends, published complete proofs of his propositions.

Stories about Archimedes.

To obtain an adequate idea of his personality and of the tremendous impress of Archimedes upon his contemporaries, we can not restrict ourselves to the few established facts, but we must also take into account the legends which have gathered around him and his fabulous discoveries.

Every one knows the story of the golden sacrificial wreath of king Hiero (Plate 26), of which Archimedes had to determine the purity[3]. Thinking about this problem, he

---

[1] This at least is the interpretation given by F. Blass, *Astron. Nachr. 104* (1883), No. 2488, p. 255, of a passage from Archimedes' "Sand-Counter".

[2] See R. C. Archibald, *The Cattle problem, Amer. Math. Monthly 25* (1918), p. 411. The equations ultimately lead to a "Pell equation":

$$t^2 - 4,729,494u^2 = 1,$$

with the subsidiary condition that u be a multiple of 9304. The least solution is so enormous that the number of cattle would require a number of more than 206,500 digits!

[3] For further particulars see E. J. Dijksterhuis, *Archimedes*, I, Groningen (Noordhoff), 1938, from which I have borrowed a great deal.

went into the bath, and promptly got the idea for determining the volume (either, as Vitruvius believed, by weighing the water that is spilled from a filled vessel when the object is immersed, or according to others, by measuring the upward pressure), and then ran home naked, shouting "Eureka, eureka!"

When the famous ship Syracosia, a wonder of technical accomplishment, equipped with the most refined luxury, which king Hiero had ordered built for his collegue Ptolemy, was to be launched, there was difficulty until Archimedes was called into consultation. He designed an apparatus which could be operated by one man alone. The king himself launched the ship and exclaimed: "From this day forward, Archimedes will be believed no matter what he says". On a similar occasion, Archimedes is believed to have spoken the winged words: "Give me a spot where I can stand, and I shall move the earth."

Polybius, Livy and Plutarch tell elaborate tales about the machines, invented by Archimedes, which, under the personal direction of the 70-years old mathematician, repelled the attack of the Romans on the city of Syracuse; powerful catapults which dropped from afar heavy rocks on the Roman legions, smaller catapults, so-called scorpions, which threw a hail of projectiles through the embrasures, cranes at the seashore which, turned outwards, dropped big rocks or heavy blocks of lead on the Roman ships, or which reached out with iron grip to lift the bow of the ships, and then to fling them down suddenly upon the water (Plate 23).

Marcellus, who personally directed the assault of the fleet stood aghast. "Shall we have to continue to fight this geometrical Briareus"[1], so he taunted his own technicians, "who empties the sea with our ships, who has knocked out our battering rams and who surpasses the mythical giants with a hundred arms in the many projectiles which he drops on us at one time?" But the Roman soldiers were scared to death: "If they only saw a rope or a piece of wood extending beyond the walls, they took to flight exclaiming that Archimedes had once again invented a new machine for their destruction", so writes Plutarch in the Life of Marcellus. In 212, after a long siege, Marcellus finally succeeded in taking the town.

For Archimedes himself, these mechanical inventions were but the "byproducts of a playful geometry, which he used to cultivate at an earlier time, when king Hiero emphatically urged him to direct his art somewhat away from the abstract and towards the concrete, and to reveal his mind to ordinary people, by occupying himself in some tangible manner with the demands of reality", so writes Plutarch. And again: "Although these discoveries had brought him the fame of superhuman sagacity, he did not want to leave behind any writing on these subjects; he considered the construction of instruments, and, in general, every skill which is exercised for its practical uses, as lowbrow and ignoble, and he only gave his efforts to matters which, in their beauty and their excellence, remain entirely outside the realm of necessity."

[1] Briarcus was a hundred-armed giant in Greek mythology.

Plutarch says that he was possessed by mathematics. "Constantly held in thrall by an ever-present Siren, he forgot to take food and he neglected the care of his body; and when, as was often the case, he was forcibly driven to the bath and to chrism, he would draw geometric figures in the ashes, and he would draw lines on his anointed body with his fingers, so possessed was he by a great enchantment, in truth a prisoner of the Muses."

At the sack of Syracuse in 212, in spite of the explicit orders of Marcellus, a Roman soldier killed the grey-haired mathematician. Archimedes, lost in the study of a figure, which he had drawn in his sandbox, asked him (according to Plutarch) to wait a few moments until he had finished the solution.

It is an open question whether the soldier had intended to kill him from the start or merely wanted to lead him to Marcellus, and slew him in anger, when he did not obey immediately. A later legend has it — if not true, it makes a good story — that Archimedes said to the soldier "Fellow, don't touch my figure".

This dramatic episode is represented in a famous mosaic, reproduced in Plate 24a. At the auction of Jerome Bonaparte's estate, it was said that this mosaic came from Herculaneum. But, in his dissertation (Zur Kunst der römischen Republik, Berlin 1931), F. W. Goethert made it highly probable, by an examination of the cuirass and the cloak of the soldier, and of the attitude of Archimedes and of the folds in his garment that the mosaic is not antique, but that it comes from the school of Raphael.

Nevertheless, it seems to me that F. Winter (Der Tod des Archimedes, 82. Winckelmannsprogramm, Berlin 1924) is right in thinking that the sandbox, which Archimedes has before him, indicates an antique prototype. Valerius Maximus and, on his authority, all the other more recent writers, have always thought that Archimedes wrote in the sand on the floor. But the ancients made their geometric drawings on a board (plinthium, abacium, abacus) covered with sand. This becomes clear from the statements in Iamblichus, Apuleius and Hieronymus, which Winter has brought together. For convenience in use, such a sandbox was placed on a table. Indeed, a scholium in Persius speaks of "the table into which the geometers describe their loci and their measures."

Apparently this fact was no longer known during the renaissance. In Raphael's Athenian school, a mathematician, in the right foreground, draws his lines on a board which lies on the floor. On other paintings, the scholars draw in the sand on the floor itself. The fact that in our mosaic, the sandbox is placed on a table nust therefore be attributed to the happy inspiration of a renaissance artist or, and this is more probable, to the fact that we have here a copy of an antique prototype.

Winter also calls attention to this, that the expression of Archimedes' head much resembles that of a head from Herculaneum, considered to be of Democritus (see Plate 24b and c) and that the attitude of the soldier strongly recalls that of a warrior on the sarcophagus of Alexander.

Marcellus rendered all honors to the relatives, and placed on the grave, in

accordance with Archimedes' own wishes, a representation of a cylinder, circumscribed about a sphere, with an inscription concerning Archimedes' greatest discovery, that the volumes of these solids have the ratio 3 : 2. When Cicero was
quaestor in Sicily, he still found the monument, with sphere and cylinder over-
grown with underbrush and thorns.

About

Archimedes as an astronomer

we know, unfortunately, but little. According to a statement of Hipparchus[1], he
made an error of ¼ day in the observation of the solstices, a small error indeed.
In the Sand-Counter, he describes an apparatus which he used to measure the
sun's apparent diameter. The principal purpose of the Sand-Counter is to explain,
that, by means of an effective notation for large numbers, it is a simple matter to
write down a number greater than the number of grains of sand, that would be
contained in the universe, if it were entirely filled with sand, even if one were to
suppose that the universe has the size assumed by Aristarchus. We recall that, in
the system of Aristarchus, the earth revolves about the sun, while the distance to
the fixed stars is much, much greater than the radius of the earth's orbit.

Archimedes seems to have favored the view of "most astronomers" that the
earth is at the center of the universe. According to Macrobius, on the whole a not
very reliable Roman writer, he calculated the number of stadia from the earth to
the moon, from the moon to Venus, thence to Mercury, to the sun, to Mars, to
Jupiter, to Saturn and finally from there to the realm of the fixed stars.

But his greatest accomplishment in the field of astronomy, especially admired in
antiquity, was the construction of a planetarium, a revolving open sphere with
internal mechanisms with which he could imitate the motions of the sun, the moon
and the 5 planets. A single turn started the entire complicated movement with the
varying periods of rotation of the different celestial bodies. "When Gallus set the
sphere in motion", writes Cicero who had himself seen the apparatus, "one could,
at every turn, see the moon rise above the earth's horizon after the sun, just as
occurs in the sky every day; and then one saw how the sun disappeared and how
the moon entered into the shadow-cone of the earth with the sun on the opposite
side ...".[2]

The book which Archimedes himself wrote about this "On the making of
spheres" has been lost.

The works of Archimedes

will not be discussed at great length; this would require a large book by itself.
And it is not necessary to do this, because any one who wants to give himself the

[1] Claudius Ptolemy, *Syntaxis* III 1.
[2] Cicero, *De republica* I 14. The text breaks off in the middle of a sentence.

pleasure can read them himself in the excellent English, German and French translations of Heath, Czwalina or Ver Eecke; moreover there exists in Dutch the
excellent work of Dijksterhuis, Archimedes, of which Volume I was published by
P. Noordhoff in 1938.[1]

I shall therefore restrict myself here to some general characterizations, to a
survey of the contents of the principal works, and a discussion of a few of the
important methods of proof.

Heath (History of Greek mathematics II, p. 20) has the following to say about
the general character of the works of Archimedes:

> The treatises are, without exception, monuments of mathematical exposition; the gradual
> revelation of the plan of attack, the masterly ordering of the propositions, the stern elimin-
> ation of everything not immediately relevant to the purpose, the finish of the whole, are so
> impressive in their perfection as to create a feeling akin to awe in the mind of the reader. As
> Plutarch said, "It is not possible to find in geometry more difficult and troublesome questions
> or proofs set out in simpler and clearer propositions". There is at the same time a certain
> mystery veiling the way in which he arrived at his results. For it is clear that they were not
> *discovered* by the steps which lead up to them in the finished treatises. If the geometrical
> treatises stood alone, Archimedes might seem, as Wallis said, "as it were of set purpose to
> have covered up the traces of his investigation, as if he had grudged posterity the secret
> of his method of inquiry, while he wished to extort from them assent to his results". And
> indeed (again in the words of Wallis) "not only Archimedes but nearly all the ancients so
> hid from posterity their method of Analysis (though it is clear that they had one) that more
> modern mathematicians found it easier to invent a new Analysis than to seek out the old".
> A partial exception is now furnished by *The Method* of Archimedes, so happily discovered
> by Heiberg. In this book Archimedes tells us how he discovered certain theorems in quadra-
> ture and cubature, namely by the use of mechanics, weighing elements of a figure against
> elements of another simpler figure the mensuration of which was already known. At the
> same time he is careful to insist on the difference between (1) the means which may be suffic-
> ient to suggest the truth of theorems, although not furnishing scientific proofs of them,
> and (2) the rigorous demonstrations of them by orthodox geometrical methods which must
> follow before they can be finally accepted as established:
>
> "Certain things", he says, "first became clear to me by a mechanical method, although they
> had to be demonstrated by geometry afterwards because their investigation by the said
> method did not furnish an actual demonstration. But it is of course easier, when we have
> previously acquired, by the method, some knowledge of the questions, to supply the proof
> than it is to find it without any previous knowledge."

We begin with a discussion of this work from which we can best acquire an
understanding of Archimedes' line of thought.

The "Method".

In 1906, the Danish philologist Heiberg, who had already provided us with
so many excellent text-editions (i.a. of Euclid and of Archimedes), went to Constantinople to study a papyrus from the library of the San Sepulchri monastery in

---

[1] It is regrettable that the book has remained incomplete; the second part was printed only in the periodical
Euclides, vols. 15—17 and 20 (1938—44).

Jerusalem. It was a so-called "palimpsest", originally covered with Greek letters, obviously a mathematical text with diagrams, afterwards scraped off by monks and written upon anew. Heiberg succeeded in restoring and deciphering nearly the whole of the old text. It contained parts of various known works of Archimedes, but in addition the extremely important work "Method", which had been believed to be lost. We shall now briefly indicate the contents of this work.

1. *Area of the parabolic segment.*

The first example of Archimedes is at the same time best suited to explain his mechanical method. Let $AB\Gamma$ be a parabolic segment, bounded by a straight line

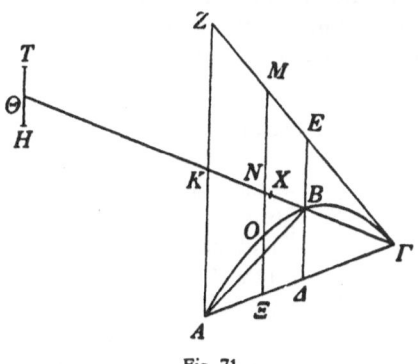

Fig. 71.

$A\Gamma$ and a parabola $AB\Gamma$. Through Δ, the midpoint of $A\Gamma$, a line ΔBE is drawn, parallel to the axis of the parabola. Now Archimedes states that the segment $AB\Gamma$ is $1\frac{1}{3}$ times as large as the triangle $AB\Gamma$.

Draw AZ, parallel to ΔB, to its point of intersection Z with the tangent line ΓEZ. Extend ΓB beyond its intersection K with AZ and make $K\Theta = K\Gamma$. Now consider $\Gamma\Theta$ as a lever with fulcrum at K. Draw also $M\Xi$, parallel to $E\Delta$ through an arbitrary point O of the parabola.

It follows from properties of the parabola, which Archimedes assumes as known that $B\Delta = BE$, so that $N\Xi = NM$ and $KA = KZ$, and furthermore

(1) $\Xi M : \Xi O = A\Gamma : A\Xi = K\Gamma : KN = K\Theta : KN.$

Now we suspend, at the other end Θ of the lever, a line segment $TH = \Xi O$. From the law of the lever, which was first obtained by Archimedes himself in his treatise on the equilibrium of plane figures, it follows then that this segment TH will be in equilibrium with the segment $M\Xi$, placed where it is. For, the proprotionality (1) states exactly that the weights of the two segments are inversely proportional to their lever arms. The same conclusion holds for all segments drawn in the triangle $A\Gamma Z$, parallel to ΔE. In the positions which they occupy, they are in equilibrium with their sections within the parabola, if transferred to the point Θ.

Thus far everything is completely rigorous. Now comes the crux of the method: *because the triangle $A\Gamma Z$ consists of all lines (like ΞM), which can be drawn in the triangle, and because the parabolic segment $AB\Gamma$ consists of all lines, like ΞO, within the parabola, therefore the triangle $A\Gamma Z$, placed where it is, will be in equilibrium with the parabolic segment, placed with its centroid at Θ, so that K is their common center of gravity.*

Now we are practically at home. For the centroid of the triangle $A\Gamma Z$ is at the point X, such that $KX = \frac{1}{3}K\Gamma$. Since now the lever arm $K\Theta$ of the parabolic segment is three times as long as the arm KX to the centroid of the triangle, and since the triangle is in equilibrium with the segment, the weight of the triangle must be equal to three times that of the segment. But triangle $A\Gamma Z$ is twice as large as $A\Gamma K$, and hence four times as large as triangle $AB\Gamma$; hence the parabolic segment is equal to $^4/_3$ of triangle $AB\Gamma$.

Archimedes remarks himself that this argument does not give a rigorous proof of the proposition; nevertheless it carries conviction.

To conceive of a parabolic segment or of a triangle as the sum of infinitely many line segments, is closely akin to the idea of Leibniz, who thought of the integral $\int y\,dx$ as the sum of infinitely many terms $y\,dx$. But, in contrast with Leibniz, Archimedes is fully aware that this conception is, as a matter of fact, incorrect and that the heuristic derivation should be supplemented by a rigorous proof.

2. *Volume of the sphere.*

Archimedes announces that the volume of the circumscribed cylinder is $1\frac{1}{2}$ times as great as the volume of the sphere. Let $AB\Gamma\varDelta$ be a great circle on the sphere. Think of a second great circle on the diameter $B\varDelta$ in a plane perpendicular to that of the first, furthermore of a cone through this second circle with vertex at A and axis $A\Gamma$, whose base is a circle with diameter EZ, and finally of a cylinder $EZH\varDelta$ with axis $A\Gamma$ on the circle EZ as a base. If now $M\varSigma N$ is an arbitrary line in the plane of the circle $AB\Gamma\varDelta$, parallel to $B\varDelta$, cutting this circle in O and $\varXi$ and the surface of the cone in $\varPi$ and P, then

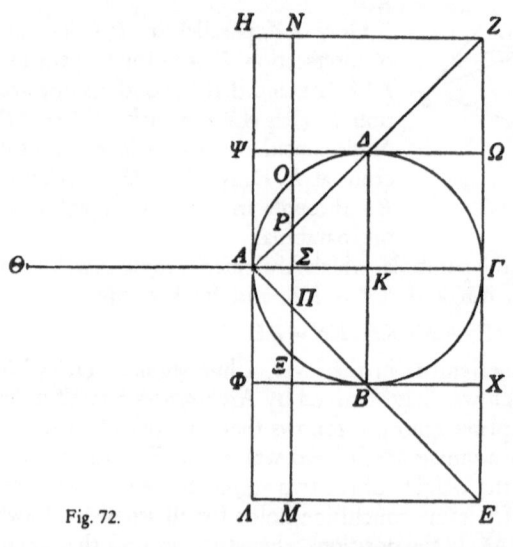

Fig. 72.

$$\varSigma P^2 + \varSigma O^2 = \varSigma A^2 + \varSigma O^2 = AO^2 = A\varSigma \cdot A\Gamma,$$

(2) $$(\varSigma P^2 + \varSigma O^2) : \varSigma N^2 = A\varSigma . A\Gamma : A\Gamma^2 = A\varSigma : A\Gamma.$$

Hence the ratio of the sum of the circles on the diameters $\varPi P$ and $\varXi O$ to the circle on the diameter MN is equal to the ratio of $A\varSigma$ to $A\Gamma$.

Now $A\Gamma$ is thought of again as one arm of a lever with fulcrum at A, the other arm $A\varTheta$ is made equal to $A\Gamma$, and the circles on the diameters $\varPi P$ and $\varXi O$ are transferred to $\varTheta$. It follows from (2) that in this position they will be in equilibrium with the circle on the diameter MN, placed at its own center $\varSigma$.

Because the cylinder *EZHΛ* is made up of these circles, the cylinder will, in its own position, be in equilibrium with the sphere and cone together, both placed at *Θ*. Since *K* is the centroid of the cylinder, the ratio of the cylinder to the sum of cone and sphere will be equal to that of *AΘ* to *AK*, i.e. as 2 : 1. Hence sphere plus cone equals half of the cylinder. But Eudoxus has shown that the cone equals ⅓ of the cylinder; therefore the sphere is ¹/₆ of this cylinder, or ²/₃ of the smaller cylinder *ΦΧΨΩ*.

The result may also be formulated as follows: the sphere is four times as large as a cone whose base is a great circle of the sphere and whose height equals the radius. This led Archimedes to the idea that the area of the sphere is equal to that of four great circles. For, he said, just as every circle is equal to a triangle whose base is the perimeter of the circle and whose height equals the radius, it will probably be true that every sphere equals a cone whose base is equal to the surface of the sphere and whose height equals the radius. In *On the sphere and the cylinder* he proved the correctness of all these results rigorously.

3. *Volume of a spheroid.*

By means of the same method it is found that a "spheroid", i.e. an ellipsoid of revolution equals ⅔ of the circumscribed cylinder.

4. *Volume of a segment of a paraboloid of revolution.*

If a paraboloid of revolution is cut by a plane perpendicular to the axis, then the volume of the finite segment is 1½ times as great as that of a cone of the same base and the same axis. By using the same method, this result is obtained even more readily than the two preceding ones.

5. *Centroid of a segment of a paraboloid of revolution*, cut off by a plane perpendicular to the axis.

To show that this centroid *K* divides the axis *AΔ* in the ratio 2 : 1, Archimedes constructs in the segment a cone *ABΓ* with vertex at *A*, produces *ΔA* to *Θ*, so that *AΘ = AΔ*, and considers *ΘΔ* again as a lever with fulcrum at *A*. Taking *OΞ* as an arbitrary line in the plane of the parabola *BAΓ*, parallel to *BΓ*, we find

$$\Sigma O^2 : \Sigma P^2 = A\Delta : A\Sigma;$$

hence the circle on *OΞ*, placed where it is, will be in equilibrium with the circle on *PΠ*, transferred to *Θ*. From this the conclusion is again drawn that the segment of the paraboloid, placed

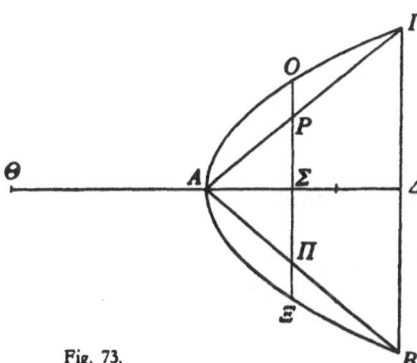

Fig. 73.

where it is, will be in equilibrium with the cone, placed at *Θ*. Since the weight of the segment equals $^3/_2$ of that of the cone, the ratio of the lever arms $A\Theta$ and AK is also equal to 3 : 2, from which the proposition follows.

6. *Centroid of a hemisphere.*

The method is the same as that used in the preceding case.

7. *Volume of a segment of a sphere.*

The method is identical with that in example 2. In entirely analogous manner, the following cases are treated:

8. *Volume of a segment of a spheroid.*

9. *Centroid of a spherical segment.*

10. *Centroid of a segment of a spheroid.*

11. *Centroid of a segment of a hyperboloid of revolution.*

Next follow the cubatures of two remarkable solids, bounded by cylindrical surfaces and planes, but equal to polyhedra bounded by planes.

12. *Cylindrical segment.*

Let a cylinder be constructed in a right prism of square base, whose base is the inscribed circle of the base of the prism; pass a plane through the center of the base and an edge of the upper base. Then the volume of the segment of the cylinder bounded by the cylindrical surface, the oblique plane and the base, is equal to $^1/_6$ of the entire prism.

Archimedes derives this proposition by his mechanical method and then gives a strictly geometrical proof. Finally he announces the following proposition:

13. *Intersection of two cylinders.*

If, in a cube, two cylinders are constructed, with mutually perpendicular axes, then the volume of the solid common to these cylinders equals $^2/_3$ of the volume of the cube.

The proof is not found on the palimpsest.

We can see very clearly how Archimedes supplemented his heuristic consider-ations with rigorous proofs, from the treatise on

The quadrature of the parabola.

Archimedes gives two strict proofs for the proposition concerning the area of the parabolic segment: one mechanical and the other geometrical. The mechanical derivation is essentially the same as that in the "Method", but Archimedes made it into a rigorous proof by means of the exhaustion method. In the case in which the segment is formed by a line $B\Gamma$ perpendicular to the axis, this proof proceeds as follows:

Draw *BΔ*, parallel to the axis, to its point of intersection *Δ* with the tangent at *Γ*. Divide *BΓ* into an arbitrary number of equal parts and, through the points of division *E, Z, H, I*, draw lines *EΛ, ZM, HN* and *IΞ* parallel to the axis, and connect *Γ* with the points in which these lines intersect the parabola. Now it is asserted

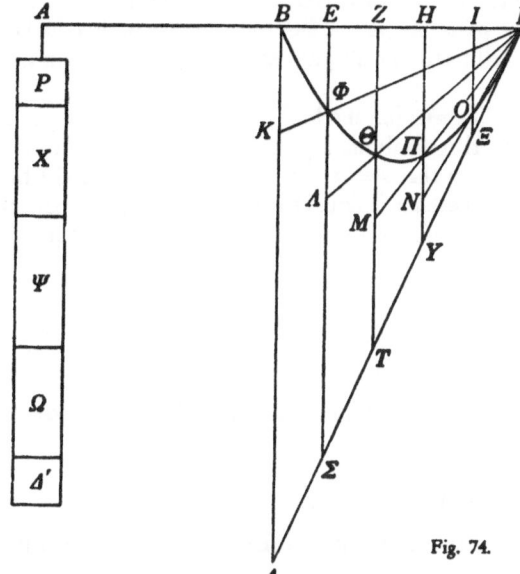

that the triangle *BΓΔ* is less than three times the sum of the trapezoids *KE, ΛZ, HM* and *NI* and the triangle *ΞIΓ*, and more than three times the sum of the trapezoids *ZΦ, HΘ* and *IΠ* and the triangle *IOΓ*.

To prove this, Archimedes extends *ΓB* to *A* such that *BA = BΓ* and he considers *AΓ* as a lever with fulcrum at *B*. At the extremity *A* of the lever he suspends the areas *P, X, Ψ, Ω, Δ'* which are respectively in equilibrium with the trapezoids *ΔE, ZΣ, TH, YI* and the triangle *ΞIΓ*. The sum of all these areas will then be in equilibrium with triangle *BΓΔ* and

Fig. 74.

therefore equal to one third of its area. From the proportion (1) of the "Method", we have

$$BA : BE = BΓ : BE = EΣ : EΦ = \text{trap. } ΔE : \text{trap. } KE,$$

and similarly

$$BA : BZ = \text{trap. } ΣZ : \text{trap. } ΛZ,$$
$$BA : BH = \text{trap. } TH : \text{trap. } MH,$$
$$BA : BI = \text{trap. } YI : \text{trap. } NI.$$

Therefore, if these trapezoids *ΔE, ΣZ*, etc. are suspended from their "right" vertices *E, Z*, etc., they will be in equilibrium with the trapezoids *KE, ΛZ*, etc. suspended from *A*. But if, instead of from these points, they are suspended from the lines *BE, EZ*, etc. they will be in equilibrium with *P, X*, etc. Therefore the areas *P, X*, etc. are smaller than the trapezoids *KE, ΛZ*, etc. The sum of the areas *P + X + Ψ + Ω + Δ'* is therefore less than the sum of the trapezoids *KE, ΛZ, MH* and *NI* and the triangle *IΞΓ*.

But the sum *P + X + Ψ + Ω + Δ'* is ⅓ of the trangle *BΓΔ*, so that ⅓ of this triangle is less than the sum of the trapezoids *KE, ΛZ*, etc. and the triangle *IΞΓ*, which extend beyond the parabola.

In an entirely analogous manner Archimedes proves that $\frac{1}{3}$ of the triangle $B\Gamma\Delta$ is greater than the sum of the trapezoids ΦZ, ΘH, $I\Pi$ and the triangle ΠO, interior to the parabolic segment. This completes the mechanical part of the proof.

The difference between the first series of trapezoids plus triangle and the second, is equal to the sum of the trapezoids $B\Phi$, $\Phi\Theta$, $\Theta\Pi$, ΠO and the triangle $\Gamma O\Xi$, through the interior of which the parabola passes. But this sum is exactly equal to the triangle $B\Gamma K$, which is an arbitrarily small part of the triangle $B\Gamma\Delta$ (in our case $^1/_8$).

Now follows the proof by the "exhaustion method", which we may present as follows. Let s denote the area of the parabolic segment and z one third of the triangle $B\Gamma\Delta$; we have to prove then that $s = z$. Archimedes has enclosed s between two areas s_1 and s_2, whose difference is equal to an arbitrarily small part of the triangle $B\Gamma\Delta$, and he has shown, by an argument from mechanics, that z is also between s_1 and s_2:

(1) $$s_1 > s > s_2.$$
(2) $$s_1 > z > s_2.$$
(3) $$s_1 - s_2 = \varepsilon.$$

Suppose now that s were greater than z. Choose ε as less than the difference $s - z$; then

$$s - z > \varepsilon = s_1 - s_2,$$
and $$s > z + (s_1 - s_2) > s_2 + (s_1 - s_2) = s_1,$$

which contradicts (1). If, on the other hand, s is less than z, take ε less than the difference $z - s$:

$$z - s > \varepsilon = s_1 - s_2.$$
Then: $$z > s + (s_1 - s_2) > s_2 + (s_1 - s_2) = s_1,$$

which contradicts (2). Therefore $z = s$.

With slight variations, this indirect method of proof recurs repeatedly in the works of Archimedes. It is a technique which originated with Eudoxus. In contrast with Archimedes, we shall not give all the steps of the indirect proof in the sequel, but we shall merely derive the inequalities (1), (2) and (3), from which the conclusion $z = s$ follows. It is clear that the inequality

$$s_1 - s_2 < \varepsilon$$

can perfectly well replace the equality (3). Sometimes Archimedes uses ratios instead of differences and proves that the ratio $s_1 : s_2$ can be made less than the ratio of the larger to the smaller of two arbitrarily given magnitudes.

This mechanical proof is followed by a geometrical derivation, at least as elegant. In the parabolic segment $AB\Gamma$, Archimedes constructs a triangle $AB\Gamma$, by drawing through the midpoint Δ of the base $A\Gamma$ a line ΔB parallel to the axis. Now he constructs, in each of the two parabolic segments determined by the arcs AB

and *BΓ*, triangles *AZB* and *BHΓ* in the same manner, and then he derives from the properties of the parabola the conclusion that the sum of these two triangles is

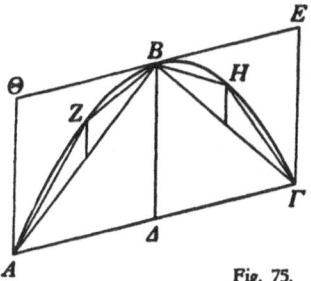

Fig. 75.

equal to ¼ of triangle *ABΓ*. If this process is continued, the next step will lead to 4 triangles, whose sum is equal to ¼ of the sum of the two triangles which have just been considered, and so forth. Moreover, triangle *ABΓ* is one half of the circumscribed parallelogram *AΓEΘ*, and therefore greater than one half of the parabolic segment. If we remove now from the parabolic segment first the triangle *ABΓ*, i.e. more than one half, and then from the remaining segments the two inscribed triangles *AZB* and *BHΓ*, i.e. again more than half, and if this process be continued, then there will ultimately remain less than any arbitrarily given area, in accordance with a well-known proposition, which we encountered in our discussion of Book X of the Elements.[1] Moreover the areas of the triangles which have been removed from a geometric progression of ratio ¼.

Next Archimedes proves the formula for the sum of such a geometric progression: if an arbitrary set of numbers *A, B, Γ, Δ, E* form a geometric progression with ratio ¼, then their sum, increased by ⅓ of the last term, is exactly equal to ⁴⁄₃ of the first term. The proof is very simple:

$$B + B/3 = 4B/3 = A/3,$$
$$\Gamma + \Gamma/3 = 4\Gamma/3 = B/3,$$
$$\Delta + \Delta/3 = 4\Delta/3 = \Gamma/3,$$
$$E + E/3 = 4E/3 = \Delta/3,$$

so that we obtain by addition

$$B + \Gamma + \Delta + E + B/3 + \Gamma/3 + \Delta/3 + E/3 = (A + B + \Gamma + \Delta)/3.$$

If $B/3 + \Gamma/3 + \Delta/3$ is subtracted from both sides and *A* added, we obtain the asserted statement:

(4) $$A + B + \Gamma + \Delta + E + E/3 = 4A/3.$$

Archimedes does not speak about the sum of the infinite geometric progression, but, although he does not know the expression "sum of an infinite series", he does have command of the concept. Indeed, the modern concept "sum of an infinite series" means nothing but the limit of a finite sum, i.e. a magnitude which differs from a finite partial sum by less than an arbitrarily given positive *ε*. Now, Archimedes proves and formulates very explicitly that the parabolic segment, although greater than every partial sum $A + B + \Gamma + \Delta + E$ of his geometric progression, differs from this sum by less than an arbitrarily given area.

[1] See p. 177.

After this, it is not much of a trick to show, by manipulating inequalities, that the parabolic segment can neither be greater nor less than $4A/3$, where A is the first inscribed triangle. For, supposing that the segment exceeds $4A/3$, a partial sum such as $A + B + \Gamma + \Delta + E$, which differs from the segment by an arbitrarily small amount, would also exceed $4A/3$, contrary to formula (4). And, if the segment Z were less than $4A/3$, we would but have to continue the series $A + B + \ldots$ to a point, at which the last term E is less than the difference $4A/3 - Z$. Then the excess of $4A/3$ over the segment would be greater than E, but the excess of $4A/3$ over $A + B + \Gamma + \Delta + E$ is $E/3$, and hence less than E; but this would require that the segment Z is less than $A + B + \Gamma + \Delta + E$, which is impossible.

All this is found in Archimedes, in essentially the same words. It shows that the estimations, which occur in the summing of infinite series and in limit operations, the "epsilontics", as the calculation with an arbitrarily small ε is sometimes called, were for Archimedes an open book. In this respect, his thinking is entirely modern.

On sphere and cylinder I.

For the measurement of plane areas and of volumes, the axioms laid down by Eudoxus, such as "The whole is greater than the part", combined with the "lemma of Archimedes", to which we shall return presently, were adequate. But for the measurement of lengths of arcs and of curved surfaces, other postulates are required. For how could one know otherwise that the perimeter of a circle is greater than that of an inscribed polygon and less than that of a circumscribed polygon? Archimedes starts therefore with some new axioms.

He considers bounded plane curves which lie entirely on one side of the line joining their endpoints, and surfaces which span plane curves and which lie entirely on one side of the plane of the bounding curve. Such a curve or such a surface he qualifies as "concave on one side", if all the line segments which connect two arbitrary points of the curve or the surface, always lie on the same side of the curve or the surface, or on them, but not on the other side. And now he postulates:

1) Of all lines with the same endpoints the straight line is the shortest.

2) If two curves in one plane have the same endpoints and are concave on the same side, and if one of them is entirely enclosed by the other (or coincides with it in part), then that one is the shorter.

3) Of all surfaces, spanning the same plane curve, the plane has the least area.

4) The analogue of 2) for surfaces.

5) If the difference between two unequal lines, areas or volumes, is added to

PLATE 23

Pl. 23a. Roman warship. Augustan relief from Palestrina. Vatican Museum, Rome. Photo Alinari. (see p. 209).

Pl. 23b. Hydraulic screw, made in accordance with the system of Archimedes, operated by a negro slave. As seen on our representation, the slave moves the cylinder round the screw with his feet, while with his hands he holds a horizontal bar, which rests on two supports. Terracotta statuette from the late Ptolemaic period, about 100 B.C., in the British Museum. Place of origin uncertain. According to Rostovtzeff (Social and Economic History of the Hellenistic World, pp. 360, 363, 1160), draining an irrigation by means of hydraulic instruments were actively carried on during the period of Alexander. The Greek historian Diodorus states positively, that the Delta was irrigated by means of a special machine, invented by Archimedes of Syracuse and called "snail" (κοχλίας) on account of its form. Rostovtzeff adds: This contrivance appears to have come into common use in the Delta in Hellenistic times and is still employed in some parts of middle Egypt.

PLATE 24

PL. 24a. Death of Archimedes. Mosaic, probably from the school of Rafael. Copy (or falsification), originally thought to be authentic Roman work. Various elements in the framework (the waterhens – porphyrions – in the four corners and the tendrils) are very successful imitations of Roman examples, see p. 210. (Städtisches Kunstinstitut, Frankfurt am Main).

PL. 24b. Democritus(?), (460–400 B.C.). Bronze copy after a Hellenistic original (3rd century B.C. in the same material. From the "Villa dei Papiri" in Herculaneum National Museum, Naples.

(Photo Alinari)

PL. 24c. Warrior on one of the gables of the Alexander Sarcophagus at Istanbul (see page 210).

itself a number of times, it will exceed every prescribed magnitude of the same kind. This is the famous "Postulate of Archimedes", already used by Eudoxus, but probably explicitly formulated for the first time by Archimedes.

After a few preliminary propositions, Archimedes proceeds now to the determination of the *curved areas of a right cylinder and of a right circular cone*, in both cases by enclosing them between an inscribed and a circumscribed prism or pyramid, on the basis of postulate 4). In both cases he constructs a circle whose area equals that of the cylinder or the cone. For example, in the case of the cylinder, the radius of this circle is a mean proportional between the height and the diameter of the cylinder.

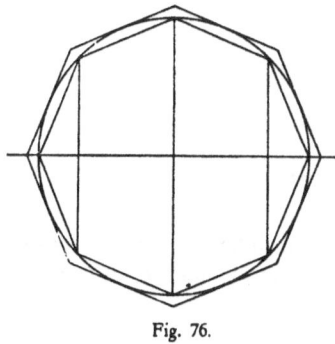

Fig. 76.

Next Archimedes takes up his famous determination of the *area and the volume of a sphere*.

He constructs an inscribed and a circumscribed regular polygon for a circle, of which the number of sides is divisible by 4. Revolution about an axis through two diametrically opposite vertices produces a solid I_n, inscribed in the sphere and bounded by sections of conical surfaces. In the same manner, a circumscribed polygon of the same number of sides generates by rotation a solid C_n, similar to I_n, which encloses the sphere. From postulate (4) we derive

(1)   area of I_n < area of the sphere < area of C_n.

Now let A be a circle whose radius is the diameter of the sphere and whose area is therefore four times as great as that of a great circle on the sphere. Then Archimedes proves the following inequality:

(2)   area of I_n < area of A < area of C_n.

Since the ratio of the areas of the similar solids I_n and C_n can be brought arbitrarily close to 1, we derive from (1) and (2) the conclusion that the sphere has the same area as the circle A, by use of an argument with which we have already become familiar.

Analogous inequalities hold for the volumes. It turns out that the volume of I_n is equal to that of a cone, whose base is a circle equal in area to the solid itself and whose height is the radius of the sphere circumscribed about I_n. The same conclusion holds for C_n and hence, by means of the exhaustion method, for the sphere as well.

The only difficulty lies therefore in the proof of inequality (2). The area of I_n is a sum of areas of sections of conical surfaces between parallel circles. From his theorems on the area of a conical surface, Archimedes deduces that the area I_n equals that of a circle whose radius R is given by

$$R^2 = a \cdot s,$$

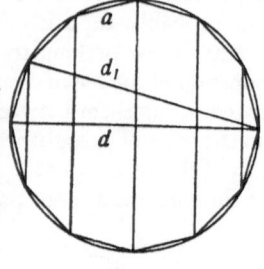

Fig. 77.

where a is the side of the revolving polygon and s the sum of all the diagonals which are perpendicular to the axis of rotation. The product $a \cdot s$ is now transformed into $d \cdot d_1$ where d is the diameter of the circumscribed circle and d_1 is the diagonal which subtends one less than half the number of sides. The ratio of d to d_1 of course approaches 1, so that in the limit R^2 tends to d^2. These are the main features of the proof.

In an entirely analogous manner, the *area of a spherical segment* and also the *volume of a spherical sector* are determined.

On sphere and cylinder II.

This treatise gives the solution of two problems which lead to cubic equations. The first, to determine a sphere equal to a given cone or cylinder, leads to a pure cubic of the form

(3) $$x^3 = b^2c,$$

which is solved in the familiar way, by determining two mean proportionals between b and c. Archimedes does not say how this is to be done. This fact led his commentator Eutocius to the elaborate report on the duplication of the cube on which we have been happy to draw earlier.

The second problem, to divide a sphere in a given ratio, by means of a plane, leads to a cubic equation of the form

(4) $$x^2(a - x) = bc^2,$$

or, in Archimedes' formulation: to divide a line segment a into two parts x and $a - x$, such that

$$(a - x) : b = c^2 : x^2,$$

Archimedes promises to give an analytical and a synthetic solution, but this part of the treatise has been lost. Eutocius has however found a manuscript in Dorian dialect, which he attributes to Archimedes and which gives first the necessary condition for the solvability of (4) and then a solution by means of the intersection of a parabola and a hyperbola. The condition for solvability is that bc^2 be at most equal to the maximum value of $x^2(a - x)$, which is attained when $x = 2a/3$.

On spirals.

When a straight line revolves uniformly about a point O, while a point P, starting from O moves uniformly along the line, then the point P describes a *spiral*.

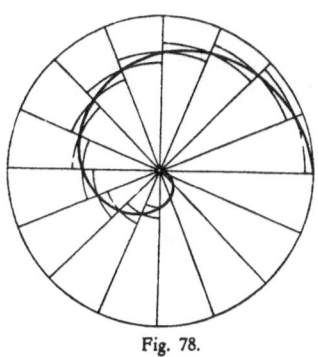

Archimedes derives the characteristic property of a point on the spiral in polar coordinates, then determines the tangent line at an arbitrary point on the spiral and finally the area contained between two arbitrary radii vectores and between two successive windings, or within the first winding. The adjoining figure indicates how this area is enclosed between sums of circular sectors. The only difficulty in the proof by the exhaustion method is the summing of the series $1^2 + 2^2 + 3^2 + \ldots + n^2$. For this sum Archimedes gives the following formula, in geometric formulation (Proposition 10):

Fig. 78.

$$3[a^2 + (2a)^2 + (3a)^2 + \ldots + (na)^2] = n(na)^2 + (na)^2 + a(a + 2a + 3a \ldots + na).$$

On conoids and spheroids.

By conoids Archimedes means paraboloids of revolution and hyperboloids of revolution of one sheet, by spheroids he means ellipsoids of revolution. He determines the volume of the spheroid and of all segments of all spheroids and conoids, which are cut off by an arbitrary plane, as well as the area of the ellipse.

The volume of the sphere is of course contained in this as a special case, but the method of proof is totally different. Archimedes might have transformed the spheroids into spheres by stretching or contracting (a trick which he himself applies to the ellipse in proposition 4), but this method is not applicable to conoids. For this reason he invented a new method, which is invariant with respect to stretching and contracting. According to his own saying he first found the results for the paraboloids, and only later on, and with great difficulty, those for the ellipsoids and the hyperboloids. The method consists in this that the segment of a conoid or a spheroid under consideration is divided into slices of equal thickness by planes parallel to the base, as shown in the adjoining figure for the paraboloid, and that each of these slices is enclosed between two cylinders, one less and the other greater than the slice.

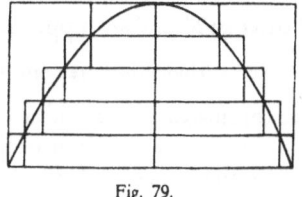

Fig. 79.

In the case of the paraboloid, the outer cylinders form an arithmetic progression whose difference is equal to the smallest term:

$$S_1 = A + 2A + 3A + \ldots + nA,$$

while the inner cylinders form the same progression except for the last term:
$$S_2 = A + 2A + \ldots + (n-1)A.$$

Now, says Archimedes, it is clear that the sum of n terms, each equal to nA, is less than $2S_1$ and greater than $2S_2$:

(1) $$2S_1 > n \cdot nA > 2S_2.$$

The geometrical meaning of $n \cdot nA$ is a pile of equal cylinders, all on the same base as the segment of the paraboloid and all of the same height. These form together a cylinder C. Thus, for (1), we can write:

(2) $$2S_1 > C > 2S_2.$$

Moreover, we have

(3) $$S_1 > S > S_2,$$

where S denotes the volume of the segment of the paraboloid.

Since the difference $S_1 - S_2 = nA$ represents the volume of the cylinder at the bottom of the pile, which can be made arbitrarily small, the familiar method leads form (2) and (3) to the result

$$S = \tfrac{1}{2}C.$$

In the case of an ellipsoid or a hyperboloid, the cylinders do not form an arithmetic progression. But their sum can be reduced to the sum of a sequence of areas of rectangles $x(a \pm x)$, whose altitudes x form an arithmetical progression $b, 2b, 3b, \ldots, nb$. If now S_1 designates again the sum of n such rectangles and S_2 the sum of the same rectangles, except for the last term B, i.e. in modern notation, if

$$S_1 = \sum_1^n kb(a+kb), \quad S_2 = \sum_1^{n-1} kb(a+kb), \quad B = nb(a+nb),$$

then Archimedes proves the inequality

(4) $$nB : S_2 > (a+nb) : (\tfrac{1}{3}a + \tfrac{1}{3}nb) > nB : S_1,$$

which replaces the inequality (1) in the proofs. In the proof of (4), Archimedes starts from the formula for the sum of squares, which he proved in "On Spirals".

The notion of integral in Archimedes.

Riemann gave a rigorous definition of the integral by enclosing it between a "lower sum" and an "upper sum", as indicated in the adjoining figure. The integral $\int_a^b y\,dx$ is then the area under the curve between the ordinates $x = a$ and $x = b$, and the X-axis; the "lower sum" is the sum of the areas of the rectangles below the curve, and the "upper sum" is the sum of rectangles, of somewhat greater height, which cover the area. The treatise on conoids and spheroids shows that Archimedes was familiar with this method of inclusion and that he used it for the determination of volumes. But, it seems to me that one can not say that he was familiar with

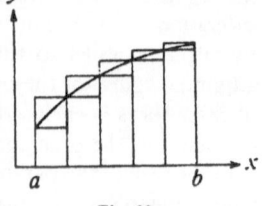

Fig. 80.

the concept of the integral. His integrals always remain tied to a definite geometric interpretation, as volumes or as areas of plane figures. We have no evidence that he understood that one single concept is the foundation of all these geometric interpretations. For instance, in the quadrature of the parabola on the one hand, and in the determination of the volume of the sphere and the spheroid on the other, he bases his rigorous proofs on totally different methods, although both lead to the same integral

$$\int_0^a x(a - x)dx,$$

so that any method effective for one of these problems could also have been used for the other.

Nevertheless, his rigorous determination of areas and volumes make Archimedes the precursor of the modern integral calculus.

The book of Lemmas

(Liber Assumptorum) has been preserved only in Arabic, not in Greek. It begins with a set of propositions concerning a figure, called the Arbelos or cobbler's

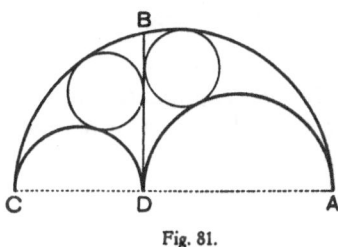

Fig. 81.

knife, bounded by three semicircles, tangent to each other at their extremities. The Arbelos has the same area as the circle which has BD as a diameter (see Fig. 81). This line divides the Arbelos into two parts, whose inscribed circles are equal. Furthermore Archimedes indicates how to express the diameter of the inscribed circle of the Arbelos itself in terms of AC, when the ratio of the parts into which D divides AC is given.

The book contains furthermore a remarkable proposition: 8. Extend a chord AB of an arbitrary circle by a segment BC equal to the radius, and draw the diameter FDE through C. Then the arc AE is three times as great as the arc BF. The proof can be given very easily, e.g. as follows:

$\angle ADE = \angle DAB + \angle ACD = \angle ABD + \angle BDC = 2 . \angle BDC + \angle BDC = 3 . \angle BDC.$

By means of this proposition, the trisection of a given arc can be accomplished as follows: draw the diameter EF and lay off

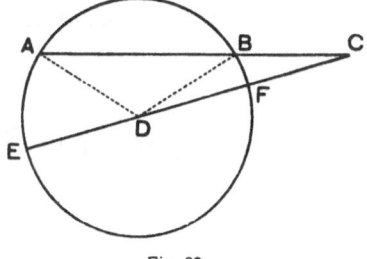

Fig. 82.

the segment BC equal to r, in such a manner that the extension of CB goes through

A, e.g. by means of a ruler on which distances equal to *r* have been marked off. Then the arc *BF* equals one third of the arc *AE*.

The Greeks frequently applied such "Neusis" constructions, in which a segment of definite length is placed between two lines in such a manner that the extension of the segment passes through a fixed point. Hippocrates used them in the quadratures of his lunules.

The construction of the regular heptagon

has also been preserved in Arabic only. C. Schoy discovered it in his study of a treatise of Al-Biruni, on the construction of the regular nonagon.[1] Tabit ben Qurra, who still had seen Archimedes' treatise in Greek, complains of the poor condition of the manuscript, so that it was only after a great deal of effort that he succeeded in unraveling the proofs.

The treatise, which Tabit has brought down to us, begins with a number of propositions on right triangles. But propositions 16 and 17, which deal with the construction of the heptagon, are independent of these.

Proposition 16. *Let the diagonal BC of the square ABDC and the transversal DTEZ be so drawn that the triangles DTC and ZAE are equal in area.* (One sees readily, by rotating the transversal, that such a position always exists.) *Drop the perpendicular TK from T on AB; and let the segments ZA, AK and KB be denoted by x, y and z respectively. Then it follows that*

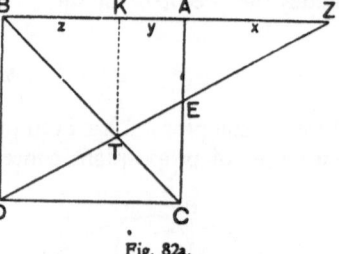

Fig. 82a.

(1) $AB \cdot KB = AZ^2$,i.e. $(y + z)z = x^2$, and

(2) $ZK \cdot AK = KB^2$,i.e. $(x + y)y = z^2$.

These statements follow easily from the equality in area of the two triangles. Archimedes does not speak about the construction of the transversal and of the segments *x* and *y*. But they can readily be carried out by the use of conic sections. For, if we set $y + z = a$, the equations (1) and (2) take the form

(3) $a(a - y) = x^2$

(4) $(x + y)y = (a - y)^2$.

Equation (3) represents a parabola, when *x* and *y* are interpreted as rectangular coordinates; and (4) is then the equation of a hyperbola. These two curves have three points of intersection in the finite part of the plane; one of these lies in the first quadrant.

[1] C. Schoy, *Die trigonometrischen Lehren des ... Al-Biruni*, Hannover, 1927. See also J. Tropfke, *Zeitschr. math. und naturw. Unterricht*, 59 (1928), p. 195, and *Die Siebeneckabhandlung des Archimedes, Osiris 1*, p. 636. I follow Tropfke in the notations.

Now Archimedes constructs a triangle AKH, of which $AK = y$, $HK = z$ and $AH = x$, he circumscribes a circle about triangle BHZ and asserts that BH is equal to the side of the inscribed regular heptagon! Is this not breath-taking? The proof can be found in Tropfke. But one can not help asking how Archimedes hit upon this scheme. Tropfke's answer is as follows:

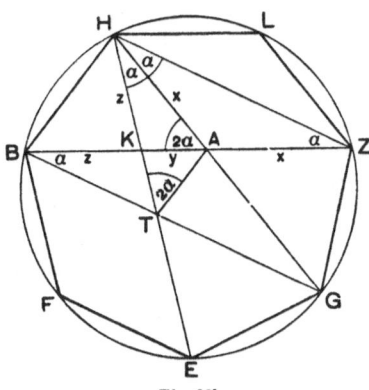

Fig. 82b.

Let the regular heptagon $BHLZGEF$ be drawn, let the diagonal BZ meet HE in K and HG in A, and let BG meet HE in T. We denote the segments ZA, AK and KB by x, y and z respectively and we designate by α the inscribed angle which intercepts one seventh of the circumference. Then we find that $AH = x$ and $KH = z$. The triangles ZHK, HAK and HTA are similar, because each of them has one angle equal to α and another equal to 2α. But from this fact we obtain immediately the proportionalities

$$z : (x + y) = y : z, \text{ and}$$
$$z : x = x : (y + z),$$

which are equivalent to (2) and (1) respectively.

The other works of Archimedes.

We have already briefly mentioned the *Measurement of the Circle* and the *Sand Counter*.

In spite of the importance of the treatises on mechanics, we can only barely mention them here. To begin with, in *On equilibria of plane figures*, the famous law of the lever is derived strictly logically from a set of axioms: *two magnitudes are in equilibrium if their distances are in inverse ratio to their weights.*[1] By means of it, the centroids of parallelogram, triangle and trapezoid are determined, and, in Book II, the centroid of a parabolic segment and of the section of a parabola between two parallel lines.

In the treatise *On floating bodies*, Archimedes first derives the laws of upward pressure for bodies, whose specific gravity is less than, equal to or greater than that of the liquid. A real masterpiece is the treatment of the stability of the equilibrium of a floating right segment of a paraboloid of revolution. One has to read this splendid piece oneself in order to appreciate its value.

In the fifth book of his Collectio, Pappus relates something about Archimedes investigation of semi-regular solids, bounded by:

[1] A careful analysis of this deduction and a discussion of E. Mach's critique is found in W. Stein, *Quellen und Studien*, B I, p. 221.

     4 regular triangles and 4 regular hexagons
or 8 triangles and 6 squares,
or 6 squares and 8 hexagons,
or 8 triangles and 6 octagons,
or 8 triangles and 18 squares,
or 12 squares, 8 hexagons and 6 octagons,
or 20 triangles and 12 pentagons,
or 12 pentagons and 20 hexagons,
or 20 triangles and 12 decagons,
or 32 triangles and 6 squares,
or 20 triangles, 30 squares and 12 pentagons,
or 30 squares, 20 hexagons and 12 decagons,
or 80 triangles and 12 pentagons.

According to an Arabic tradition[1], Archimedes is also the discoverer of the well-known formula for the area of a triangle, usually attributed to Heron:

$$\sqrt{s(s-a)(s-b)(s-c)}.$$

Somewhat younger than Archimedes is

Eratosthenes of Cyrene.

He was learned, industrious, conscientious and he had artistic tastes, but he lacked the genius of Archimedes. This is perhaps the reason why his friends called him Beta — a "second string man". Another nickname for him was Pentathlos, the five-sports-athlete, the allround sportsman, who accomplished excellent results in a great diversity of fields, without being at the top in any one. Indeed he was eminent in various domains, as a mathematician, a geographer, a historian, a philologist and a poet. His epigram on the duplication of the cube, which was mentioned at an earlier point[2], is a sample of his subtle poetic art. His dialogue Platonicus deals not only with the Delian problem, but also with philosophical questions and with matters relating to the theory of music. He described the starry heavens in a poem, taking the form of a walk through the sky of Hermes, in which he encounters various pretty adventures. He also collected myths about the constellations. He made an estimate of the inclination of the ecliptic, of the distances of sun and moon, and of the earth's perimeter. He designed a new map of the world, based on the view that the earth is round, and he wrote a great work on ancient Greek comedy. He was the founder of critical chronology; it was from him that man learned to determine scientifically the dates of historical events.

Life.

Around the year 260, young Eratosthenes travelled from his birthplace on the

---

[1] Al-Biruni, *The book concerning the chords*; see Bibliotheca Mathematica XI 3, p. 11–78.
See p. 160–1.

coast of Africa to Athens, in order to study philosophy. In his own opinion he
arrived there at the most favorable moment, because the coryphees of philosophy
were then in Athens. It was the period in which the schools were constantly
wrangling with each other. Following the example of Plato's Academia and Aristotle's Lyceum, Epicurus had established the "garden" and Zeno the ,,stoa", the
colonnade. Neither Epicurus nor Zeno were great philosophers; they were dog-
matists and their views of nature especially were long antiquated; nevertheless
Stoicism and Epicureanism continued to be the fashionable tendencies, even
lasting into the Roman period.

Very popular were also the Kynics, who took this designation after Diogenes,
"the dog" (κύων). The doctrine of Diogenes has little connection with our concept
of "cynicism"; he taught that man can only attain real liberty by liberating himself
from his desires. There are numerous anecdotes about Diogenes, how he walked
through the city, wearing nothing but a cloak and sandals, carrying his wooden
watercup, how he sometimes spent the night in a big urn (the famous vat), how
he ridiculed all conventions and all imagined needs unmercifully.

Zeno and Chrysippus, the first stoics, were indeed influenced by the kynic
dogma, but they tried to make it into a moral code for reliable functionaries; we
must remember that they lived in the time of the great Hellenistic monarchies with
their staffs of government officials. But Ariston of Chios pitched into the compromi-
ses and the moral casuistry of the Stoa. The wise man, he said, who has experienced
justice as the inner truth, does not need a rule of behavior for every possible case,
he will know for himself how to act. "The wise man is like a good actor, who plays
the parts of Agamemnon and of Thersites equally well", i.e. he accepts life as it
comes, but stands above it inwardly.

On account of his magnetic oratorical talents, Ariston was called "the siren".
But Eratosthenes admired him above everything else for his uncompromising
ethics.

He equally admired Archesilaus who had reorganized the Academy. Plato's
successors, Xenocrates and Polemon, were very decent and well-meaning people,
but under their direction, the Academy had come into the blind alley of sterile
speculations on numbers. Archesilaus had put a stop to this and had enthusiasti-
cally taken up the cudgels against the dogmatism of the Stoa. He disputed relent-
lessly the naively-empirical epistemology of the Stoics. He did not indoctrinate
his pupils, but he trained them in the dialectical method; they had to defend
propositions against his attacks. He also maintained uncompromisingly the re-
quirement that every philosopher had to begin by learning mathematics. He himself had an excellent teacher of mathematics in Autolycus.

According to Schwarz[1], Eratosthenes must have been around fifty when he was
called to the court of Ptolemy III to educate the crown prince. At the same time
he became the head of the world-famous library. It is probable that he had been

[1] E. Schwarz, Charakterköpfe aus der Antike, Leipzig 1943.

in Alexandria for some time before this, because his life-work is unthinkable without constant use of a great library. Where else would he have found the material for his map of the world, for his chronography, his philological work and his celestial travelogue? Neither could he learn mathematics and astronomy in Athens, but he could in Alexandria.

Indeed, Eratosthenes was a typically Alexandrian scholar. The refined culture of the Hellenistic royal court, where poets, philosophers and grammarians competed in the purification of the language, in learned arguments and in artfully made poetry, was his element. The subtle compliment to the king and his son, which occurs in his epigram, betrays the well-versed courtier.

In his later years, Eratosthenes grew blind. He died the "philosopher's death" by suicide.

Chronography and measurement of a degree.

The chronological work of Eratosthenes excels in critical precision. Is was his principle to eliminate all unverifiable legends and to date the events exclusively on the basis of authentic documents (such as the list of winners in the Olympic games) and reasonable estimates. It is true that, in connection with the Trojan war, he could not avoid deviating from this principle.

His measurement of the earth was equally conscientious. An older estimate, mentioned by Archimedes, started from the assumption that the distance from Lysimachia on the Hellespont to Syene in Egypt is 20,000 stadia. But this line went over land and sea, so that it was impossible to verify the distance. For this reason Eratosthenes preferred to take a smaller distance, which could be measured accurately, viz. from Alexandria to Syene, situated practically due south of Alexandria. Probably on the authority of professional counters-of-steps, he put the distance from Syene to Alexandria at 5000 stadia. Then he established the fact that at the time of the summer solstice, the sun is exactly in the zenith in Syene, but at an angle of $1/50$ of 4 right angles from the zenith in Alexandria. This gave him $50 \times 5000 = 250,000$ stadia for the perimeter of the earth. Since we do not know the exact length of a stadium, we can say little more than that the order of magnitude is about right.

Duplication of the cube.

Recalling what has already been said about the problem (see p. 160), we shall now discuss the mechanical solution given by Eratosthenes which he had engraved in stone in the temple of the king-god Ptolemy. I shall reproduce in translation the text, which appeared on the stone tablet, without the commentary (superfluous, in my judgment) of the forged Eratosthenes letter. To understand it, one has to know that at the top there was a model in bronze, consisting of 3 tri-

angular or rectangular plates, which could be pushed back and forth between a
fixed and a rotating ruler. Under this were the following figure and text.

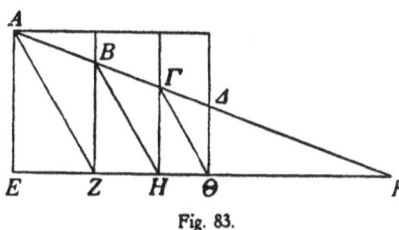

Fig. 83.

To determine two mean proportionals in
continued proportion to two given lines. Let
the lines AE and $\Delta\Theta$ be given. Then I move the
plates of the instrument towards each other
until the points $AB\Gamma\Delta$ lie on one line. Now,
since AE and BZ are parallel, KA and KB
have the same ratio as KE and KZ. And since
AZ and BH are also parallel, this ratio also
equals that of KZ to KH. Hence KE is to KZ
as KZ is to KH. But AE and BZ also have the
same ratio, as well as BZ and ΓH. In the same way we prove that BZ is to ΓH as ΓH is to
$\Theta\Delta$. Therefore AE, BZ, ΓH and $\Delta\Theta$ form a (continued) proportion. Thus two mean proportionals between two given (lines) have been found.

If now the given lines are not equal to AE and $\Delta\Theta$, we shall obtain the mean proportionals
by making AE and $\Delta\Theta$ proportional to them; we reduce them to these and thus the task
will have been carried out.

In case more mean proportionals have to be found, we take every time one more plate in
the instrument than the number of mean proportionals to be constructed. The proof remains the same.

Theory of numbers.

The best known discovery of Eratosthenes in the theory of numbers is the
sieve of Eratosthenes, a method for sorting out the prime numbers among the odd
numbers, transmitted by Nicomachus of Gerasa (Introductio arithmetica, chapter
13). The odd numbers, beginning with 3, are written down, going as far as one
wishes. Now one takes the first number 3 and crosses off all multiples of 3. There
are every time two places between two successive ones. Now one takes the second
number 5 and crosses off all its multiples, which now have 4 places between
them. Next one takes the third number 7, etc. In the end only the prime numbers
are left.

Medieties.

Another arithmetical theory which has to be ascribed to Eratosthenes is the
generation of all kinds of medieties (or mean proportionals) from the geometric
mean proportional, and of all proportionalities from equality, as they are explained
by Nicomachus, Theon of Smyrna, and Pappus.

The explanations of Theon and of Pappus begin with a philosophical introduction:

(Theon): "*Eratosthenes* says that ratio is the source of proportionality, and the
origin for the generation of everything which is produced in an ordered way. For all proportionalities arise from ratios and the source of all ratios is equality."

(Pappus): "Proportionality is composed from ratio and equality is the origin of all
ratios. Geometric mediety indeed has its first origin in equality; it establishes itself and also
the other medieties. It shows us, as says divine *Plato*, that proportionality is the source of

all harmonies and of all rational and ordered existence. For he says that the only connecting link between all sciences, and the cause of all existence, and the tie between everything that has become, is the divine nature of proportionality."

The words which are here attributed to Plato, do not occur in the dialogues, and Theon ascribes almost the same words to Eratosthenes. The explanation is very simple: it is not the writer Plato who is speaking here, but the Plato of the dialogue Platonicus of Eratosthenes.

In the same way, Aristotle says sometimes: "Socrates says in The Republic . . ." and at other times "Plato says in The Republic". Of course, the reference is always to the person Socrates who appears in the dialogue of Plato's Republic.

In Theon, Eratosthenes discusses elaborately the elements from which various kinds of entities can be evolved: numbers from the unit, magnitudes from the point, ratios from equality.

But what is the meaning of the words "Geometric mean proportionality produces itself and also the other medieties"?

In Pappus, the 10 medieties ($\mu\varepsilon\sigma\sigma\tau\tilde{\eta}\tau\varepsilon\sigma$) are defined as follows:

1) $A - B = B - \Gamma$ or $A + \Gamma = 2B$: arithmetical mediety;
2) $A : B = B : \Gamma$ or $A\Gamma = B^2$: geometrical mediety;
3) $(A - B) : (B - \Gamma) = A : \Gamma$: harmonic mediety;
4) $(A - B) : (B - \Gamma) = \Gamma : A$: subcontrary of the harmonic mediety;
5) $(A - B) : (B - \Gamma) = \Gamma : B$: fifth mediety;
6) $(A - B) : (B - \Gamma) = B : A$: sixth mediety;
7) $(A - \Gamma) : (A - B) = B : \Gamma$, or $A = B + \Gamma$: seventh mediety;
8) $(A - \Gamma) : (A - B) = A : B$: eighth mediety;
9) $(A - \Gamma) : (A - B) = A : \Gamma$: ninth mediety;
10) $(A - \Gamma) : (B - \Gamma) = B : \Gamma$: tenth mediety.

The first three of these are old-Pythagorean, the next three were discovered by Eudoxus, the last four by "later" writers. After having quoted the passage from "Plato", cited above, Pappus continues as follows:

"We shall now show how to generate the 10 medieties from the geometrical one. First now following theorem:

Proposition 17. Let A, B and Γ be three proportionals and set

$$\Delta = A + 2B + \Gamma,$$
$$E = B + \Gamma,$$
$$Z = \Gamma,$$

Now Δ, E, Z are again three proportional terms".

Pappus proves this proposition by transforming proportionalities; in proposition 18, he shows briefly, by three examples, how all proportionalities in three terms (A, B, Γ) arise, by repeated application of proposition 17, from the equality $(1, 1, 1)$. This is developed more fully by Theon, and by Nicomachus. Starting from the proportionality in 3 terms $(1, 1, 1)$, one obtains by application of proposition 17, successively

$$\begin{array}{ccc} 1 & 2 & 4, \\ 1 & 3 & 9, \text{ etc.;} \end{array}$$

and by inversion

$$\begin{array}{ccc} 4 & 2 & 1, \text{ etc.;} \end{array}$$

then, by applying proposition 17 again,

$$\begin{array}{ccc} 4 & 6 & 9, \text{ etc., etc.} \end{array}$$

The inverse process makes it possible to transform every proportionality in three integers into the equality $(1, 1, 1)$. None of this is very profound, but it is rather nice.

How the geometric proportionality, thus having generated itself, produces also the other medieties, is shown in the succeeding set of propositions 20—27, of which we present one by way of example.

     Proposition 20. If A, B, Γ are proportional, then

$$\Delta = 2A + 3B + \Gamma$$
$$E = 2B + \Gamma$$
$$Z = B + \Gamma.$$

constitute a harmonic mediety.

These propositions mean that the quadratic equations, which are equivalent to the various medieties, such as, e.g.,

$$E(\Delta + Z) = 2\Delta Z$$

for the harmonic mediety for Δ, E and Z, are all reducible to the normal form

$$B^2 = A\Gamma,$$

by means of linear substitutions such as

$$\Delta = 2A + 3B + \Gamma, \quad E = 2B + \Gamma, \quad Z = B + \Gamma.[1]$$

The propositions which have been mentioned have without doubt been taken from Eratosthenes' treatise on medieties, referred to twice by Pappus in Book VII. In the Platonicus, Eratosthenes evidently hinted at this theory and developed some of its philosophical implications, but in his mathematical treatise he has added the proofs.

Theon brings his explanation, taken from Adrastus, to a close in the following words:

> "Eratosthenes proves that all figures are composed of certain proportionalities, so that they originate from equalities and are again transmuted into equalities; but it is not necessary to speak about this now".

If the word "figures" were here replaced by "medieties", everything would be

[1] Pappus naturally does not consider the arithmetical and the seventh medieties which lead not to quadratic but to linear equations. The Latin commentator Commandinus and Hultsch did not adhere to this sensible restriction, and have inserted two propositions, 19 and 24, which do not appear in Pappus, Hultsch has even "translated back" these interpolations into Greek.

perfectly clear; but what does Eratosthenes mean by "all figures", τὰ σχήματα
πάντα?

It is my conjecture that the "figures" are intended to be conic sections, given by
quadratic equations in homogeneous coordinates, and that Eratosthenes wants to
say that it is possible to transform all such equations by means of linear substitutions to the form $B^2 = A\Gamma$ or $A : B = B : \Gamma$.

A few years ago, when I casually suggested this explanation, it appeared to me
as very risky and as having little chance of being confirmed. But recently I discovered that Tannery had made a similar conjecture on the basis of entirely different sources. His starting point is the following:

In Chapter VII, enumerating books on geometric analysis and on geometric
loci, Pappus mentions, along with books on those subjects by Euclid and by
Apollonius, a work of Eratosthenes in two volumes: Περὶ μεσοτήτων, "On medieties". A bit further on, he talks about certain geometrical loci (τόποι), which
Eratosthenes called τόποι πρὸσ μεσότητασ, i.e. "geometrical loci referring to medieties". Of these loci he says that they "belong to the types mentioned before,
but differ from them by the peculiarity of the hypothesis". The types mentioned
before are straight lines, circles, conic sections, higher plane curves and surfaces.
It becomes clear from another place (p. 652), where the τόποι πρὸσ μεσότητασ
appear along with other types of curves, that they are not surfaces, and at a still
different place (p. 672) they are placed in contrast to the straight line and the
circle. Hence these τόποι have to be conics or higher plane curves.

It appears from various sources that Aristaeus and Apollonius were much interested in "geometrical loci on three or four lines", defined as the loci of
points, whose distances from 3 or 4 given lines form a proportionality; for 3
lines this would be a proportionality of 3 terms

$$A : B = B : \Gamma,$$

and for 4 lines one of 4 terms

$$A : B = \Gamma : \Delta.$$

Such loci are conics, as shown partly by Aristaeus, and completely by Apollonius.
Now one can define for all the other medieties loci of points, whose distances (perpendicular or oblique) to 3 given lines form a mediety (harmonic, etc.). Such
curves are again conics and τόποι πρὸσ μεσότητασ would be a good name for them.
For this reason Tannery surmises that these are the curves which Eratosthenes has
defined in his work on medieties. If this surmise is correct, it would at the same
time account for the passage in Theon.

Before coming to the last geometrician of genius of antiquity, to Apollonius of
Perga, we must say a word about a competitor of Eratosthenes, the geometer

Nicomedes.

Chronologically, Nicomedes comes between Eratosthenes and Apollonius, since he criticised Eratosthenes' duplication of the cube, while Apollonius refers somewhere to the cochloid of Nicomedes. Apparently he belonged to the same group of Alexandrian mathematicians as Eratosthenes and Apollonius.

The *cochloid* or *conchoid of Nicomedes* is described in the following manner by Pappus (Book IV, proposition 20): Let $\Gamma\Delta E$ be perpendicular to AB. Let the line

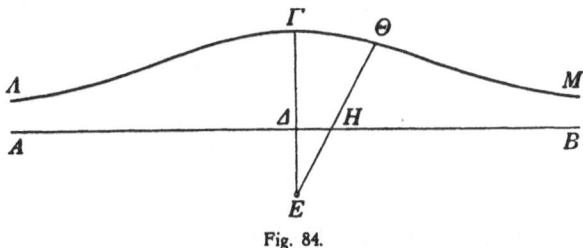

Fig. 84.

$\Gamma\Delta E$ revolve about E, so that Δ always remains on AB while $\Delta\Gamma$ preserves its length. Then Γ describes a curve $\Lambda\Gamma M$ which has the following property: If a line $EH\Theta$ is drawn to the curve through E, then the curve and the line AB determine on this line a segment $H\Theta$ which is always equal to $\Gamma\Delta$. The point E is called the *pole* of the conchoid, AB its *ruler* and $\Gamma\Delta$ its *distance*.

Pappus states that Nicomedes described an instrument, with which the curve can be drawn. He proved that on both sides of E, it approaches the line AB, and that every line drawn from a point on AB to the side on which the curve lies, will cut the curve somewhere. From this follows: If two lines AB and AH are given, and

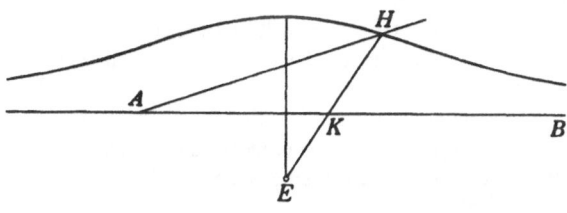

Fig. 85.

if E is a point external to the angle BAH, then it will always be possible to con-struct a line segment of given length HK in such a manner that H lies on AH, and K on AB, while the extension of HK passes through E. This "neusis-construction" can be performed by determining the point of intersection of the line AH with the cochloid whose "pole" is E, whose "ruler" is AB and for which the given length is the "distance".

Nicomedes took great pride in the discovery of this curve. He applied his neusis-construction to two famous problems, viz. the trisection of the angle[1] and the duplication of the cube.

The trisection of the angle.

Let $AB\Gamma$ be the given angle, and let $A\Gamma$ be perpendicular to $B\Gamma$. Draw

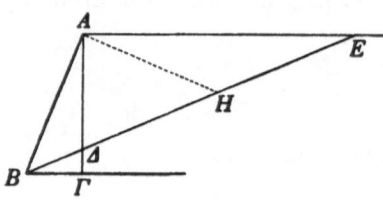

Fig. 86.

$AE \mathbin{/\mkern-5mu/} B\Gamma$. Now construct between the lines $A\Gamma$ and AE a line segment $E\Delta$, twice as long as AB, and such that its extension passes through B. As we have seen, this neusis-construction can be carried out with the aid of the cochloid. Now the angle ΓBE is exactly ⅓ of the given angle $AB\Gamma$.

Proof: Bisect ΔE in H and draw AH. Then the line segments $H\Delta$, HA and HE are all three equal to one half of ΔE (because the right triangle $A\Delta E$ can be inscribed in a semicircle) and therefore equal to AB. In the isosceles triangle ABH we have

$$\angle AB\Delta = \angle AH\Delta,$$

and, since $\angle AH\Delta$ is an exterior angle of the triangle AHE which is also isosceles,

$$\angle AH\Delta = 2 \cdot \angle E.$$

Furthermore, since $AE \mathbin{/\mkern-5mu/} B\Gamma$,

$$\angle \Gamma B\Delta = \angle E,$$

so that finally

$$\angle AB\Gamma = \angle \Gamma B\Delta + \angle AB\Delta = \angle E + 2 \cdot \angle E = 3 \cdot \angle E.$$

The underlying idea of the trisection will be seen to be the same as that of the trisection of Archimedes. But Nicomedes only uses a neusis between two straight lines, while Archimedes calls for a neusis between a circle and a straight line.

Pappus shows that the neusis of Nicomedes can also be obtained by the intersection of a circle and a hyperbola. Nicomedes used the intersection of his cochloid with a straight line, every bit as simple, since a cochloid is defined and drawn more easily than a hyperbola.

The duplication of the cube in Nicomedes.

Let AB and $B\Gamma$ be two given line segments, between which we wish to construct two mean proportionals. Complete the parallelogram $AB\Gamma\Delta$. Bisect AB in Λ, and $B\Gamma$ in E. Extend $\Delta\Lambda$ until it meets the extension of ΓB. (This is the way Pappus has

---

[1] At least according to Proclus, *Comm. in Eucl.* p. 272. But it looks as if Pappus claims the honor of this construction for himself.

it; the construction would be much simplified by making $BH = B\Gamma$, and then drawing $\varDelta H$, which will then bisect AB). Draw $EZ \perp B\Gamma$ and determine Z so as to make $\Gamma Z = A\varDelta$ (Again simpler: determine Z so that BZ and ΓZ are both equal to $B\varLambda_1$. Draw $\Gamma\Theta \mathbin{//} HZ$. Now draw a line $Z\Theta K$ through Z to the extension of $B\Gamma$ in such a way that the segment ΘK will be equal to $A\varLambda$, or to ΓZ. This "neusis-construction" can be carried out by intersecting the line ΓK with a cochloid, which has Z as "pole", $\Gamma\Theta$ as "ruler" and ΓZ as "distance".

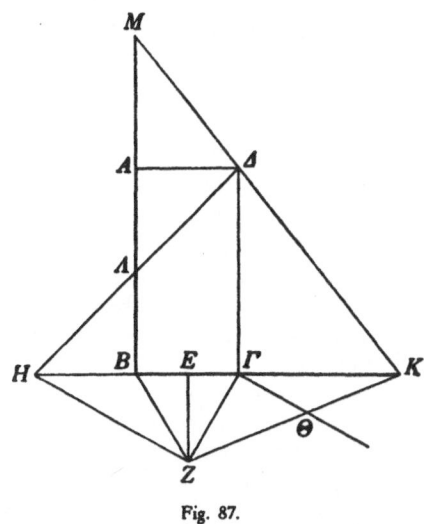

Fig. 87.

Finally, extend $K\varDelta$ and BA to their point of intersection M. Then ΓK and AM are the two required mean proportionals between AB and $B\Gamma$.

Strictly speaking, the line $K\varDelta M$ is superfluous. From the proportions

$$AM : 2 \cdot A\varLambda = AM : AB = \varDelta M : \varDelta K = \Gamma B : \Gamma K, \text{ and}$$
$$\Theta Z : A\varLambda = \Theta Z : \Theta K = \Gamma H : \Gamma K = 2 \cdot \Gamma B : \Gamma K,$$

it follows at once that $AM = \Theta Z$, so that AM can be replaced by ΘZ and we can say: ΓK and ΘZ are the two required mean proportionals between AB and $B\Gamma$. The proof requires only a consideration of the part of the figure below HK.

Let the reader try to find the proof for himself. The more he worries over it, the more he will admire the ingenuity of Nicomedes, who discovered not only the proof but the construction itself. A single hint: determine KZ from the Pythagorean theorem and make use of the second proportion derived above.

If you do not succeed, you can look up the proof in Heath I, p. 261, or in the third book of Pappus, translated by Ver Eecke (I, p. 43).

We come now to the last really great geometer of antiquity:

Apollonius of Perga.

Apollonius' period of florescence falls around 210 B.C., under Ptolemy Philopator. As a young man, he came to Alexandria, and there he learned mathematics from the pupils of Euclid. At the museum he had the nickname Epsilon, as it is said, because he established a theory of the moon and the crescent of the moon has the shape of an ε. It is of little importance whether this story is genuine or not, but it is certain that he was a great astronomer as well as a great mathematician.

Our first witness to this fact is the astrologer Vettius Valens, who informs us that he used tables of Apollonius to determine the position of sun and moon at the time of eclipses. It is not certain whether he referred here to Apollonius of Perga or to another Apollonius; neither do we know how these tables for the sun and moon were arranged, nor the theory on which they were based. Let us therefore rather go to more reliable sources to learn something about the astronomy of Apollonius.

The theory of the epicycle and of the excenter

was the foundation of the theoretical astronomy of the Alexandrian mathematicians, and it continued as such until Claudius Ptolemy, who crowned the work of the astronomers of antiquity with his masterly "Almagest".

In the theory of the epicycle, each planet describes a small circle, called *epicycle*, whose center travels at the same time on a larger circle about the observer. In the theory of the excenter, the roles are reversed; the planet traverses a large circle, the *excenter*, whose center describes a small circle about the observer.

Ptolemy states that Apollonius proved two important propositions which show us how to determine the points on the planetary orbit, at which the direct motion, as seen from the earth, changes to a retrograde motion, or vice versa. The first of these propositions deals with the epicycle hypothesis; it is as follows:

If a straight line ZHB is drawn from our eye Z, which cuts the epicycle in such a way that one half of the segment BH interior to the epicycle has the same ratio to the segment ZH from the eye to the proximate point of intersection, as the angular velocity of the epicycle has to the angular velocity of the planet in the epicycle, then the point H lies precisely at the boundary between the direct and the retrograde motion, i.e., arrived at H, the planet will appear to be standing still. Here it is supposed that both motions, that of the epicycle and that of the planet in the epicycle have the same sense of rotation.

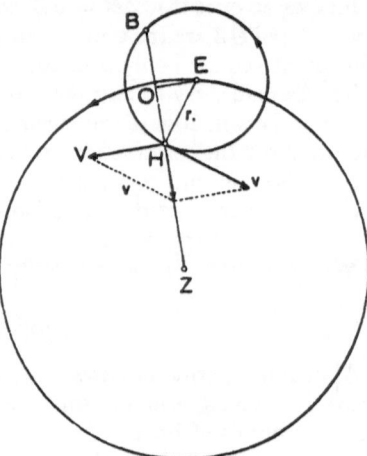

Fig. 88.

The second proposition, analogous to the first, is concerned with the excenter hypothesis.

If we had to prove such a proposition, we should argue about as follows:

Let r be the radius of the epicycle, Ω the angular velocity of the center E of the epicycle, and ω the angular velocity of the planet H on the epicycle. Then the linear velocity of the planet on the epicycle is

$$v = r\omega.$$

In the same way, we find for the linear velocity of H, resulting from the rotation

of the entire epicycle about Z

$$V = ZH \cdot \Omega.$$

As viewed from Z, the planet will then appear to be standing still, when the resultant of these two velocities passes through Z. The parallelogram of the velocities will then consist of two right triangles. If the perpendicular EO is dropped from E to ZH, then these right triangles are similar to the triangle EOH. This gives the proportion

$$V : v = HO : r,$$
$$V \cdot r = HO \cdot v,$$
$$ZH \cdot \Omega r = HO \cdot \omega r,$$
$$ZH : HO = \omega : \Omega.$$

This is exactly the result of Apollonius.

This reasoning is of course entirely non-classical; the composition and resolution of velocities do not occur anywhere in classical writings. Nevertheless I consider Apollonius capable of having obtained his conclusion by an argument of this kind. In the Mechanics of Heron, the composition of uniform rectilinear motions does occur; the parallelogram of velocities was therefore not entirely foreign to the Greeks.

I think that Apollonius did not give a simple deduction of this type, not because he was not capable of it, but because he wanted more. He wanted not merely to determine the instant at which the apparent angular velocity is zero, but he wanted to show incontrovertibly that before this instant the motion is direct, and retrograde after.

According to Ptolemy, Apollonius based his proof on the following lemma: If in triangle $AB\Gamma$, in which $B\Gamma > A\Gamma$, a segment $\Gamma\varDelta$ is laid off on ΓB, greater than or equal to ΓA, then

$$\Gamma\varDelta : \varDelta B > \angle B : \angle \Gamma.$$

Proof: Complete the parallelogram $A\varDelta\Gamma E$ and extend BA and ΓE to their point of intersection Z. Consider first the case $\Gamma\varDelta = \Gamma A$. Then, a circle of radius AE about A as a center will pass through Γ. Now

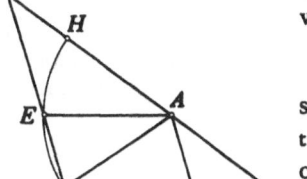

Fig. 89.

triangle $AEZ >$ sector AEH,
triangle $AE\Gamma <$ sector $AE\Gamma$,

so that

triangle AEZ : triangle $AE\Gamma >$ sector AEH : sector $AE\Gamma$.

or

$$ZE : E\Gamma > \angle EAZ : \angle EA\Gamma.$$

But

$$ZE : E\Gamma = ZA : AB = \Gamma\varDelta : \varDelta B.$$

and hence

$$\Gamma\varDelta : \varDelta B > \angle B : \angle \Gamma.$$

It is obvious that the ratio is increased further if $\Gamma\Delta$ is greater than $A\Gamma$, instead of being equal to it.

Next comes the proof of the proposition. The diameter ZE intersects the epicycle in the "perigee" Γ, and its extension meets it in the "apogee" A. Suppose now that ZHB has been determined so that

$$\tfrac{1}{2}BH : ZH = \text{angular velocity of the epicycle : angular velocity of the planet}$$

as stated in the proposition. Lay off from H an arbitrary arc HK towards the apogee A, or towards the perigee Γ. We have then to show that, in the first case the motion of the planet in this arc, as seen from Z, is direct, and in the second case retrograde.

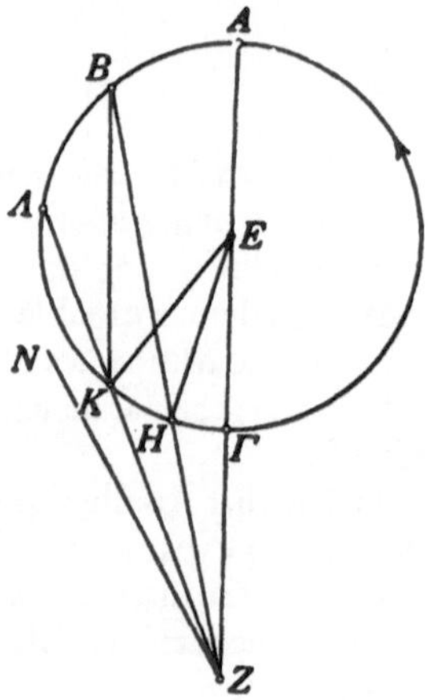

Suppose that the arc HK is directed towards A. Draw $ZK\Lambda$, BK, EK and EH. In triangle BKZ, we have

$$BH > BK,$$

and hence, in accordance with the lemma

$$BH : HZ > \angle HZK : \angle B,$$
$$\tfrac{1}{2}BH : HZ > \angle HZK : 2\angle B.$$

But, twice the inscribed angle B is equal to the corresponding central angle HEK; therefore

$$\tfrac{1}{2}BH : HZ > \angle HZK : \angle HEK.$$

On the other hand, the ratio of $\tfrac{1}{2}BH$ to HZ is equal to that of the angular velocities of the epicycle and of the planet; therefore the ratio of these angular velocities is greater than $\angle HZK : \angle HEK$.

It follows that the angle, which has to $\angle HEK$ a ratio equal to that of the angular velocities, is greater than $\angle HZK$; let this angle be $\angle HZN$.

During the time in which the planet describes the arc KH of the epicycle, the center of the epicycle has travelled, in direct motion, through a distance whose angular measure is equal to $\angle HZN$. Therefore motion through the arc KH of the epicycle has moved the planet backwards by a smaller amount than the amount by which the motion of the epicycle through the angle HZN has moved it forward. Thus the resultant motion of the planet is a forward one.

Is not this proof a masterpiece of classical reasoning?

The proof that the apparent motion of the planet along the arc HK is retrograde, when this arc is laid off towards the perigee, is carried through in similar manner.

Fig. 90.

Conica

i.e. *conic sections*, is the title of Apollonius' most famous work. This masterpiece has rightly aroused the utmost admiration among all mathematicians of antiquity and of modern times.[1]

A complete discussion of the Conics is out of the question here. It would fill an entire book; but this book is already in existence, it is the excellent work of Zeuthen, Die Lehre von den Kegelschnitten im Altertum, supplemented in a few important points by the "Apollonius-Studien" of Neugebauer.[2] I shall therefore restrict myself to giving an impression of the structure and the style of Apollonius' work, and to sketching in broad outline what was known about conic sections before him and what he has added himself.

The conic sections before Apollonius.

Menaechmus already knew the parabola and the equilateral hyperbola; indeed he used their intersection for the duplication of the cube. About 300 the theory of conic sections was developed far enough to enable Euclid to write a textbook on this subject. These "elements of conic sections" have been lost, but we can get an idea of the work because Archimedes frequently quotes propositions from it. The interested reader is referred to Dijksterhuis, Archimedes I, p. 51—100 for further details.

A short time before Euclid, Aristaeus wrote a book on "spatial loci", i.e. on conics as geometrical loci. And Archimedes proves incidentally some propositions on conic sections, which he needed in his determinations of areas and volumes.

Thus Apollonius had important precursors, but, as he says himself, he put the theory on a more complete and more general foundation than his predecessors. In what does this generalization consist?

Archimedes and the more ancient writers already represent the conics systematically by means of "symptoms", i.e. by means of equations, usually referred to rectangular coordinate axes, but occasionally also to oblique axes. In Menaechmus we already met with the equations of the parabola

(1) $$y^2 = px$$

and of the equilateral hyperbola

(2) $$xy = A.$$

Archimedes always gives the equations of ellipse and hyperbola in the "two-abscissas form", which is as follows. Let $AB = a$ be the major axis of the conic. The perpendicular $PQ = y$ from a point on the conic to AB, is called the "ordinate", and the distances $AQ = x$ and $BQ = x_1$ are called the "abscissas". In the case of the ellipse we have therefore $x_1 = a - x$, for the hyperbola $x_1 = a + x$. The "symptom" of the curve, the condition to be satisfied by every point

[1] English translation: Heath, Apollonius of Perga, Treatise on Conic Sections (1896). French translation: Ver Eecke, Les Coniques d'Apollonius (1924).
[2] O. Neugebauer, Studien zur antiken Algebra II, Quellen und Studien B 2, p. 215.

P of the curve, is then in both cases:

(3) $$y^2 : xx_1 = \alpha.$$

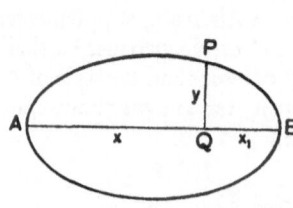

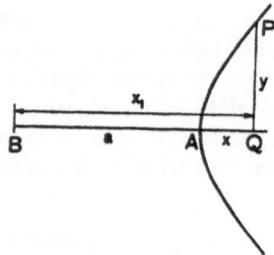

Fig. 91.   Fig. 92.

were α is a given ratio. For the circle $\alpha = 1$, and therefore $y^2 = xx_1$. If $\bar{x}$, $\bar{x}_1$ and y are the abscissas and the ordinate of another point on the curve, then (3) can be replaced by

(4) $$y^2 : xx_1 = \bar{y}^2 : \bar{x}\bar{x}_1.$$

It is in this form that Archimedes always writes the symptom. He chooses the second point $(\bar{x}, \bar{y})$ in various ways, depending upon the problem with which he is concerned. In I 21, Apollonius gives the symptom in the same form (4) and in I 20 the analogous form for the parabola:

$$y^2 : \bar{y}^2 = x : \bar{x}$$

In the discussion of the Data, we were quite naturally led to introduce a new symbol αB for a magnitude which has a given ratio α to B. When α is given as the ratio $c : d$, we can also write, instead of αB

$$(c : d)B.$$

In this very convenient notation, which makes it possible to put the elaborate formulations of the ancients in a very compact form, we can write instead of (3)

(5) $$y^2 = \alpha xx_1$$

and thus, for ellipse or hyperbola

$$y^2 = \alpha x(a - x) \quad \text{or} \quad y^2 = \alpha x(x + a).$$

This already comes very close to modern analytic geometry. But how were the symptoms (3), (4) and (5) derived from the definitions of the conics?

We have already pointed out that, before Archimedes, each of the conic sections was obtained from one type of cone of revolution, by cutting it with a plane perpendicular to a generator. This is the reason why the parabola was called "section of the rectangular cone", the hyperbola "section of an obtuse-angled cone", and ellipse "section of an acute-angled cone".

The ellipse as section of a cone according to Archimedes.

In Archimedes [1] we find a proof that every ellipse can be considered as a section of a circular cone, whose vertex may even be chosen arbitrarily in a plane of symmetry of the ellipse. Archimedes starts from the symptom of the ellipse in the form (5) and proves that the curve represented by this symptom does indeed lie on a right or on an oblique circular cone. But exactly the same proof can be used to derive the symptom if it has not yet been obtained. Let us therefore look a little more closely at this proof.

Archimedes first considers the case in which the vertex of the cone lies on a perpendicular to the plane of the ellipse and through its center (proposition 7) and then he reduces the general case to this special case (proposition 8). The method of proof is exactly the same in both cases, but the proof of proposition 8 is more concise. It proceeds as follows:

> Let AB be a principal axis of the given ellipse, Δ the center, N one half of the other axis, and let the point Γ lie in a plane through AB perpendicular to the plane of the ellipse. We have to find then a cone with vertex Γ, on which the ellipse lies.
>
> Make $\Gamma E = \Gamma B$ and draw through Δ a line ZH parallel to EB. Construct now in the plane through EB, perpendicular to the plane $AB\Gamma$, a circle, in case N^2 is equal to $\Delta Z \cdot \Delta H$,

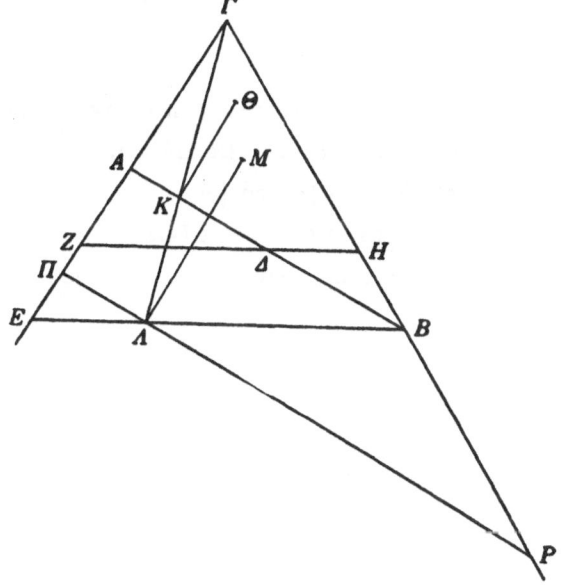

Fig. 93.

[1] *On Conoids and Spheroids*, propositions 7, 8.

but otherwise an ellipse of which EB is one principal axis, while the square on the other, Σ, is determined by

(6) $N^2 : \Delta Z \cdot \Delta H = EB^2 : \Sigma^2$.

From proposition 7, already proved, it follows that this circle or ellipse will lie on a right or on an oblique cone with vertex Γ. We have then to show that the ellipse, of which AB is a principal axis, also lies on the surface of this cone.

If this were not the case, there would be a point Θ on the ellipse, which does not lie on this conical surface. From Θ drop the perpendicular ΘK to AB. Extend ΓK to its intersection Λ with EB. Draw ΛM, perpendicular to BE (the intention here is, perpendicular to the plane $B\Gamma E$), so that M lies on the surface of the cone. Finally draw ΠP through Λ parallel to AB.

Then it follows from (6) that

(7) $N^2 : \Delta Z \cdot \Delta H = \Lambda M^2 : E\Lambda \cdot \Lambda B$.

Here Archimedes uses the symptom of the ellipse on EB, viz. the ratio $\Lambda M^2 : E\Lambda \cdot \Lambda B$ is the same for every point on the ellipse and therefore equal to the ratio of the squares of the semi-axes,

Now he uses a lemma from plane geometry, which he obviously supposes to be known and which may be formulated as follows: *If through a given point Δ two lines, such as AB and ZH, are drawn to the sides of an angle $A\Gamma B$, and through another point Λ two lines, such as ΠP and EB, parallel to the first, then the ratio of the products of the intercepts is the same for Δ as for Λ:*

(8) $\Delta Z \cdot \Delta H : \Delta A \cdot \Delta B = \Lambda E \cdot \Lambda B : \Lambda \Pi \cdot \Lambda P$.

The lemma is readily proved by multiplication of the two proportions

$\Delta Z : \Delta A = \Lambda E : \Lambda \Pi$, and $\Delta H : \Delta B = \Lambda B : \Lambda P$.

By composition we obtain from (7) and (8)

$N^2 : \Delta A \cdot \Delta B = \Lambda M^2 : \Lambda \Pi \cdot \Lambda P$.

The symptom of the given conic allows us to replace the left member by $\Theta K^2 : AK \cdot BK$, so that

(9) $\Lambda M^2 : \Lambda \Pi \cdot \Lambda P = \Theta K^2 : AK \cdot BK$.

Moreover

(10) $\Lambda \Pi \cdot \Lambda P : \Gamma \Lambda^2 = AK \cdot BK : \Gamma K^2$.

From (9) and (10) we get, again by composition

$\Lambda M^2 : \Gamma \Lambda^2 = \Theta K^2 : \Gamma K^2$,

and hence

$M\Lambda : \Gamma \Lambda = \Theta K : \Gamma K$,

which leads to the conclusion that Γ, Θ and M are collinear, and therefore that Θ is on the surface of the cone.

How were the symptoms derived originally?

It is not much of a trick to reverse the order of the steps in the above proof, and thus to derive the symptom of the conic section from the assumption that Γ, Θ, M are collinear. It is therefore possible to obtain the symptom of the curve $A\Theta B$ from that of the conic EMB by means of simple proportions from plane geometry.

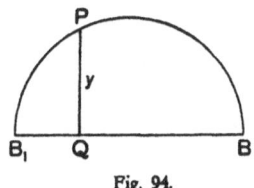

Fig. 94.

The proof becomes still simpler, when EMB is a circle, whose symptom is

$$y^2 = xx_1.$$

For in that case, one obtains, instead of (6)

$$N^2 = \Delta Z \cdot \Delta H \quad \text{and} \quad EB = \Sigma,$$

and, in place of (7)

$$\Lambda M^2 = E\Lambda \cdot \Lambda B.$$

The cone is then a right circular cone. And, this was exactly the case which Menaechmus and Euclid needed for the derivation of the symptom of their "section of the acute-angled cone". One can imagine, that these "ancients" had already used such a method of proof and that Archimedes only generalized and inverted a well-known proof. This would also account for the conciseness of his statement.

It is easy to devise analogous proofs for the parabola and the hyperbola. Such proofs have been worked out by Dijksterhuis, in Archimedes I, p. 53—54.

The supposition that Menaechmus actually proceeded in some such manner is made still more plausible by the fact that similar proofs are found in Apollonius. But before passing on to this matter, we have to answer another question.

A question and an answer.

Why did "the ancients", Menaechmus first of all, restrict themselves to sections made by planes perpendicular to a generator of the cone? Was the general case perhaps too difficult for them? This can hardly be supposed.

Rather do I look for the explanation, with Zeuthen, in the fact that they were chiefly concerned with proving that a curve, whose equation is given in the forms (1) or (5), can always be obtained as a conic, i.e. as a section of a cone. For it is exactly this proof which is particularly simple when the plane of section is perpendicular to a generator. There is an infinite number of cones of revolution, which are cut by a given plane in a given conic, but of all these cones, the two whose vertex lies in a perpendicular to the plane of this section at one of its vertices, are by far the easiest to construct.

The further development of this line of thought is found in Chapter 21 of Zeuthen's book.

The derivation of the symptoms according to Apollonius, [1]

is entirely analogous to the one which we dug out of Archimedes, only simpler and more general.

The axis of an oblique circular cone is for Apollonius the line which joins the vertex T with the center of the base. Suppose now that an arbitrary plane cuts the base in the line EZ; then he draws in the base a diameter perpendicular to EZ. If the plane $TT\Delta$ is taken as the plane of drawing, the plane of the conic will intersect this plane in a line AB. This line is taken as the X-axis. The ordinate $K\Theta$ of an arbitrary point of the conic is drawn parallel to EZ. This line will then lie in the plane of the conic AB and also in the plane of the circle ΠKP, through K, parallel to the circle of the base. In the plane of this circle, ΘK is perpendicular to the diameter $P\Pi$, as a result of the construction; hence the equation of the circle is again

$$\Theta K^2 = \Theta\Pi \cdot \Theta P.$$

Now Apollonius draws $T\Lambda \ // \ AB$, and he multiplies the proportions

$$\Theta\Pi : \Theta B = \Lambda\Gamma : \Lambda T,$$

and

$$\Theta P : \Theta A = \Lambda\Delta : \Lambda T;$$

thus he obtains

$$\Theta\Pi \cdot \Theta P : \Theta B \cdot \Theta A = \Lambda\Gamma \cdot \Lambda\Delta : \Lambda T^2.$$

This is a special case of the very same lemma which we have already met in

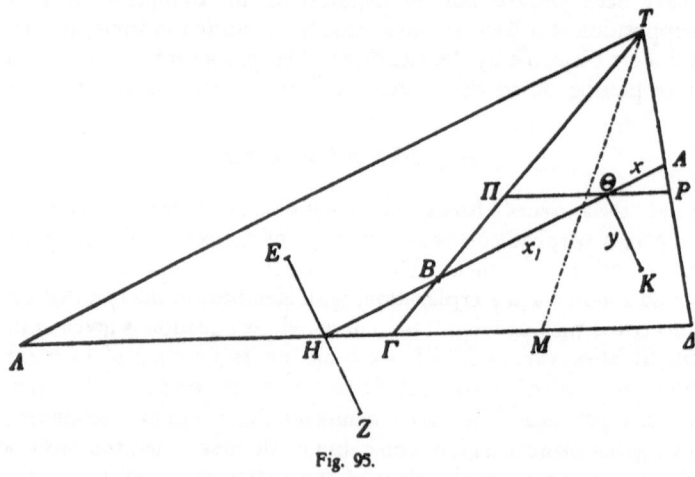

Fig. 95.

Archimedes. If we replace $\Theta\Pi \cdot \Theta P$ by ΘK^2 and if we designate the ratio in the right member by α, we find as the symptom of the ellipse

[1] *Conica* I, 11–13.

$$K\Theta^2 = \alpha \cdot (\Theta A \cdot \Theta B),$$

or, in our earlier notation

(11) $$y^2 = \alpha x x_1.$$

For the parabola, Apollonius finds in a similar manner

(12) $$y^2 = px.$$

The equations are exactly the same as the earlier ones, but the difference lies here, that the ordinate $K\Theta$ in the plane of the conic, which we shall henceforth designate by $PQ = y$, is no longer perpendicular to the diameter AB, from now on called AA_1. *The new element in Apollonius is therefore that the cone and the plane of the conic no longer have a common plane of symmetry, so that the x- and y- directions are no longer mutually perpendicular.*

But there is still more. By the use of geometric algebra, Apollonius reduces equations (11) and (12) to a new and very useful form. Equation (12) simply states that y^2 is equal to a rectangle of base x and constant height p. In both other cases, y^2 is also equal to a rectangle of base x and height αx_1 but this altitude is

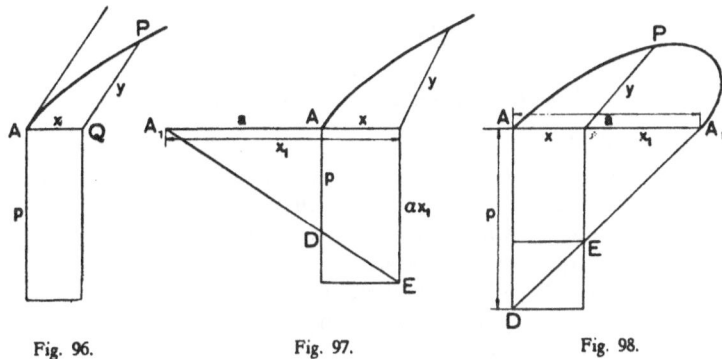

Fig. 96.　　　　Fig. 97.　　　　Fig. 98.

now no longer constant. For the hyperbola (Fig. 97) $x_1 = a + x$, and therefore $\alpha x_1 = \alpha a + \alpha x$; the rectangle is therefore "applied" to a constant line segment $p = \alpha a$, in such a way that there is a rectangle in "excess", whose base x and height αx have a constant ratio $\alpha = p : a$. In the ellipse (Fig. 98), we have $x_1 = a - x$, so that $\alpha x_1 = \alpha a - \alpha x$; the rectangle is therefore "applied" to the line segment $p = a\alpha$ in such a manner that there is a rectangle in "defect" whose base and height have again the constant ratio $\alpha = p : a$.

As we know, the terminology of such applications of areas comes from geometric algebra. The names parabola, hyperbola and ellipse have their origin the Greek words for "application", "excess" and "defect".

The diagrams of Apollonius, here reproduced, consist of two unequal parts; one could speak of a geometric and an algebraic diagram. The geometric diagram consists of the conic section, in which we are actually interested, the oblique axes, the abscissa x and the ordinate y. The algebraic diagram is not oblique, but rectangular; it expresses the algebraic relation between a, p, x and y. The areas which are here added and subtracted correspond to the terms of an equation in modern analytical geometry, and Apollonius proves geometrically all the algebraic transformations performed on the equation. The line of thought is mostly purely algebraic and much more "modern" than the abstract geometric formulation would lead one to think. Apollonius is a virtuoso in dealing with geometric algebra, and also a virtuoso in hiding his original line of thought. This is what makes his work hard to understand; his reasoning is elegant and crystal clear, but one has to guess at what led him to reason in this way, rather than in some other way.

The relation between the areas of the applied rectangles and the square y^2, just stated in geometric language, can be expressed as follows in algebraical formulas:

(13) $$y^2 = x\,\alpha x_1 = x \cdot (a\alpha \pm \alpha x) = x(p \pm \alpha x) = x\{p \pm (p : a)x\},$$

in which the $+$ sign applies to the hyperbola and the $-$ sign to the ellipse. The magnitudes a and p, which occur in these equations, are sometimes called *latus transversum* and *latus rectum*. The proportion $p : a = \alpha x_1 : x_1$ can also be expressed geometrically by drawing a straight line A_1DE.

Conjugate diameters and conjugate hyperbolas.

Apollonius calls the midpoint of AA_1 the *center of the conic section* and the line through the center in the direction of the ordinate, the *diameter conjugate to* AA_1.

In the ellipse the conjugate diameter cuts the curve in two points D and D_1, whose ordinates are obtained immediately from equation (11) by setting $x = x_1 = \frac{1}{2}a$. The length $\bar{a} = DD_1$ of the conjugate diameter is obviously the mean proportional between a and p,

(14) $$\bar{a}^2 = \alpha a^2 = ap.$$

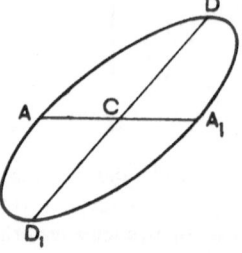

Fig. 99.

Apollonius proves (I, 15) that the two conjugate diameters are interchangeable, i.e. that the ellipse which has DD_1 as the direction of the abscissas, and CA_1 as that of the ordinates, is given by a symptom of exactly the same form as the original symptom (13), the new latus transversum $\bar{a}$ being defined by (14) and the new latus rectum $\bar{p}$ by the proportion

(15) $$\bar{a} : a = a : \bar{p}, \quad \text{i.e. by } a^2 = \bar{a}\bar{p}.$$

For Apollonius, the word "hyperbola" means what we call a branch of a hyperbola. The two branches which we consider as forming together one hyperbola, he calls "opposite hyperbolas". This concept is not original with Apollonius, for he says in the preface to Book IV that, in his pamphlet against Conon, Nicoteles has already stated (without proof), in how many points at most a circle or a conic section can meet a pair of opposite sections. What probably is new, is the introduction of the concept of *pairs of conjugate hyperbolas*. These are defined as follows at the end of the first book:

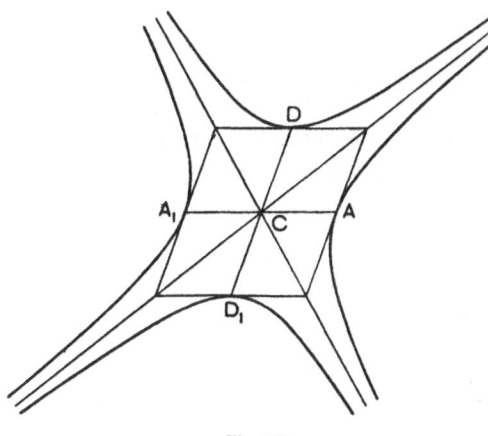

Fig. 100.

Apollonius starts with a pair of "opposite hyperbolas" and draws through the center C a line DD_1 in the direction of the ordinates. On this line equal distances CD and CD_1 are laid off on either side of C, such that $DD_1 = \bar{a}$ is, in accordance with (14), a mean proportional between a and p. Next $\bar{p}$ is determined by (15), so that a is a mean proportional between $\bar{a}$ and $\bar{p}$. Then two new "opposite hyperbolas" are constructed on the diameter DD_1, the ordinates having the direction of CA, the latus rectum being equal to $\bar{a}$ and the latus transversum to $\bar{p}$; these are the conjugate hyperbolas. It is proved in Book II that they have the same asymptotes. By introduction of the conjugate hyperbolas, Apollonius succeeds in proving all the propositions concerning conjugate diameters etc. for hyperbolas as elegantly and as simply as for ellipses.

Tangent lines.

A superficial reading of the first book of Apollonius' Conica may well create the impression that the author does not follow a definite method, but "writes down whatever happens to come to his mind", as Descartes expresses it. [1] Zeuthen has shown however that Apollonius did have a definite purpose in mind, realized at the end of the first book, and that he included precisely those propositions which were needed for the accomplishment of this purpose. This aim was the following: to prove that the sections which he had initially defined by means of an arbitrary oblique circular cone and which he had characterized by a "symptom", can also be obtained as sections of right circular cones and were therefore identical with the types that were known of old.

For this purpose he has to show first that the diameter AA_1 with the corresponding direction of the ordinates, which appeared naturally in the space-derivation of

[1] Descartes, *Géométrie* (ed. van Schooten), p. 7.

the symptom, can be replaced by any other diameter PP_1, the corresponding direction of the ordinates always being that of the tangent line at P. He has to begin therefore by determining the tangent line at an arbitrary point of the curve.

Apollonius conceives of a tangent line as a line which has one point in common with the conic section, but which lies outside it everywhere else. In I 17, it is shown to begin with, that the line through A in the direction of the ordinates, is a tangent line in that sense. Then follow a number of simple propositions about lines which intersect a conic. In I 32, we find a repetition of the statement that the line through A, in the direction of the ordinates, is tangent to the curve, but the statement is sharpened by the additional remark that between this line AC and the curve no other line is possible. He proves this by showing that every other line AD meets the conic in a second point H.

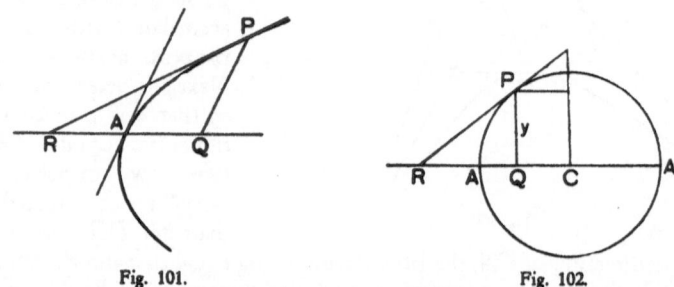

Fig. 101.   Fig. 102.

Now he constructs the tangent line at an arbitrary point P. For the parabola, the construction is very simple: The ordinate PQ is drawn through P, QA is extended by an amount AR equal to QA; then PR is tangent at P (Proposition 33).

For the hyperbola, the ellipse and the circle, R is determined by use of the proportion

(16) $$RA : RA_1 = QA : QA_1$$

(I 34). At a later time, this relation among 4 points received the name "harmonic position".

In I 35, 36, it is proved that the tangent line at P, constructed in this way, is the only one, and that no other line through P is possible between the conic and this tangent line.

In I 37, the proportion (16) is transformed into a relation between segments, issuing from the center C:

$$CQ \cdot CR = CA^2.$$

Moreover a relation is indicated by means of which the position of the point of contact P can be determined when R, and hence Q, are given:

(17) $$CQ \cdot CR : y^2 = a : p.$$

Similar relations hold for the conjugate diameter (I 38).

The equation referred to the center.

The two-abscissas-form of the equation of the ellipse and the hyperbola was found to be

$$y^2 : xx_1 = p : a,$$

or, equivalently,

(18) $$xx_1 = (a : p)y^2.$$

When the origin of coordinates is translated to the center C, i.e. if we take $x_0 =$

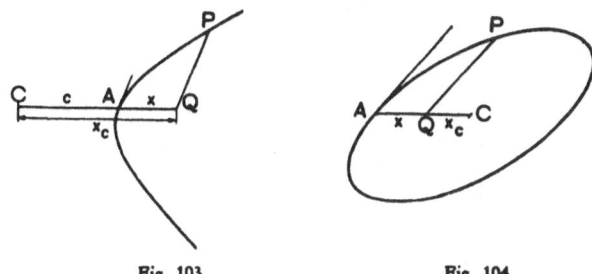

Fig. 103.   Fig. 104.

CQ as the new abscissa, and if we set $AC = \tfrac{1}{2}a = c$, then we obtain for the hyperbola

$$xx_1 = (x_0 - c)(x_0 + c) = x_0^2 - c^2.$$

and for the ellipse or the circle

$$xx_1 = (c - x_0)(c + x_0) = c^2 - x_0^2.$$

so that equation (13) now becomes

(19) $$x_0^2 - c^2 = (a : p)y^2 \quad \text{(hyperbola)},$$
(20) $$c^2 - x_0^2 = (a : p)y^2 \quad \text{(ellipse and circle)}.$$

This "central equation" is formulated in a somewhat different form by Apollonius in Proposition 41. First he replaces the squares in the left members by two similar rectangles $c \cdot \beta c$ and $x_0 \cdot \beta x_0$ in which β is an arbitrary ratio. Then, γ representing the composite ratio of $a : p$ and β, (19) becomes

$$x_0 \cdot \beta x_0 - c \cdot \beta c = y \cdot \gamma y, \quad \text{or}$$
(21) $$x_0 \cdot \beta x_0 - y \cdot \gamma y = c \cdot \beta c.$$

In the same way, one finds for the ellipse or the circle

(22) $$x_0 \cdot \beta x_0 + y \cdot \gamma y = c \cdot \beta c.$$

Next the three rectangles occurring in the left and in the right members of (21) and (22) are replaced by three parallelograms, having equal angles, and proportional to these rectangles. Let Π be a parallelogram with sides c and βc, Π_1 a similar parallelogram with sides x_0 and βx_0, and Π_2 a parallelogram with sides y and γy with the same angles as Π and Π_1. Then equations (21) and (22) can be expressed as follows:

(23) $$\Pi_1 - \Pi_2 = \Pi \quad \text{(hyperbola)}$$
(24) $$\Pi_1 + \Pi_2 = \Pi \quad \text{(ellipse or circle)}.$$

In the applications, Apollonius always chooses as the angle of the parallelograms, the angle between the directions of the abscissas and the ordinates. The diagrams have been drawn for this case:

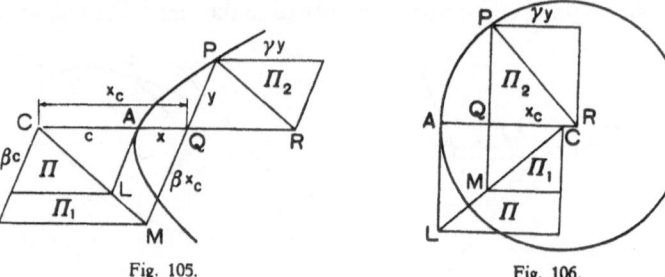

Fig. 105.   Fig. 106.

Bisection of the parallelograms gives a relation between the areas of triangles:

(25) $$\triangle CQM \pm \triangle PQR = \triangle CAL.$$

The ratio β can still be chosen at will. The form (25) of the central equation is the formulation best adapted to the transformation of coordinates, to which Apollonius now proceeds.

The two-tangents theorem and the transformation to new axes.

Apollonius wishes to demonstrate that all diameters are equivalent, i.e. that, starting from an arbitrary point E on the curve, taking as axis of abscissas the diameter EC and as direction of the ordinates that of the tangent line at E, a symptom is found of exactly the same form as that referred to the original diameter and the corresponding direction of the ordinates.

To accomplish this, he transforms the symptom of the conic, by means of area-calculation, into a form in which the two diameters AC and EC, and their corresponding ordinate-directions, appear symmetrically. In its original form, the symptom is an equation involving rectangles and squares. In proposition 41, already discussed, he changes the first to parallelograms, then to triangles, next to a trapezoid and a triangle, and finally (in Book III) to oblique quadrangles, whose sides are parallel to the two diameters and their ordinates.

In all of this, the leading part is played by two propositions, which Neugebauer has called the "two-tangents propositions". The first of these, referring to a parabola, is the following

I 42. *Let a parabola be given through the origin A with diameter AD, and draw, at an arbitrary point E of the parabola, the tangent line ED and the ordinate EZ. Through an*

PLATE 25

PL. 25. Battle of Issus (334 B.C.) or of Gaugamela (333 B.C.), showing the victory of Alexander the Great over Darius, king of the Persians. A mosaic (5.92 m by 3.42 m) found in the Casa del Fauno, a large patrician house in Pompeii. National Museum, Naples. Copy from about 100 B.C. after a famous Greek painting probably of Philoxenus of Eretria (end of 4th century B.C.). The mosaic is in subdued brownish tints, probably in imitation of the "four-colors technique" of the original. In every other respect this work must have been a "novum" in the art world. The style is very realistic, almost naturalistic. Although the means of expression appear to us as very sober (the representation of the background is restricted to a single feature, the bare tree), space is represented very strongly. Everywhere, especially in the representation of the horses, there is foreshortening. Only the principal figure, that of Alexander, is seen in its full width. As a result of this foreshortening, which is worked into the close crowding of the conflict, the tumult of battle is shown very suggestively. The almost cruel rendering of the pitiless battle is realistic, the exact representation of the armour, clothing and equipment, of the incidence and reflection of light are naturalistic. The face of the Persian falling backwards under Darius' battle-wagon, is reflected in his shield. We realize that this work of art originated in a period when applied mathematics and experimental natural science were being developed vigorously. Mechanical and acoustic problems were discussed in the school of Aristotle, as seen from the collection of problems ascribed to Aristotle. Strato of Lampsacus made physical experiments, Timocharis and Aristyllus made observations on the moon and the stars (Ptolemy, Almagest 7, Chapter 3). Euclid wrote an Optica and a Catoptrica which discussed perspective and the theory of optical images (see p. 200). These optical theories and the tendency towards careful observation of nature are reflected in this painting.

(Photo Alinari)

PLATE 26

PL. 26a. Portrait of Ptolemy II Philadelphus, (283–247 B.C.) and his wife
Arsinoe. Cameo (sardonyx) in the
Kunsthistorisches Museum in Vienna
(early Hellenistic). A cameo achieves its
peculiar effect, through the stone, consisting of two or more layers of different colors, being fashioned in such a
manner, that the upper layer or layers
are preserved for the representation,
while the rest is cut away down to
the darker layer which serves as a
background.

PL. 26b.
Hiero II of Syracuse
269/8–214 B.C. Silver
Syracuse coin. (British
Museum).

PL. 26c. Idealised portrait of Alexander the
Great, marble; herma, with name inscribed,
from Tivoli, Louvre, Paris. This, as well as
numerous other portraits of Alexander, derive
from the famous Alexander-portraits of his
court-sculptor Lysippus.

(Photo Alinari)

arbitrary point P on the parabola, draw the lines PR and PQ, parallel to the tangent line

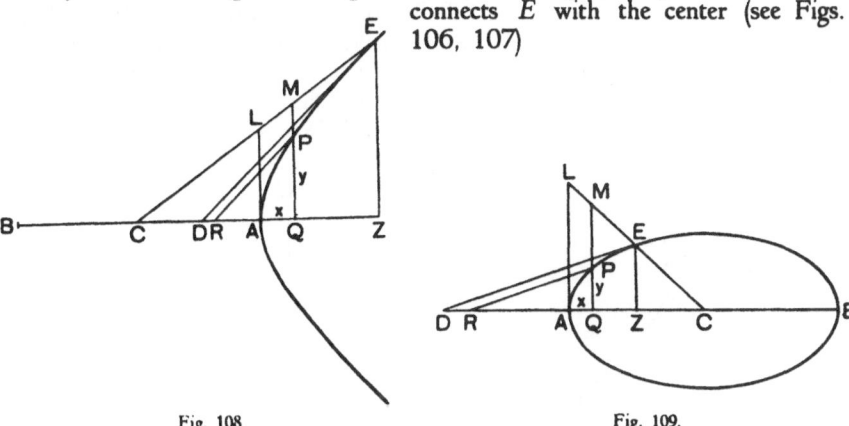

ED and the ordinate EZ, to their intersections with the diameter AD. Complete the parallelogram AZEL. Then we have

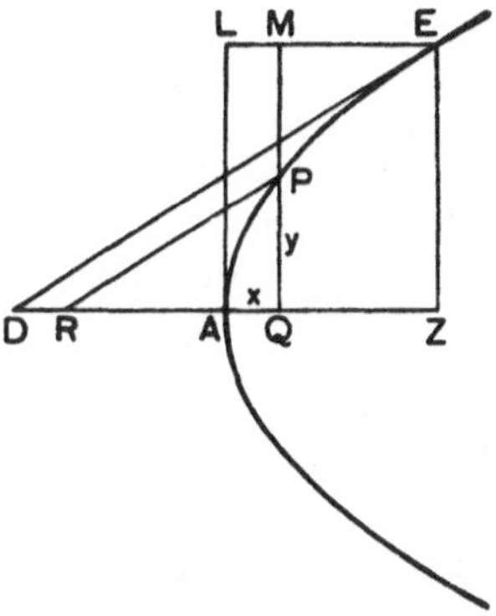

(26)   triangle PQR = parallelogram $ALMQ$.

The proof is very simple.

The similar triangles PQR and EZD are proportional to the squares on their altitudes

$$\triangle PQR : \triangle EZD = PQ^2 : EZ^2 = AQ : QZ;$$

the last equality follows from the symptom of the parabola.

Since $AZ = AD$. $\triangle EZD$ = par. $ALEZ$. so that

$$\triangle PQR : \text{par. } ALEZ = AQ : AZ.$$

But we have also

par. $ALMQ$: par. $ALEZ = AQ : AZ$;

hence $\triangle PQR$ = par. $ALMQ$.[1]

Fig. 107.

Thus we see that I 42 is nothing but a transformation of the equation of the parabola, analogous to the transformation of the central equations of ellipse and hyperbola in I 41. But the formulation is already of such a character, that it contains two diameters, AD and EL, and two tangent lines AL and ED. This prepares for interchanging the roles played by the points A and E.

The two-tangents proposition for the hyperbola, the ellipse and the circle is exactly the same, except that the parallel EL is now replaced by a line EC which connects E with the center (see Figs. 106, 107)

Fig. 108.   Fig. 109.

[1] Following Neugebauer, the line of thought of the proof can also be formulated as follows:

As P varies, $\triangle PQR$ is proportional to $PQ^2 = y^2$ and par. $ALMQ$ to $AQ = x$. But y^2 is proportional to x, therefore $\triangle PQR$ is proportional to par. $ALMQ$. By letting P coincide with E, the factor of proportionality is found to be 1, because $\triangle EZD$ = par. $ALEZ$.

The proposition (I 43) is now, analogous to (26):

(27)   triangle PQR = trapezoid $ALMQ$.

This equality follows at once from (25), because the trapezoid $ALMQ$ equals the difference between the two similar triangles CAL and CQM. It only remains to show that the hypothesis, from which the conclusion (25) follows, is fulfilled, i.e. that the ratio $\gamma = QR : QP$ is obtained by composition from the ratio $\beta = ZE : ZC$ and the ratio $a : p$, of the diameter to the latus rectum. But this is an immediate consequence of a property of the tangent line, which Apollonius had obtained earlier (I 37).

The form (27) does not yet show immediately that the roles of A and E, with their diameters and tangent lines, are interchangeable, But a slight transformation suffices to gain this result. For if the trapezoid $ALMQ$ is represented as the difference between the constant triangle CAL and the variable triangle CQM, then (27) can be replaced by

(28) $\triangle CQM - \triangle PQR = \triangle CAL$ (hyperbola), and
(29) $\triangle CQM + \triangle PQR = \triangle CAL$ (ellipse).

In the cases drawn in our figures, the left member of (28) or (29) is the quadrangle $CMPR$, of which two sides are the diameters CM and CR, while the two other sides are lines through P parallel to the tangent lines at A and E.

In these cases the two-tangent proposition can therefore be formulated very simply as a *proposition on areas*: *the area of the quadrangle CMPR remains constant when P moves along the curve* (see figs. 110 and 111). But there are conditions under which the same formulation of the proposition would require the introduction of the concept "improper quadrangle", as Zeuthen shows. In Fig. 112, for example, the "area $CMPR$" would have to be understood to mean the difference in area of the triangles CMS and PRS.

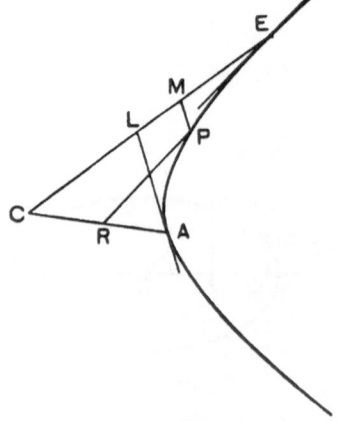

Fig. 110.

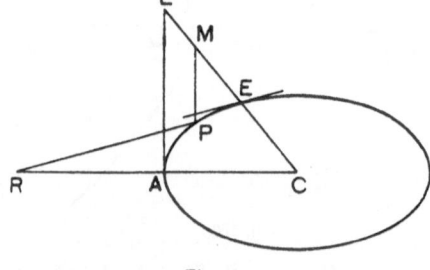

Fig. 111.

Apollonius does not use improper quadrangles; this is the reason why the formulates his area propositions in a somewhat different manner. In III 3, e.g., it is stated that for two arbitrary points P, P', the areas $SPMM'$ and $SP'R'R$ are equal (see Figs. 113—115).

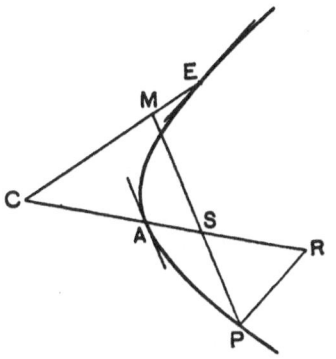

This formulation has the further advantage of being valid also for the parabola, for which a center does not exist. In the two other cases, III 3 is evidently equivalent to Zeuthen's formulation

$$CMPR = CM'P'R'.$$

In the third book, Apollonius treats the area-propositions in a systematic sequence:

III 1: Limiting case $P' = A$, $P = E$.

III 2: Limiting case $P' = A$, P arbitrary.

Fig. 112.   III 3: P and P' both arbitrary.

III 4—12: The same for two opposite branches of a hyperbola.

III 13—15: The same for two conjugate hyperbolas.

This is a clear indication of the great value which he attached to these propositions![1]

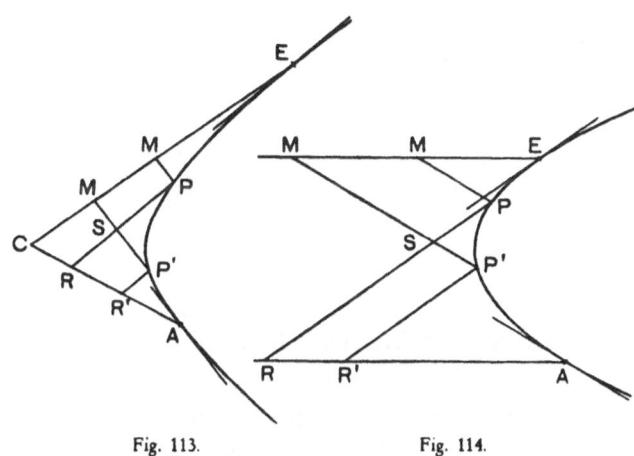

Fig. 113.   Fig. 114.

According to Zeuthen, the significance of the area-propositions lies in this that they can be considered as *symptoms of conic sections referred to two arbitrary diameters as axes of coordinates*. If, e.g., in the case of Fig. 116, one divides the quadrangle

---

[1] The second endpoint of the line $P'M$ in Figs. 113, 114 and 115 should be changed to M'.

CMPR by means of the dotted lines in a parallelogram *CQPT* and the two triangles *PQR* and *PTM*, and if one designates the sides of the parallelogram by
x and y, then it becomes clear that the paral-
lelogram is proportional to xy and the two
triangles to x^2 and y^2. The symptom of the
ellipse then takes the form

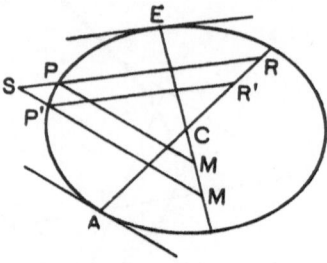

$$\alpha x^2 + \beta xy + \gamma y^2 = \text{constant},$$

with analogous results in all other cases. The
consideration of the conjugate hyperbolas in
III 13—15 is necessary to make possible a
treatment of the cases in which the axes do
not intersect the conic.

Fig. 115.

In the first book, Apollonius treats almost
exclusively those area-propositions which are needed for his immediate purpose,
viz. the interchange of the roles of the points *A* and *E*. In I 46—48, he proves
that the new diameter bisects all the chords parallel to the tangent line at *E*. In
I 49, 50 he demonstrates that the two-tangents proposition is equally valid with
respect to the new diameter as with regard to the original diameter *AC*; after

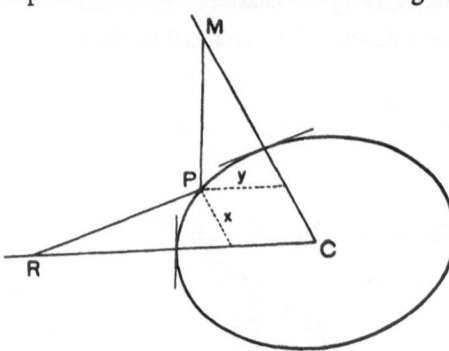

that there is little difficulty in de-
riving the symptom of the curve
referred to the new diameter and
the corresponding direction of the
ordinates.

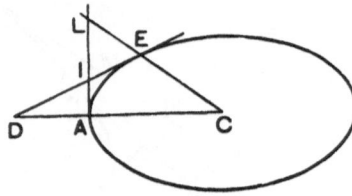

Fig. 116.   Fig. 117.

The new latus rectum p' is determined from

(30) $$p' = (EI : EL) \cdot 2ED,$$

where *I* is the point of intersection of the tangents *AL* and *ED*, while *D* is on the
diameter *AC* and *L* on the diameter *EC*.

Cones of revolution through a given conic.

Apollonius now proposes (I 52, 53) the following problem: to "find" a conic
section when a fixed point *A* is given, a diameter and a direction of the ordinates,
the latus rectum p and (in the case of a hyperbola or an ellipse) the latus transver-

sum a. From the construction it becomes clear that "finding" means to obtain the curve, whose symptom is given, as the section of a cone of revolution.

The constructions are first carried out for the case in which the ordinate direction is perpendicular to the given diameter, i.e. that the given diameter is a *principal*

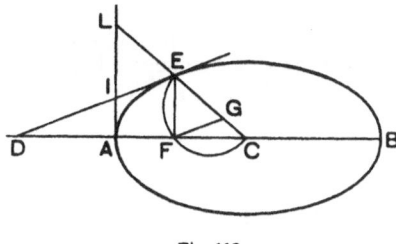

axis. The general case is reduced to this one by first constructing a principal axis. We shall begin here with the latter construction because it connects directly with the figure just drawn and with formula (30). Let E be the given fixed point, EC the given diameter, $a' = 2EC$ the latus transversum and p' the latus rectum. It is required to find a point A such that the tangent AL is perpen-

Fig. 118.

dicular to the diameter AC. Draw the ordinate $EF//LA$; then F must lie on a semicircle of diameter EC. Draw also $FG//DE$. Then we have from (30)

$$p' : 2ED = EI : EL = GF : GE,$$

and also

$$2ED : a' = ED : EC = GF : GC.$$

If now, from the two ratios $p' : 2ED$ and $2ED : a'$, one forms the composite ratio $p' : a'$, we obtain

(31) $$p' : a' = GF^2 : GE \cdot GC.$$

This analysis is not found in Apollonius. Without preamble, Apollonius says: draw, parallel to DE, a line FG between the semicircle and the diameter EC, in such a way that the proportion (31) holds. He does not even say how this is to be done; but it is not hard. [1] Now he draws CFD and determines A on this line in such a way that $CA^2 = CD \cdot CF$; as shown in I 34, this is indeed the characteristic property of the tangent line in E. AC is then the required principal axis; the latus transversum $AB = a$ is found by doubling AC, the latus rectum is determined in such a way that the point E satisfies the symptom of the curve.

Finally, to construct, with a given principal axis AB, a cone of revolution whose intersection with the plane of drawing is the required conic, Apollonius passes through AB a plane perpendicular to the plane of drawing, and constructs in this plane an arbitrary circle through A and B, which must contain the vertex T of the cone. This amounts to choosing the vertical angle of the cone arbitrarily. It only remains to select T on this circle in such a way that the conic will have the desired ratio $p : a$ of the latus rectum to the latus transversum. This is a simple problem of plane geometry, which we shall not enter into here, because our only purpose is to indicate the main structural outline of the first book.

[1] Neugebauer has shown that a general method, which Apollonius applies over and over again to proportions of the form (31), suits the purpose also in this case.

The second book

gives a detailed treatment of the theory of conjugate diameters and principal axes, of asymptotes and of conjugate hyperbolas. Of course, the "asymptotic equation"

$$xy = \text{const.}$$

of the hyperbola also comes up for discussion; furthermore propositions concerning the segments determined on an arbitrary secant by two conjugate hyperbolas and their asymptotes, the construction of tangents through a given point, etc.

Of greater importance is

The third book

We have already discussed the area-propositions III 1—15.

In III 16—23, a *power-proposition* is developed, sometimes called the Proposition of Newton. It is a well-established custom to name propositions after mathematicians, who did not discover them. Newton himself says explicitly that he has taken the proposition from "the ancients". The power-proposition is stated as follows: *If through a variable point Z two lines are drawn in given directions, determining on a conic two chords EK and DT, then the ratio*

$$ZD \cdot ZT : ZE \cdot ZK$$

is constant.

One or both lines may be tangents, exactly as in the well-known case of the circle. Apollonius obtains the power-proposition directly from the area-propositions III 1—3, and then treats a few special cases and applications (III 24—29).

Propositions 30—40 refer to what is nowadays called the *theory of poles and polars*. Apollonius associates, by definition, with every point D, internal to the conic or external, a line l, conjugate to the diameter through D, such that for every line through D, which meets the conic in E and F and the line l in G, the proportion

$$DE : DF = GE : GF$$

is valid. When D is external to the conic, then l is the join of the points of contact of the tangents through D. The special cases which arise when, in contemporary terminology, one of the four points "is at infinity", are dealt with as separate propositions.

III 41—44 are propositions about tangents; relations are indicated which hold between two sets of points, determined on two fixed tangents, or on two asymptotes, by a variable tangent. This topic is closely related to modern projective geometry. These propositions can be, and probably have been, used to construct the tangents through a given point. Propositions 41—43 reduce these constructions to the following problem: to draw a line through a given point, which intercepts, on two given lines, segments with given initial points, such that either the ratio of these segments (III 41: parabola), or their product (III 42, 43: ellipse and hyperbola) have a given value. The first problem, in which the ratio is given, was solved by Apollonius himself in his work *On cutting off a ratio*, the second in his work *On cutting off a product*.

The tangent propositions III 41—44 are applied in *the theory of foci* (III 45—52). It was shown by Zeuthen (Lehre von den Kegelschnitten im Altertum, p. 213) that Euclid already knew the proposition, according to which the locus of points for which the distances to a fixed point and to a fixed line have a given ratio, is a parabola, an hyperbola or an ellipse, according as this ratio is equal to, greater than or less than 1. It would have been an inconceivable denseness not to have observed that every ellipse, hyperbola and parabola can be obtained in this way. The foci and directrices of all three conics were therefore known. [1] However Apollonius treats only the focal properties of the ellipse and the hyperbola. He proves that the lines, which join a point P on the curve to the foci F_1 and F_2, make equal angles with the tangent line at P (III 48), and that their sum or their difference is constant, indeed equal to the major axis (III 51, 52).

The last propositions, 53—56, state the relations between the two pencils of rays, generated when a variable point P on the conic is joined to two fixed points A and B on the same conic. The rays AP are brought to intersection with a line through B parallel to the tangent line at A, and the rays BP with a line through A parallel to the tangent line at B. Apollonius proves that the products of the segments from A to the points determined on the rays AP, and of the segments from B to the points determined on the rays BP, have a constant value.

This proposition is also very closely related to modern projective geometry.

Loci involving 3 or 4 straight lines.

In the preface to his Conica, Apollonius says that the propositions of Book III can be used to solve completely the problem of the "loci involving 3 or 4 straight lines", not completely disposed of by Euclid.

This famous problem, mentioned also by Pappus, is the following: to determine the locus of points whose distances x, y, z, u from four given lines satisfy an equation of the form

(32) $$xz = \alpha \cdot yu,$$

when α is a given ratio. If the last two lines coincide, we have a "locus involving 3 straight lines".

Pappus already knew that these loci are conics. Descartes prides himself on the fact that his analytical geometry enables him to determine these conics quite simply. But the preface just cited shows that Apollonius could solve the problem quite easily by his method and Zeuthen has made it clear (Abschnitt 7, 8) that the propositions of Book III are indeed sufficient for this purpose.

The conic determined by (32) obviously passes through the 4 vertices of the

[1] Neugebauer has shown in his Apollonios-Studien that those for the parabola follow from III 41 exactly as those for the ellipse and hyperbola follow from III 42.

quadrangle formed by the four given lines. The constant α can be chosen so as to make the conic pass also through a fifth given point. The solution of the problem of the "locus involving 4 lines" therefore solves at the same time the problem of passing a conic (i.e. a circle, an ellipse, a parabola, or a pair of opposite hyperbolas) through five given points.

Apollonius does not take up the problem of the conic through five given points, probably because the necessary preparations and the complicated distinction of cases would have carried him too far afield. But he does prove that the problem can have only one solution. Indeed the fourth book is chiefly occupied with the proof of the proposition that two conics can never have more than 4 points in common. In Apollonius this proposition is broken down into 3 propositions:

III 25. Two conics can not meet in more than four points.

III 36. A conic or a circle can not have more than four points in common with a pair of opposite hyperbolas.

III 53. A pair of opposite hyperbolas meets another such pair in at most four points,

In his preface to Book IV, he says that Conon of Samos had dealt with the first case (III 25), but that his proofs are not correct. "Nicoteles of Cyrene has justly attacked Conon on this point. Nicoteles said also that he could prove the second propositions (III 36), but he did not publish the proof, and the third proposition (III 53) has not yet been thought of by anybody", thus writes Apollonius.

The fifth book

treats, in the words of Apollonius in the introduction, the problem of the shortest and the longest line segments from a point O to a conic. But he gives even more than he promises: he determines all the lines through O, which intersect the conic at right angles (nowadays, we call them normals), he investigates the positions of O for which there are two, three or four solutions, and he studies carefully the maximum and minimum properties of these lines.

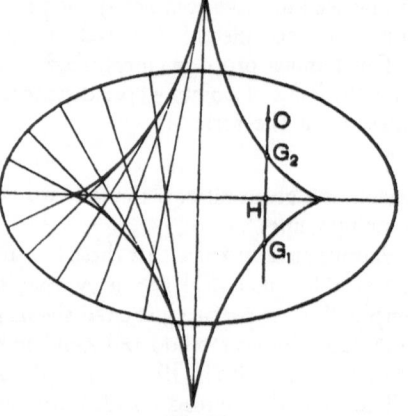

He starts by taking the point O on one of the principal axes. In that case there are either no, or two maximal or minimal lines through O, situated symmetrically, besides the principal axis itself. He proves that these maximal or

Fig. 119.

minimal lines OM are always perpendicular to the tangent line at M. Next he allows O to move along the ordinate HO, perpendicular to the principal axis, and he determines on all these ordinates the limiting positions G_1 and G_2, at which the number of normals through O jumps from 2 to 4, or inversely. These limiting points lie on a curve which envelops all these normals and which we call the evolute of the conic. Apollonius does not introduce this curve; he restricts himself to determining the limiting positions G_1 and G_2 fore very position of H.

The sixth, seventh and eighth books.

The sixth is of minor importance.

The principal propositions of Book VII are the following:

VII 12. The sum of the squares on two conjugate diameters of an ellipse is equal to the sum of the squares on the principal axes.

VII 13. The difference of the squares on two conjugate diameters of a hyperbola is equal to the difference of the squares on the principal axes.

VII 31. In the ellipse, and also in two conjugate hyperbolas, the parallelogram on two conjugate diameters, under the angle at which they intersect at the center, is equal to the rectangle on the axes.

According to Apollonius' own statement, the seventh book contains furthermore propositions which are useful for the diorism (i.e. the distinction of the possible cases as to the number of solutions) of problems which are taken up in the eighth book. This eighth book has been lost, but Halley succeeded in reconstructing a considerable number of its problems on the basis of a set of lemmas of the eighth book, which are found in Pappus. Most of the modern editions of Apollonius contain Halley's very successful reconstructions.

Further works of Apollonius.

There is only one other work of Apollonius, besides the Conica, that has been preserved in its entirety, namely:

1. On cutting off a ratio.

This work, in two books, treats the following problem: given two straight lines and a point on each of them, to draw a straight line through a given point, such that the segments cut off on the given lines, starting at the given points, have a given ratio.

Apollonius gives first an analysis of the problem, which leads to a quadratic equation, and then a synthesis consisting of the solution of the quadratic and the proof that this actually solves the problem.

The contents of the following works are known to us only from excerpts in the geometrical compendium of Pappus:

2. *On cutting off areas.*

This treats an entirely analogous problem in which the two segments which are cut off are required to have a given product, or, as expressed by Apollonius of course, to determine a given rectangle. This problem also leads to a quadratic equation, as does the next:

3. *On a fixed division.*

Let the points A, B, C, D be given on a straight line. Determine on the same line a point P, such that the rectangle $AP \cdot CP$ has a given ratio to the rectangle $BP \cdot DP$.

4. *On contacts.*

These two books contain the famous tangency problem of Apollonius: Given three things, each of which may be a point, a line or a circle. Determine a circle which passes through each of the given points and which is tangent to the given lines and circles.

For a reconstruction of the contents of this book, see Th. Heath, History of Greek Mathematics II, p. 182.

5. *On plane loci.*

A number of propositions on loci which turn out to be circles or straight lines. The most interesting proposition deals with a type of transformation of the plane, which carries circles and straight lines into circles and straight lines. In Pappus, this proposition is formulated as follows:

If two lines are drawn, from one given point, or from two, which

(*a*) are extensions of one another, or

(*b*) are parallel, or

(*c*) include a given angle
and which either (1) have a given ratio, or
                   (2) have a given product,
and if the endpoint of one of these lines describes a circle or a straight line, then the endpoint of the other will also describe a circle or a straight line.

Observe that this proposition contains as a special case the "inversion with respect to a point O": If (*a*) OP and OQ have the same direction, and if (2) $OP \cdot OQ$ is constant, and if P describes a straight line or a circle, then Q will also describe a straight line or a circle. Furthermore, all similarity transformations are special cases. But still more: All transformations of the plane, which carry circles and straight lines into circles or straight lines (i.e. the so-called circular transformations) are contained as special cases in Apollonius' general proposition. Was Apollonius aware of this?

6. *On Neuses.*

The reader will recall that a "neusis construction" (neusis = inclination) was for the Greeks the construction of a line segment of given length, which passes through a given point, extended if necessary, and of which the endpoints have to lie on two given straight lines or circles. In this book, Apollonius indicates a number of cases in which the construction can be carried out by means of compasses and straight edge. [1]

The following book is mentioned by Hypsicles in the introduction to the so-called 14th book of Euclid:

7. *Comparison of the dodecahedron and the icosahedron.*

If a dodecahedron and an icosahedron are inscribed in the same sphere, their areas have the same ratio as their volumes.

Entirely lost are:

8. *A general treatise.*

9. *On the helix.*

10. *On unordered irrationalities.*

11. *The rapid delivery.*

According to Eutocius, Apollonius gave, in this last work, a closer approximation to π than the one given by Archimedes. We can only conjecture whether this approximation has any connection with the origins of trigonometry and whether Apollonius knew or made tables of chords.

---

[1] The idea, sometimes expressed, that the Greeks only permitted constructions by means of compasses and straight edge, is inadmissible. It is contradicted by the numerous constructions, which have been handed down, for the duplication of the cube and the trisection of the angle. It is true however that such constructions were considered to be more elementary, and somewhere Pappus says that, whenever a construction is possible by means of compasses and straight edge, more advanced means should not be used. See in this connection A. D. Steele, *Quellen und Studien*, B 3, p. 287.

CHAPTER VIII

THE DECAY OF GREEK MATHEMATICS

External causes of decay

After Apollonius Greek mathematics comes to a dead stop. It is true that there were some epigones, such as *Diocles* and *Zenodorus*, who, now and then, solved some small problem, which Archimedes and Apollonius had left for them, crumbs from the board of the great. It is also true that compendia were written, such as that of Pappus of Alexandria (300 A.D.); and geometry was applied to practical and to astronomical problems, which led to the development of plane and spherical trigonometry. But apart from trigonometry, nothing great nothing new appeared. The geometry of the conics remained in the form Apollonius gave it, until Descartes. Indeed the works of Apollonius were but little read and were even partly lost. The "Method" of Archimedes was lost sight of, and the problem of integration remained where it was, until it was attacked anew in the 17th century, particularly in England. Germs of a projective geometry were present, but it remained for Desargues and Pascal to bring these to fruition. Higher plane curves were studied only sporadically; the algebraic tools for a systematic investigation were lacking. Geometric algebra and the theory of proportions were carried over into modern times as inert traditions, of which the inner meaning was no longer understood. The Arabs started algebra anew, from a much more primitive point of view. The theory of irrationals was interpreted by commentators, but it was not truly understood. Briefly, Greek geometry had run into a blind alley.

How did this come about?

Sometimes it is said: "Greek civilization got old and lost the spark of life." Very true, but this is only a general summary, not an explanation.

Political and economic conditions of course play a very important role. We have already seen that science had become the concern of courtiers, depending upon libraries and upon royal subsidies. When it happened that the kings did not wish to spend money on science, the subsidies shriveled. Wars, crushing taxes and, later on, the Roman hegemony with the accompanying exploitation, brought about the end of prosperity in the Hellenistic countries. When Caesar was besieged in Alexandria, a large part of the famous library was burned. The Roman emperors had very little use for pure science, and the wealthy Romans, while importing Greek sculptors, pedagogues and historians, did not invite mathematicians.

All of this is certainly very important, but it does not furnish a complete explanation. For it only accounts for periods of scientific inactivity, but not for retrogression and actual decay. For in astronomy the development was quite different. It is true that there also shorter and longer periods of inactivity occurred, but after such periods the work was always continued from the point at which it had stopped. Hipparchus (150 B.C.) continued the work of Apollonius and also took Babylonian observations into consideration. Around 150 A.D., Ptolemy resumed the work of Hipparchus and carried theoretical astronomy to a truly admirable point of development. And the 300 years between Hipparchus and Ptolemy had not been a period of complete idleness, for Ptolemy himself refers to other authors, who had attempted to explain the motion of the planets by means of epicycles and excenters. The Indian Surya-Siddhânta is to a large extent based on Greek astronomy before Ptolemy. The great Arabic astronomers, such as al-Battâni, improved the Ptolemaic system. Even Copernicus started from the Ptolemaic system, and so did Kepler, the father of modern astronomy. Each of these great astronomers stood on the shoulders of his precursors, and no essential progress was ever lost.

How totally differently things went in mathematics! Euclid was studied diligently, but not Archimedes, nor Apollonius. Many of the writings of the great mathematicians were lost, others were reputed to be extremely difficult and were practically not read at all. Al-Khwârizmi, the father of Arabic algebra, purposely avoided "Greek erudition". He wanted to write a book intelligible to simple folks, who were concerned, e.g., with inheritance problems, and who needed simple little rules for setting up and solving algebraic equations. It is on his work, rather then on Greek learning, that Arabic algebra is based, from which in turn the algebra of the Renaissance was derived.

Thus, in astronomy we have a progressive development, even though not uninterrupted, but in mathematics a long-continued decline, followed by a new growth on a quite different foundation. What is the cause of this?

Obviously, general political and economic factors do not adequately explain the decline, since these causes should have had the same effect in astronomy. There must be inner grounds for the decay of antique mathematics.

It was Zeuthen, who uncovered these inner causes most clearly in his "Lehre von den Kegelschnitten im Altertum".

The inner causes of decay.

1. The difficulty of geometric algebra.

We have already frequently pointed out the importance of the algebraic element in Greek geometry. Theaetetus and Apollonius were at bottom algebraists, they thought algebraically even though they put their reasoning in a geometric dress.

Greek algebra was a geometric algebra, a theory of line segments and of areas, not of numbers. And this was unavoidable as long as the requirements of strict logic were maintained. For "numbers" were integral or, at most, fractional, but

at any rate rational numbers, while the ratio of two incommensurable line segments can not be represented by rational numbers. It does honor to Greek mathematics that it adhered inexorably to such logical consistency.

But, at the same time, this set bounds for Hellenic algebra. Equations of the
first and second degree can be expressed clearly in the language of geometric
algebra and, if necessary, also those of the third degree. But to get beyond this
point, one has to have recourse to the bothersome tool of proportions.

Hippocrates, for instance, reduced the cubic equation $x^3 = V$ to the proportion

$$a : x = x : y = y : b,$$

and Archimedes wrote the cubic

$$x^2(a - x) = bc^2$$

in the form

$$(a - x) : b = c^2 : x^2.$$

In this manner one can get to equations of the fourth degree; examples can be
found in Apollonius (e.g. in Book V). But one can not get any farther; besides,
one has to be a mathematician of genius, thoroughly versed in transforming proportions with the aid of geometric figures, to obtain results by this extremely
cumbersome method. Any one can use our algebraic notation, but only a gifted
mathematician can deal with the Greek theory of proportions and with geometric
algebra.

Something has to be added; that is

2. The difficulty of the written tradition.

Reading a proof in Apollonius requires extended and concentrated study.
Instead of a concise algebraic formula, one finds a long sentence, in which each
line segment is indicated by two letters which have to be located in the figure. To
understand the line of thought, one is compelled to transcribe these sentences in
modern concise formulas. The ancients did not have this tool; instead they had
the oral tradition.

An oral explanation makes it possible to indicate the line segments with the
fingers; one can emphasize essentials and point out how the proof was found. All
of this disappears in the written formulation of the strictly classical style. The
proofs are logically sound, but they are not suggestive. One feels caught as in a
logical mousetrap, but one fails to see the guiding line of thought.

As long as there was no interruption, as long as each generation could hand over
its method to the next, everything went well and the science flourished. But as
soon as some external cause brought about an interruption in the oral tradition,
and only books remained, it became extremely difficult to assimilate the work of
the great precursors and next to impossible to pass beyond it.

This is seen very clearly in

The commentaries of Pappus of Alexandria.

Pappus (320 A.D.) had the magnificent library of Alexandria at his disposal, all the works of the foremost mathematicians and astronomers. He was gifted, diligent and enthusiastic, but he had to study the written works and there he encountered the same difficulties that we meet. In order to assist in overcoming these difficulties and to simplify matters for those who were to come after him, he wrote extended commentaries, e.g. a commentary on Ptolemy's Almagest and one on the tenth book of Euclid. And his great compendium, to which we have referred repeatedly, consists largely of commentaries on the classics. Whenever Pappus found a proof difficult or incomplete, he wrote an explanatory "lemma". Frequently Ptolemy, or whoever it was, had treated only one of the possible cases; Pappus would then add the analogous proofs for the other cases. Sometimes it happens that Apollonius uses, e.g., a relation between line segments, perhaps a proportion or a relation between products, without giving an explicit proof. Usually such relations can be taken from the figure by any one who has had the necessary practice in the transformation of products and ratios. In such cases Pappus would deduce the relation, little step by little step, from propositions which occur explicitly in Euclid. This illustrates clearly that, already in the days of Pappus, people found it difficult to understand matters which would become at once clear in an oral exposition.

A further development of mathematics urgently required a concise algebraic notation; further development of the rigorous Greek methods could not lead to such a notation. It was necessary to return to the primitive Babylonian point of view, to multiply and divide magnitudes unconcernedly, to treat expressions like $2 + \sqrt{5}$ simply as numbers without worrying about their irrationality. To progress, it was necessary first to take a step backwards. And this is what the Arabs did.

The Algebra of the Italian Renaissance did not derive from Greek geometric algebra, but from Arabic algebra. The notation used by the Arabs and by the Italians was still rather clumsy; it was simplified by the Frenchmen Vieta and Descartes [1], and it is from this source that our algebra was developed.

We come now to

The epigones of the great mathematicians

1. Diocles.

Diocles probably lived during the 2nd century B.C. He has written a book on burning-mirrors. By the use of conics he solved the "problem of Archimedes", the division of a sphere into two spherical segments whose volumes have a given ratio; and he solved the Delian problem by means of new curve, the "cissoid".

[1] A more complete description would demand also mention of the Italian Bombelli and of the Dutchman Simon Stevin.

Eutocius transmitted both constructions in his commentary on Archimedes. We shall discuss only the second of these.

The cissoid

was defined as follows:

Let AB and $\Gamma\varDelta$ be two mutually perpendicular diameters of a circle. Let E and Z lie on the circle, on either side of and at equal distances from B. Drop the perpendicular ZH form Z to $\Gamma\varDelta$ and let $\varDelta E$ meet this perpendicular in Θ. Then the point Θ will describe the cissoid.

The "symptom" of the cissoid, i.e. its equation in rectangular coordinates ΓH and $H\Theta$, is found readily by drawing $EK \perp \Gamma\varDelta$ and then writing the proportionality

$$\varDelta H : H\Theta = \varDelta K : KE;$$

this can be replaced by

$$H\varDelta : H\Theta = \Gamma H : HZ.$$

But HZ is a mean proportional between ΓH and $H\varDelta$; thus one obtains

(1) $\Gamma H : HZ = HZ : H\varDelta = H\varDelta : H\Theta.$

This says that HZ and $H\varDelta$ are two mean proportionals between ΓH and $H\Theta$. If one takes care furthermore, that the extremes ΓH and $H\Theta$ have a given ratio, $a : b$, then one has obtained a solution of the Delian problem. This is accomplished by locating the intersection of the cissoid and the line $\Gamma\varPi$, which determines on the two mutually perpendicular radii $\varLambda\Gamma$ and $\varLambda B$, segments $\varLambda\Gamma$ and $\varLambda\varPi$ in the given ratio $a : b$. It remains then only to multiply the four terms of the proportion (1) by a proportionality factor, in order to

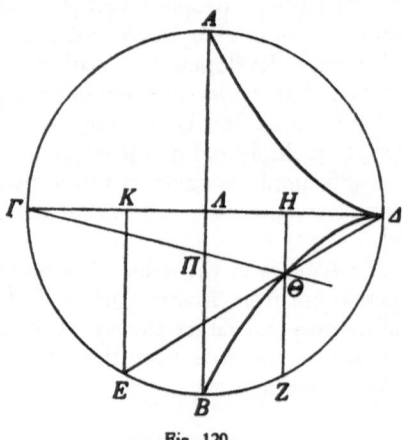

Fig. 120.

obtain two mean proportionals between given line segments a and b.

2. Zenodorus,

the author of a very interesting book on isoperimetric figures, lived between Archimedes (250 B.C.), whom he mentions, and Quintilian (75 A.D.) who mentions him. In view of the excellent quality and style of his work he is probably closer to Archimedes than to Quintilian.

Isoperimetric figures

are figures of equal perimeter. The questions investigated and partly solved by Zenodorus are the following: which plane figure of given perimeter has the largest area, and which solid of given area has the largest volume?

It is easy to guess at the answers to these questions, but very difficult to prove them. It was not until 1884 that Hermann Amandus Schwarz gave a perfectly rigorous proof of the isoperimetric properties of circle and sphere, by using methods of Weierstrass. But, for his time, Zenodorus had come a long way. His treatise, known to us from excerpts found in Pappus and in Theon of Alexandria, contains the proofs of 14 propositions, the most important of which are the following:

1. Of two regular polygons with the same perimeter, the one which has the larger number of angles is the larger.

3. If a circle and a regular polygon have the same perimeter, the circle is larger.

7. Among triangles of equal base and equal perimeter, the isosceles is the largest.

10. The sum of two similar isosceles triangles is larger than the sum of two non-similar triangles with the same bases and with the same sum of their perimeters.

11. Among all polygons of equal perimeter and equal numbers of sides, the regular polygon is the largest.

In the proof of 11, it is taken to be self-evident that a largest polygon of given perimeter exists. It then follows readily from 7 and 10 that this largest polygon must have equal sides and equal angles.

From 3 and 11, Zenodorus concludes that of all figures of equal perimeter, the circle is the largest. This conclusion is justified only if "figures" are understood to be only circles and polygons.

Next Zenodorus states that among all solids of equal surface area, the sphere is the largest. But he only proves the following propositions:

13. If a regular polygon (with an even number of sides) revolves about one of the longest diagonals, a solid is generated, bounded by conical surfaces, of less volume than the sphere of the same area.

14. Each of the five Platonic polyhedra is smaller than the sphere with equal area.

3. Hypsicles

probably lived in Alexandria around 180 B.C., not long after Apollonius. He wrote on polygonal numbers and on the harmony of the spheres; this places him in the Pythagorean tradition. Two of his writings have been preserved; the first is the so-called

Fourteenth Book of the Elements,

which is included as such in most editions of Euclid. In this short treatise, Hypsicles compares the areas and volumes of a dodecahedron and an icosahedron, inscribed in the same sphere. The circles which circumscribe the faces of these two solids are equal to each other, according to a proposition of Aristaeus. From this it follows readily that

the volume of the dodecahedron: the volume of the icosahedron =
the area of the dodecahedron: the area of the icosahedron.

In addition, Hypsicles proves now that this ratio is also equal to

edge of the cube: edge of the icosahedron

and he gives a construction of this ratio by means of plane geometry.

Hypsicles also wrote a work, called

Anaphorai (Risings)

dealing with the times required for rising and setting by the various signs and degrees of the ecliptic. His method is not based on exact trigonometric calculations, but on a rough approximation. He simply assumes that the times of rising for the signs from Aries to Virgo form an ascending arithmetic progression, and those of the signs from Libra to Pisces a descending one. If the times for Aries to Virgo are

$$A \quad B \quad C \quad D \quad E \quad F,$$

then those for Libra to Pisces are

$$F \quad E \quad D \quad C \quad B \quad A.$$

Like the Babylonians, Hypsicles divides the day into 360 "degrees of time". The sum of each of the arithmetic progressions is therefore 180 degrees:

$$A + B + C + D + E + F = 180.$$

During the light part of each day, exactly 6 signs rise above the horizon, because at the beginning of the day, the sun rises and at the end the opposite point. The longest daylight period is therefore the day on which the first three signs and the last three rise. Half of the longest day is therefore $A + B + C$.

Hypsicles assumes furthermore that, for Alexandria, the longest day lasts 210 degrees of time. Hence

$$A + B + C = 105,$$

and, because the sum is equal to three times its middle term,

$$B = 35.$$

In the same way, one finds from $D + E + F = 75$ that $E = 25$.

The remaining times can now be calculated without difficulty. Hypsicles uses the same method to determine the initial moments of the separate degrees of the ecliptic.

This primitive method of calculation is really not worthy of a Greek mathematician, although it is sufficiently accurate to serve for practical, e.g. astrological, applications. Similar arithmetic progressions are found in the writings of the Roman astrologers Vettius Valens, Manilius and Firmicus Maternus. Neugebauer has shown that these methods of calculation are directly related to the Babylonian lunar calculations. [1]

Apparently Hypsicles was not yet up to an accurate trigonometrical calculation of the times of rising and setting of the signs of the zodiac. But Hipparchus, who lived at a not much later period, was able to carry through such calculations. To understand this, we have to take a look at the

History of trigonometry.

In place of sine, cosine and tangent, the Greek astronomers Hipparchus (150 B.C.) and Ptolemy (150 A.D.) always used *chords of arcs of circles*. In fact it makes little difference whether one operates with chords or with sines, since what we now call the sine of an angle is the quotient by the radius of one half of the chord of twice the intercepted arc [2]:

$$\sin \alpha = \frac{1}{2R} \, chd(2\alpha)$$

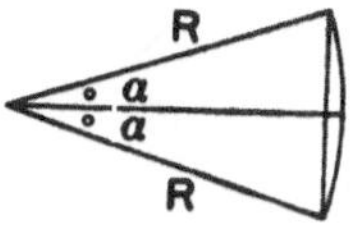

Fig. 121.

As early as the fifth century A.D., the Indian astronomers changed from the chords to the sines; in their great standard work Sûrya Siddhânta and in the Āryabhatiya of Āryabhata, tables of sines appear. After that the Arabs introduced the tangent and the cotangent, which greatly simplified the computations.

The Greeks therefore worked with tables of chords. We have discussed already how these were calculated.

We do not know whether Hipparchus was himself the inventor of trigonometry, nor whether he used only plane triangles, or only spherical triangles, or both, for his writings are lost for the greater part. For this reason we are compelled to draw chiefly on the works of Menelaus and of Ptolemy for the earliest history of plane and spherical trigonometry.

Plane trigonometry.

Ptolemy does not know any rules for oblique triangles, such as our laws of sines and cosines. When he is dealing with oblique triangles, he always divides them into right triangles.

An interesting example:

[1] See my article *Babylonian Astronomy* III, *Journal Near Eastern Studies* 10 (1951), p. 32, where further references to the literature are given.

[2] It is only since around 1800 that we make the division by the radius. Up to that time the sine was one half the chord of twice the arc, for an arbitrarily chosen radius R.

In Chapter 6 of Book 4 of the Almagest, in order to determine the moon's epicycle, Ptolemy has to solve the following problem:

The points A, B, Γ lie on a circle. The arcs $BA = 53°35'$ and $BA\Gamma = 150°26'$ are known from observations. From a point Δ, outside the circle, these arcs are seen under angles $B\Delta A = 3°24'$ and $B\Delta\Gamma = 0°37'$. Setting the radius of the circle equal to 60, how far is Δ from the center?

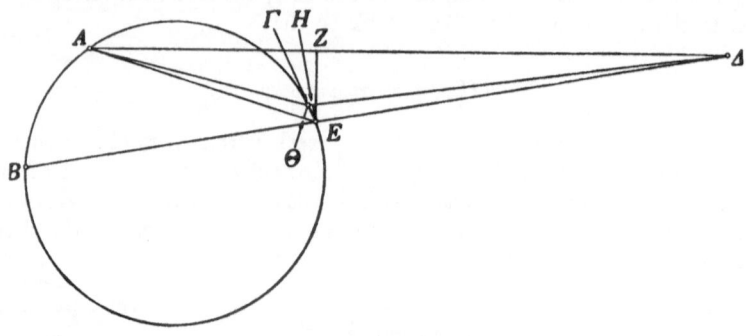

Fig. 122.

Ptolemy reasons as follows: Let E be the second intersection of ΔB with the circle. Draw EA and $E\Gamma$, and drop perpendiculars EZ and EH from E to ΔA and $\Delta \Gamma$ respectively. Finally drop the perpendicular $\Gamma\Theta$ from Γ to EA.

In the semicircle on $E\Delta$, the peripheral angle $E\Delta Z = 3°24'$; hence the arc $EZ = 6°48'$, and the chord EZ is, according to the table of chords, equal to $7^p7'$, when the diameter $E\Delta$ is set equal to 120^p.

In the same way: $\angle EAZ = \angle BEA - \angle B\Delta A$ is known, hence the arc EZ in the semicircle on AE as diameter is known, and also, from the table of chords, the chord EZ, the diameter AE being taken as equal to 120^p. If, by means of a proportionality factor, the line EZ is given the measure $7^p7'$, one finds $AE = 17^p55'32''$ in the same unit in which $\Delta E = 120^p$.

By the same method, Ptolemy obtains $EH = 1^p17'30''$ and $\Gamma E = 1^p20'23''$ when $\Delta E = 120^p$.

We see therefore that every time an oblique triangle such as $A\Delta E$ or $\Gamma\Delta E$, whose angles are known, is divided into two right triangles, the ratios of the altitude-line to the two including sides are then determined, and thus the ratio of the sides follows. The result is of course equivalent to our law of sines.

In triangle $A\Gamma E$ the angle at E is known as a peripheral angle with a given arc. The altitude-line $\Gamma\Theta$ again divides this triangle into two right triangles. In the right triangle $E\Theta\Gamma$ both angles are known, so that the ratios of $E\Theta$ and of $\Gamma\Theta$ to $E\Gamma$ can be calculated from the table of chords. Again setting $E\Gamma = 1^p20'23''$, we find $\Gamma\Theta = 1^p0'8''$ and $E\Theta = 0^p53'21''$.

In the same units we had found $AE = 17°55'32''$, therefore $A\Theta = AE - E\Theta$
$= 17°2'11''$. The theorem of Pythagoras now gives

$$A\Gamma^2 = A\Theta^2 + \Gamma\Theta^2 = 291°14'36'',$$

so that

$$A\Gamma = 17°3'57'', \quad \text{when} \quad \Delta E = 120°.$$

But, in the unit in which the diameter of the circle is $120°$, $A\Gamma$, being the chord
of an arc of $96°51'$, is, according to the table of chords, equal to $89'46°14''$.
By a change of unit for $A\Gamma$, ΓE and ΔE, which makes $A\Gamma = 89°46'14''$, one
finds $\Delta E = 631°13'48''$ and $\Gamma E = 7°2'50''$, in terms of the unit which makes
the diameter of the circle equal to $120°$. By using the table of chords backwards,
one finds arc $\Gamma E = 6°44'1''$. Addition of the arc $B\Gamma A = 150°26'$, makes arc

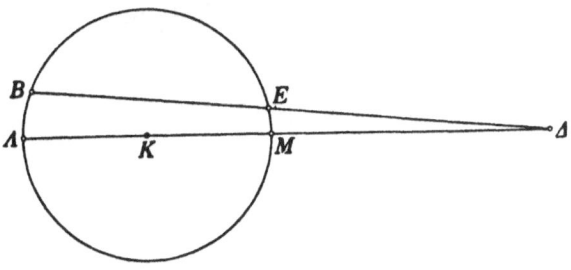

Fig. 123.

$B\Gamma E = 157°10'1''$. By use of the table of chords, the corresponding chord BE
is found to be $117°37'2''$.

The only problem now remaining is the following: In a circle of diameter $120°$,
a chord $BE = 117°37'32''$ is extended by a segment $\Delta E = 631°13'48''$. How
far is Δ from the center K?

This is easy, even without trigonometry. Draw the diameter ΛKM through Δ.
Then

$$\Delta \Lambda \cdot \Delta M = \Delta B \cdot \Delta E = (\Delta E + EB) \cdot \Delta E = 472 \cdot 700°5'32''.$$

But from geometric algebra, we know that

$$\Delta K^2 = \Delta \Lambda \cdot \Delta M + KM^2,$$

so that ΔK^2 and hence ΔK are known:

$$\Delta K = 690°8'42''.$$

This completes the solution of the proposed problem.
The words of Ptolemy suggest that this method was taken from Hipparchus.

For, in the previous chapter, he says: "We shall follow a theoretical method, which we already see applied by Hipparchus. Indeed, starting from three selected lunar eclipses, we shall also . . ."

We now pass on to

Spherical trigonometry

and take up first the work of

Menelaus.

In 98 A.D. Menelaus made astronomical observations in Rome. Besides a work on chords, that has been lost and that probably contained tables of chords, he has

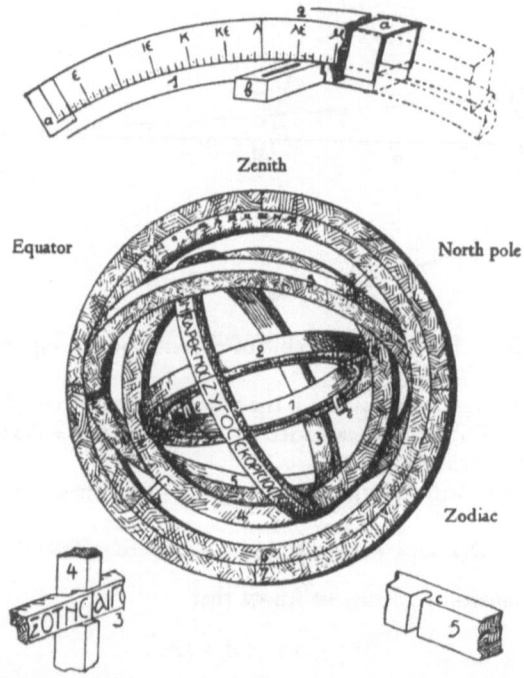

Fig. 124. Ptolemy's astrolabic instrument for observing the moon, stars and planets, as reconstructed by Mr. Paul Rome, engineer and architect, from Ptolemy's own description.

written a Spherics, which has been preserved. In the first two books, various propositions are proved on spherical triangles (i.a. congruence propositions). The third book opens with the famous

Transversal proposition

also repeatedly applied by Ptolemy; it runs as follows

Let Δ lie on AB, E on $A\Gamma$, and let $\Gamma\Delta$ and BE meet in Z. Then the ratio of ΓE to AE is equal to the ratio, compounded from the ratios $\Gamma Z : \Delta Z$ and $\Delta B : AB$. And the ratio $\Gamma A : AE$ is obtained by composition of the ratios $\Gamma\Delta : \Delta Z$ and $ZB : BE$. In modern notation:

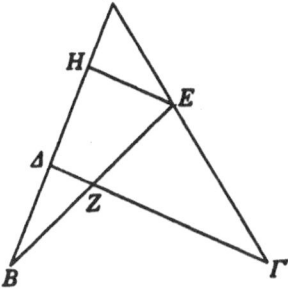

Fig. 125.

$$\frac{\Gamma E}{AE} = \frac{\Gamma Z}{\Delta Z} \cdot \frac{\Delta B}{AB} \text{ and } \frac{\Gamma A}{AE} = \frac{\Gamma\Delta}{\Delta Z} \cdot \frac{ZB}{BE}.$$

To prove, for instance, the second relation, one draws the line EH through E, $// \Gamma\Delta$, meeting AB in H. Then:

$$\frac{\Gamma A}{AE} = \frac{\Gamma\Delta}{EH} = \frac{\Gamma\Delta}{\Delta Z} \cdot \frac{\Delta Z}{EH} = \frac{\Gamma\Delta}{\Delta Z} \cdot \frac{ZB}{BE};$$

By projection from the center, Menelaus derives from these ratios, relations on the sphere. If $A\Delta B$, $AE\Gamma$, $\Gamma Z\Delta$ and BZE are great circles on the sphere, he finds the relations

$$\frac{\text{chd}\,(2\Gamma E)}{\text{chd}\,(2AE)} = \frac{\text{chd}\,(2\Gamma Z)}{\text{chd}\,(2\Delta Z)} \cdot \frac{\text{chd}\,(2\Delta B)}{\text{chd}\,(2BA)},$$

and

$$\frac{\text{chd}\,(2\Gamma A)}{\text{chd}\,(2AE)} = \frac{\text{chd}\,(2\Gamma\Delta)}{\text{chd}\,(2\Delta Z)} \cdot \frac{\text{chd}\,(2ZB)}{\text{chd}\,(2BE)}.$$

From the transversal proposition, Menelaus derives other propositions which have played a role in spherical trigonometry.[1] We shall not enter into this further, but rather take up the applications of the transversal proposition in the Almagest.

In Book I, Chapter 14, Ptolemy wants to determine the declination δ of the sun

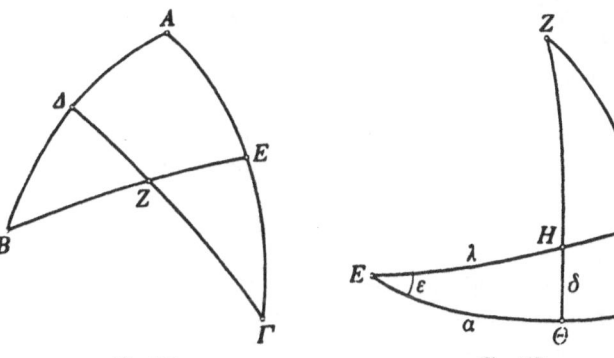

Fig. 126.   Fig. 127.

[1] See A. von Braunmühl, *Geschichte der Trigonometrie*, Leipzig 1900, p. 17.

for a given longitude λ. The inclination of the ecliptic, ε, i.e. the angle at E between the ecliptic and the equator, must of course also be given. Let the sun be in H, so that $EH = \lambda$ is the given longitude. Let Z be the pole of the equator and let the great circle ZH meet the equator in Θ; then $H\Theta = \delta$ is the required declination. Ptolemy draws the circle $\acute{Z}BA$, which passes through the poles of the equator and of the ecliptic and is therefore perpendicular to both. According to the transversal proposition, we have now

$$\frac{\text{chd}\,(2ZA)}{\text{chd}\,(2AB)} = \frac{\text{chd}\,(2Z\Theta)}{\text{chd}\,(2\Theta H)} \cdot \frac{\text{chd}\,(2HE)}{\text{chd}\,(2EB)},$$

i.e.

$$\frac{\text{chd}\,(180^\circ)}{\text{chd}\,(2\varepsilon)} = \frac{\text{chd}\,(180^\circ)}{\text{chd}\,(2\delta)} \cdot \frac{\text{chd}\,(2\lambda)}{\text{chd}\,(180^\circ)},$$

or, since the chord of the arc of 180° is equal to 120 in Ptolemy's units,

$$\text{chd}\,(2\delta) = \frac{\text{chd}\,(2\lambda)\,\text{chd}\,(2\varepsilon)}{120}.$$

This agrees with our modern formula

$$\sin \delta = \sin \lambda \sin \varepsilon.$$

These methods enable Ptolemy to solve all right spherical triangles, of which two sides, or an angle and one side are given. And this appears to be sufficient for spherical astronomy.

The period of Menelaus and of Ptolemy was one of revival in mathematics. Astrology was tremendously popular, the astrologers needed astronomical calculations, and for these mathematics was indispensable.

But during the two centuries before Menelaus, during the period in which the power of the Roman empire was spread, mathematics had sunk to a pitifully low level.

During the period of splendor in Roman literature, in the time of Vergil, of Horace and of Ovid, one hardly finds a single allusion to mathematics. Cicero knew a good deal of philosophy and something of astronomy, but he had no interest at all in mathematics. Even the architect Vitruvius, who loves to display his learning in various fields, says next to nothing about mathematics.

There is a beginning of a new revival of interest in

Heron of Alexandria

around 60 A.D.

How can we establish this date? In Heron's Dioptra a method is developed (unnecessarily complicated) to determine the difference in time between Rome and Alexandria by means of a lunar eclipse and Neugebauer observed that the eclipse which is used there as an example is probably the one of the year 62. [1]

---

[1] O. Neugebauer, Kgl. Danske Vid. Selsk., *Hist.-fil. Meddel.* 26 no. 2 (1938).

Heron's work is in the nature of an encyclopaedia of elementary geometry, applied geometry and applied mechanics. In a series of works, he treats the construction and the use of various measuring instruments, such as water clocks and diopters, and of machines, such as pneumatic machines, automatic machines, war engines, lifting machines, etc. He also wrote a commentary on Euclid and a work on definitions. Finally a number of works on areas and volumes, the best known of which is called

Metrics.

Book I of this treatise treats the calculation of areas, book II the calculation of volumes, book III the division of areas and volumes in a given proportion.

In book I, most remarkable is the calculation of the area of a triangle with given sides. Heron assumes the sides to be 13, 14, 15, and gives two methods. The first consists in first calculating the height of the triangle, by a method well-known from our schoolbooks. The second method avoids the calculation of the height. It is explained as follows:

$$13 + 14 + 15 = 42$$
$$42/2 = 21$$
$$21 - 13 = 8$$
$$21 - 14 = 7$$
$$21 - 15 = 6$$
$$21 \times 8 \times 7 \times 6 = 7056$$
$$\sqrt{7056} = 84.$$

This prescription is equivalent to the well-known formula

$$A = \sqrt{s(s - a)(s - b)(s - a)}, \qquad s = \frac{a + b + c}{2}.$$

Cantor ascribed this formula to Heron, because it appears in Heron without indication of a source. This is as if one were to attribute to Hütte all formulae in Hütte's Pocket Book for Engineers, for which no source is indicated! However, we now know from an Arabic manuscript that "Heron's formula" is due to Archimedes.

Heron next gives a geometrical proof of this formula. The proof makes use of the inscribed circle of the triangle.

Several formulas occurring in Heron, particularly those for volumes, are taken from Archimedes. As to the others, the history of the rules of calculation and of the numerical examples is almost impossible to trace. Usually such things are transmitted practically unchanged, century after century. Occasionally something is added, sometimes the name of the inventor of a formula is mentioned, but usually the source is anonymous. Some of the numerical examples in Heron's treatise are already found in cuneiform texts.

Heron's examples are repeated endlessly in the Roman era and during the Middle Ages. The oldest Chinese collection of problems on applied proportions [1] looks like an ancient Babylonian text, but it is next to impossible to prove their dependence or to trace the road along which they were transmitted.

And, after all, it is not very important.

It is mankind's really great thoughts that are of importance, not their dilution in popularizations and in collections of problems with solutions. Let us rejoice in the masterworks of Archimedes and of Apollonius and not mourn the loss of numberless little books on calculation.

By far the most original mathematician of late antiquity (the Roman era) is

Diophantus of Alexandria.

It is uncertain when Diophantus lived; according to Bombelli he lived during the reign of Antoninus Pius (150 A.D.), according to Abulfaradj during that of Julian the Apostate (350). But Psellus says that the learned Anatolius of Alexandria, who became bishop of Laodicea in 270, dedicated a book to Diophantus. This leads me to think that most probably Diophantus lived about 250.

In an algebraic puzzle rhyme, found in the Anthologia Palatina, his life is described as follows (I have taken the translation from Cantor, Geschichte der Mathematik):

> Hier dies Grabmal deckt Diophantus. Schauet das Wunder!
> Durch des Entschlafenen Kunst lehret sein Alter der Stein.
> Knabe zu sein gewährte ihm Gott ein Sechstel des Lebens;
> Noch ein Zwölftel dazu, sprosst' auf der Wange der Bart;
> Dazu ein Siebentel noch, da schloss er das Bündnis der Ehe;
> Nach fünf Jahren entsprang aus der Verbindung ein Sohn.
> Wehe das Kind, das vielgeliebte, die Hälfte der Jahre
> Hatt' es des Vaters erreicht, als es dem Schicksal erlag.
> Drauf vier Jahre hindurch durch der Grössen Betrachtung den Kummer
> Von sich scheuchend, auch er kam an das irdische Ziel.

We leave it to the reader to determine from this, which age Diophantus reached. His principal work, the

Arithmetica

is dedicated to the "highly honored Dionysius". This may have been Saint Dionysius, who became bishop of Alexandria in 247. The dedication starts as follows:

> Knowing, highly honored Dionysius, that you exert yourself to learn the study of number problems, I have begun with an explanation of the nature and the power of numbers, starting with the foundations on which the matters are based. This may appear to be more difficult than it is, because it is not yet known. Beginners are inclined quickly to lose courage. But it will be easy for you to understand it, thanks to your enthusiasm and to my explanation.

[1] P. L. van Hee, Le classique de l'île maritime, ouvrage chinois du IIIe siècle, Quellen und Studien B 2, p. 255.

The Arithmetica contains 189 problems with their solutions. Nearly all the problems of the first book lead to determinate equations such as

$$as = b, \quad \text{or} \quad as^2 = b,$$

which have only one positive solution. In the remaining 5 books, one finds especially indeterminate equations, also called

Diophantine equations,

which usually have infinitely many solutions.

Diophantus admits only *positive rational* solutions. He is usually satisfied when he has found one solution; it makes no difference to him whether the solution is integral or fractional.

His method varies from case to case. There is not a trace in his work of a systematic theory of diophantine equations. The general solution of diophantine equations of the first degree

$$ax \pm by = c$$

is first indicated by Āryabhaṭa (500) and elaborated by Brahmagupta (625). As far as diophantine equations of the second degree are concerned, we find in Diophantus the general solution of the equation of the "Pythagorean triangles"

(1) $$x^2 + y^2 = z^2$$

and also special methods for indeterminate equation of the form

(2) $$Ax^2 + Bx + C = y^2,$$

which are effective, when either A or C is zero, when A or C is a square, or when $A + C$ is a square and B is zero. Diophantus also has a method for "double equations" of the form

(3) $$ax^2 + bx + c = y^2, \qquad dx^2 + ex + f = z^2.$$

We shall see presently, with the aid of solved examples, how Diophantus proceeded in solving such equations.

Diophantus knows also that a number of the form $8n + 7$ can not be written as the sum of three squares. This about completes the general ideas to be found in Diophantus.

About

The precursors of Diophantus

in the field of indeterminate equations, we know precious little. At an earlier point we have discussed the methods ascribed to Pythagoras and to Plato for the solution of the diophantine equation (1). We also know that the Babylonians were acquainted with the general solution

$$x = p^2 - q^2, \qquad y = 2pq, \qquad z = p^2 + q^2,$$

which we shall encounter again in Diophantus.

The Pythagoreans also knew infinitely many solutions of the diophantine equation

$$x^2 - 2y^2 = \pm 1$$

(see "lateral and diagonal" numbers, p. 125).

I also refer to the "cattle problem of Archimedes". Not much more is known.

On the other hand, we do have an idea with respect to *determinate equations*, as to the tradition to which Diophantus belongs. In a papyrus from the second century A.D. [1], systems of linear equations are found with two or more unknowns. I also remind the reader of the "Flower of Thymaridas" and of Babylonian algebra. As early as the time of Archimedes, little algebraic problems, dressed up in verse form, such as the epitaph cited above, were very popular in Greece. A large number of them is found under the name of Metrodorus in the Greek Anthology. [2] In contrast to these popular collections, Diophantus gives nearly always a purely arithmetical formulation, without stories.

Connections with Babylonian and Arabic algebra.

It is probable that the tradition of these algebraic methods was never interrupted so that, along with the scholarly tradition of Greek geometry, there has always existed a more popular tradition of small algebraic problems and methods of solution, a tradition which originates in Babylonian algebra, and which ends in Arabic algebra, with radiations into Greek culture, into China and India.

We have no real proofs for the existence of such an uninterrupted tradition; too many connecting links are missing for this. It is rather a general impression of relatedness which makes itself felt when one knows the cuneiform texts and then looks through Heron or Diophantus, or the Chinese "classic of the maritime isle", or the Aryabhayta of Aryabhata or the Algebra of Alkhwarizmi. According to all Arabic sources, Alkhwarizmi was the first writer on algebra, but his algebra is so mature that we can not assume that he discovered everything himself. The algebra of Alkhwarizmi can hardly be accounted for on the basis of the Greek and Indian sources which we know; one gets more and more the impression that he has drawn on older sources which in some way or other are connected with Babylonian algebra.

Certainly problems 27—30 in the first book of the Arithmetica are closely related to Babylonian problems, and so are the solutions. They are:

To find two numbers, of which

---

[1] Karpinski and Robbins, *Michigan papyrus* 620, *Science 70* (1929), p. 311.
[2] Anthologia Palatina. See, e.g., Wertheim's German translation of Diophantus.

27) the sum and product are given;
28) the sum and the sum of the squares are given;
29) the sum and the difference of the squares are given;
30) the difference and the product are given.

The method of solution agrees completely with the Babylonian method. When the sum $2a$ is given, Diophantus sets one of the numbers equal to $a + s$, the other to $a — s$ and then obtains an equation in s. When the difference $2b$ is given, he sets the numbers equal to $s + b$ and $s — b$.

Diophantus reduces all his problems, no matter how complicated, to equations in one unknown. Either he expresses the other unknowns in terms of that one, or he gives them arbitrary values to be modified afterwards, if necessary, in order to satisfy the other conditions of the problem. Alkhwarizmi also reduces all his problems systematically to equations in one unknown. And yet Alkhwarizmi is independent of Diophantus.

The algebraic symbolism

of Diophantus is still rather primitive. He had a special symbol for the unknown (which we shall always denote by s), which looks somewhat like an s and which is also found in the Michigan papyrus 620. Perhaps this symbol was developed from the first letters, or from the final σ of Arithmos. [1]

Just as Alkwharizmi, Diophantus uses special names for the powers of the unknown s; for them he uses the following abbreviating symbols:

(1) $\overset{\circ}{M} = Μόνας$ = unity;
(s) $\varsigma = Ἀριθμός$ = number [2];
(s^2) $Δ^Y = Δύναμις$ = square;
(s^3) $K^Y = Κύβος$ = cube;
(s^6) $K^YK = Κυβόκυβος$ = cubocube.

Moreover he has names for reciprocals of powers such as s^{-1}, s^{-2}, etc.
For the addition of terms, Diophantus simply writes them in a row, e.g.

$$Δ^Y\bar{γ}\overset{\circ}{M}\overline{ιβ} = 3s^2 + 12.$$

Thus he does not simply write 12, but 12 units. In a similar way, Alkhwarizmi usually writes 12 dirhems.

Diophantus has a definite sign for subtraction, something like an inverted $Ψ$. Everything else, multiplication, equality, greater than and less than, he expresses in words, just as the Arabs did later on, and, following their example, the Italians. Alkhwarizmi does not even use the abbreviations of Diophantus; his algebra is

---

[1] For a detailed discussion see Th. Heath, *Diophantus of Alexandria*, 2nd ed., Cambridge 1910, or the same author's *History of Greek Mathematics*, p. 456.
[2] If, besides the "number" s, the square s^2 occurs also, Diophantus sometimes calls s "the side"; so does Alkhwarizmi.

still entirely "rhetorical" and in this respect more primitive than that of Diophantus. [1]

In his introduction, Diophantus explains how one multiplies and divides powers of the unknown, how polynomials are multiplied and how terms with the same power of s can be combined. Next follows what the Arabs call aljabr and almukâbala, and later writers restauratio and oppositio. Aljabr (hence our word algebra) is the transposition of negatives to the other side, and almukâbala is subtracting equal terms from the left and right members.

In the simplest case, these operations will leave only one term on each side of the equation; it then takes e.g. the form

$$as = b, \quad \text{or} \quad as^2 = bs, \quad \text{or} \quad as^2 = b$$

and it is easy to solve.

Diophantus promises to discuss at a later point the case, in which two terms remain on one side. In the part of his work that has been preserved, this promise is not redeemed, although it becomes evident that he knew how to solve quadratic equations like

$$as^2 + bs = c, \quad \text{or} \quad as^2 + c = bs, \quad \text{or} \quad bs + c = as^2,$$

as well as the Babylonians, the Indians and the Arabs.

The 6 forms of equations of the first and second degree, mentioned here, are exactly the 6 standard forms of Alkhwarizmi.

It looks therefore as if the solution of the definite equations is a traditional topic for Diophantus. Much more original is his virtuoso treatment of indeterminate problems. We shall attempt to throw some light on Diophantus' procedure by means of a few typical examples

From Book II.

Problem 9. *To divide a given number which is the sum of two squares into two other squares.*

In free translation, the solution is as follows:

Let the given number be 13, the sum of the squares of 2 and 3.

Let the side of one square be $s + 2$ and that of the other 3 units less than an arbitrarily selected multiple of s, e.g. $2s - 3$. Then one of the squares is $s^2 + 4s + 4$ and the other $4s^2 + 9 - 12s$, together $5s^2 + 13 - 8s$. This has to equal 13, hence $s = 8/5$.

The squares are therefore $(s + 2)^2 = 324/25$ and $(2s - 3)^2 = 1/25$. Their sum is indeed 13.

---

[1] A typical example from the algebra of Alkhwarizmi, in the translation of Rosen (*The algebra of Mohammed ben Musa*, ed. Rosen, London, 1831, p. 5) is the following: "A *square* and ten of its *roots* are equal to nine and thirty dirhems, that is if you add ten *roots* to one *square*, the sum is equal to nine and thirty. The solution is as follows. Take half the number of *roots*, that is in this case five, then multiply this by itself, and the result is five and twenty. Add this to the nine and thirty, which gives sixty-four; take the square root, or eight and subtract from it half the number of roots, namely five, and there remains three. This is the root."

Problem 20. To find two numbers such that the square of either added to the other gives a square.

Choose for the two number the expressions s and $2s + 1$. The square of the first plus the second then becomes automatically a square viz.

$$s^2 + 2s + 1 = (s + 1)^2.$$

The square of the second plus the first becomes $4s^2 + 5s + 1$. To make this into a square, one sets it equal to $(2s - 2)^2$. Thus one finds the equation

$$4s^2 + 5s + 1 = 4s^2 + 4 - 8s,$$

which leads without difficulty to the solution

$$s = {}^3/_{13}.$$

The identical method is used in both examples. Diophantus sets the required numbers equal to simple expressions in s, such as $s + 2$ and $2s - 3$ in the first example. He obtains then a quadratic equation in s and he knows how to arrange matters cleverly so that either the term in s^2 drops out, or the constant term, so that he can solve for s rationally.

Most of the problems in Book II are solved by this method.

Another typical ingenious device, the "method of double equality", is derived from the rule

$$x^2 - y^2 = (x + y)(x - y).$$

The method is applied whenever each of two expressions in s has to be a square. In such a case their difference has to be equal to the difference of two squares. This is accomplished by factoring the difference and equating the factors to $x + y$ and $x - y$. Half the sum of these factors is x, and half the difference is y.

The method is applied for the first time in

Problem 11. To add the same (required) number to two given numbers so as to make each of them a square.

Let the given numbers be 2 and 3; we add s to each. Therefore $s + 2$ and $s + 3$ must be squares; we call something like this a "double equality". Take their difference and look for two numbers whose product is equal to this, e.g. 4 and $\frac{1}{4}$. Set half the difference of these numbers, multiplied by itself, equal to $s + 2$, or set half their sum, multiplied by itself, equal to $s + 3 \ldots$ in each case one finds $s = {}^{97}/_{64}$. The number to be added is therefore ${}^{97}/_{64}$.

From Book III.

Problem 10. To find three numbers such that the product of two of them, increased by 12, produces every time a square.

Set the product of the first two plus 12 equal to 25; then the product of the first two is 13. Therefore set the first number equal to $13s$, the second to s^{-1}.

Set the product of the second and third numbers plus 12 equal to 16; then this product itself is 4. But the second number is s^{-1}, hence the third is $4s$.

Now the product of the first and third plus 12 must also be a square. The product of the first and third is $52s^2$, hence $52s^2 + 12$ must be a square.

If the factor 52 were a square, it would be easy to satisfy this condition. Now 52 was found as the product of 13 and 4; if these were both squares their product would be a square. Thus we have to find two squares, each of which would produce a square if increased by 12. This is easy [1]; for example, we can choose 4 for one of the squares and $\frac{1}{4}$ for the other, since each of them, increased by 12, is a square.

Let us therefore start all over again and set the first number equal to $4s$, the second to s^{-1} and the third to $\frac{1}{4}s$. We still have to make sure that that the product of the first and third plus 12 is a square. The product of the first and third is s^2, hence $s^2 + 12$ must be a square. Choose as the side of the square, e.g., $s + 3$, then the square becomes $s^2 + 6s + 9$ and we find $s = \frac{1}{2}$. This solves the problem.

What Diophantus applies here is the "method of the wrong hypothesis"; he starts by assuming certain expressions for his unknown numbers, and, when this brings him into trouble, he looks for the causes of the difficulty and for the changes in the assumptions which will enable him to solve the problem.

From Book IV.

A very pretty example of the method of the wrong assumption is the following:

Problem 31. To divide unity into two parts, such that, if given numbers are added to them respectively, the product of the two sums gives a square.

Let 3 and 5 be the added numbers. Let one part be s, the other $1 - s$. Addition of 3 and 5 to these parts gives $s + 3$ and $6 - s$. Their product is therefore $3s + 18 - s^2$. This has to be equal to a square. Let it be equal, e.g. to $4s^2$, so that one finds

$$3s + 18 = 5s^2.$$

This equation can not be solved in rational numbers. The coefficient 5 was a square plus 1. In order that the equation may have rational solutions, this coefficient multiplied by 18, increased by the square of one half of 3, must be a square. Thus we have to look for a square which, increased by 1, and multiplied by 18, and then increased by $2\frac{1}{4}$, produces a square. (Introduction of a new number s:)

Let s^2 again be the required square. Then $s^2 + 1$ times 18 plus $2\frac{1}{4}$, or $18s^2 + 20\frac{1}{4}$ must be a square. Take this four times: $72s^2 + 81$ has to be a square. Set it equal to $(8s+9)^2$; one finds $s = 18$. The required square s^2 is therefore 324.

Now we return to the original question: $3s + 18 - s^2$ must become a square. Take for it $324s^2$; then we find $s = {}^{78}/_{325} = {}^{6}/_{25}$. The first part becomes ${}^{6}/_{25}$, the second ${}^{19}/_{25}$.

Apparently Diophantus knew how to solve the general quadratic equation of the form

$$as^2 = 2bs + c.$$

He wrote the solution as a fraction with denominator a and he knew that the solution is rational if $b^2 + ac$ is a square. His formula was evidently

$$s = \frac{b + \sqrt{b^2 + ac}}{a}.$$

---

[1] e.g., by using the method of the "double inequality" and factoring 12 into two factors $12 = 2 \times 6 = (4 - 2)(4 + 2) = 4^2 - 2^2$, and again $12 = 3 \times 4 = (3\frac{1}{2} - \frac{1}{2})(3\frac{1}{2} + \frac{1}{2}) = (3\frac{1}{2})^2 - (\frac{1}{2})^2$. Thus we find $2^2 + 12 = 4^2$ and $(\frac{1}{2})^2 + 12 = (3\frac{1}{2})^2$.

PLATE 27

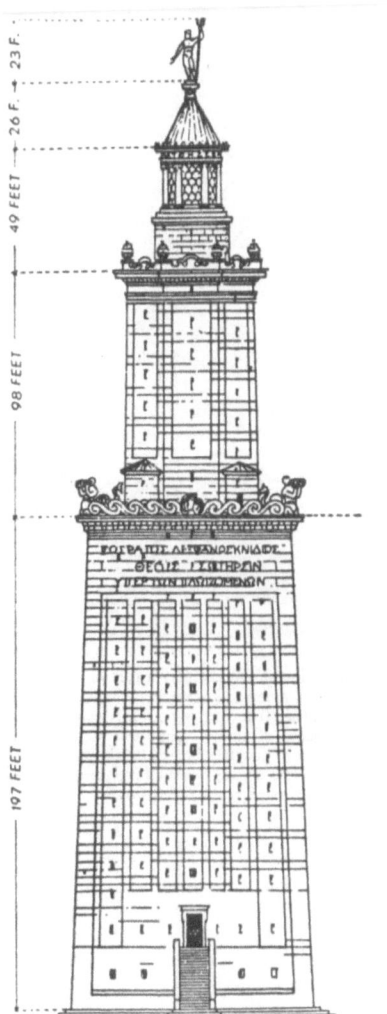

49 FEET — 26 F. — 23 F.

98 FEET

107 FEET

PL. 27a. Lighthouse on the island Pharos near Alexandria, built by Sostratus of Cnidus during the time of Ptolemy Soter (323–283 B.C.), the prototype of all lighthouses (notice the French name "phare"). More than 100 meters high. Completed about 280 B.C. The structure survived until the 14th century A.D. Reconstruction by A. Thiersch, based chiefly on representations on coins (compare Plate 27b). Our plate is taken from his book "Der Pharos von Alexandrien", 1909. The tower is a splendid example of the collaboration of science, technology and art in the Hellenistic period. The lighthouse stood at the entrance of the port of Alexandria, "the most famous harbor of the ancient world". It is still open to question whether this and similar towers served as lighthouses from the start; this is certain, only from about 50 A.D. An older form is the fire-column (harbor of Piraeus). High towers, nothing unusual for us, were quite uncommon for the Greeks. The line of development, starts in the Orient (Tower of Babel and cognate monuments, Lycian graves, Pyramids). With Alexander the Great the situation changed. From that time the Greeks also began to show on many occasions a marked love of the colossal. Before, the Mausoleum of Halicarnassus (about 350 B.C.) shows a notable tendency to verticalism. It seems to be connected with older towerlike graves in Asia minor. There is perhaps, in the Alexandrian Pharos, also a reminiscence of Persian watch towers.

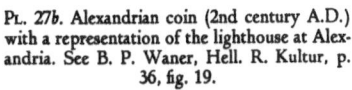

PL. 27b. Alexandrian coin (2nd century A.D.) with a representation of the lighthouse at Alexandria. See B. P. Waner, Hell. R. Kultur, p. 36, fig. 19.

PLATE 28

Pl. 28. The "Hagia Sophia". The construction of arches, vaults and cupolas did not play a role of importance in Greek architecture. Already in the second half of the 5th century B.C., Democritus studied the theory of the arch, yet the Greeks applied it but little in practice and not until late in the Hellenistic period. It is quite different with the Romans. Connecting with early Roman and Italic constructions of arches, vaults and cupolas, these forms have been developed enormously, helped also by the introduction of new building materials (bricks, lime, concrete); one has but to think of the Pantheon. Oriental influences are also active, but the Hagia Sophia is not conceivable without the flourishing of the Roman construction of vaults and cupolas.

Diophantus gave a second solution for the same problem, in which he operated with inequalities:

> *Differently*. Let the first part be s less the 3 units which have to be added to it. Then the second part is $4 - s$ and the condition becomes: $9s - s^2$ must be a square. Set this square equal to $4s^2$; then one finds $s = {}^9/_5$. But we can not subtract 3 units from this; s has to be made greater than 3, but less than 4.
>
> Now s was found by dividing 9 by 5, i.e. by a square plus 1. If s lies between 3 and 4, then the square plus 1 must lie between 3 and $2\frac{1}{4}$, and the square between 2 and $1\frac{1}{4}$. Therefore we have to look for a square that is greater than $1\frac{1}{4}$ but less than 2.
>
> Reduce these numbers to the denominator 64, so that the numerators become 80 and 128. Now it is easy: the square is $^{100}/_{64}$, i.e. $^{25}/_{16}$.
>
> Returning to the original problem, we find the equations

$$9s - s^2 = ({}^{25}/_{16})s^2,$$

from which follows $s = {}^{144}/_{41}$. Therefore the first number will be $^{21}/_{41}$, the second $^{20}/_{41}$.

From Book V.

A simple example of the "method of the double equation":

Problem 1. To find three numbers in geometrical progression such that each of them decreased by 12 gives a square.

Look for a square which, decreased by 12, again produces a square.

> This is easy: the square is $42\frac{1}{4}$.
>
> Now let the first term of the progression be $42\frac{1}{4}$, the third term s^2; then the middle term is $6\frac{1}{2}s$. Now each of the two expressions $s^2 - 12$ and $6\frac{1}{2}s - 12$ has to be a square. Their difference is $s^2 - 6\frac{1}{2}s$. Resolve this in two factors: s and $s - 6\frac{1}{2}$. Half their difference multiplied by itself is $^{169}/_{16}$; we equate this to the smallest expression $6\frac{1}{2}s - 12$. Thus we get $s = {}^{361}/_{104}$, etc.

From Book VI.

Book VI takes up right triangles with rational sides, i.e. solutions of the diophantine equation

$$x^2 + y^2 = z^2.$$

Diophantus knows that he can obtain such triangles "with the aid of" two numbers p and q, according to the formulas

$$x = p^2 - q^2, \qquad y = 2pq, \qquad z = p^2 + q^2.$$

In each of the problems of Book VI, Diophantus imposes further conditions on the sides or the area. An example:

Problem 1. To find a (rational) right triangle such that the hypothenuse minus each of the sides gives a cube.

> Let the triangle be formed with the aid of two numbers, e.g. s and 3. The hypothenuse is then $s^2 + 9$, the height $6s$ and the base $s^2 - 9$.
>
> The hypothenuse less the base is now 18. This is not a cube. Where does the number 18 come from? It is twice the square of 3. One has to take therefore (in place of 3) a number

whose square, taken twice, is a cube. Call this number again s; then $2s^2$ has to be a cube, e.g.
s^3, and s becomes 2.

Now form again a triangle with the aid of the numbers s and 2. The hypotenuse becomes
$s^2 + 4$, the height $4s$ and the base $s^2 - 4$. The difference of hypotenuse and base is a cube.

The other condition requires that the hypotenuse less the height, i.e. $s^2 + 4 - 4s$, be a
cube. Now this expression is a square of side $s - 2$. Therefore, if we set $s - 2$ equal to a cube,
the work is done.

Take for this cube 8; then we get $s = 10$.

The triangle is therefore formed from 10 and 2. The hypotenuse is 104, the height 40,
and the base 96.

Whenever, as in this case, Diophantus has more than one condition which the
unknown numbers have to satisfy, he frequently selects, as he does here, for the
unknowns such expressions in s, that they satisfy all conditions save one, "in
general", i.e. they satisfy them no matter how s is chosen. The last condition then
leads to an equation in s from which s can be determined.

Any one in whom these examples have developed a taste for it, should get hold
of the work of Diophantus himself, either in Heath's English translation, or in that
of Ver Eecke, in French.

In one more case has the Alexandrian school produced a mathematician of
calibre, namely

Pappus of Alexandria.

We happen to know when Pappus lived: on October 18, 320 he observed a
solar eclipse and he tells about it in his commentary on the Almagest. [1]

The reader will gradually have gotten some idea of his principal work, the
"Mathematical Collection", on which we have already drawn several times. In
this work, Pappus brought together everything in the work of his predecessors
which interested him: on higher plane curves, on the quadrature of the circle, the
duplication of the cube, the trisection of the angle, the method of analysis, etc.
Wherever explanations of or supplements to the works of the great geometers
seemed to him necessary, he formulated them as lemmas. This gives us a great deal
of varied information about what was contained in the lost works of Euclid and of
Apollonius.

But furthermore, Pappus supplemented and extended in several points the
work of his predecessors. In Book 3 for example, he gives a new construction for
the five regular solids inscribed in a given sphere.

Book 4 contains an interesting generalization of the Theorem of Pythagoras and
a number of nice propositions on circles inscribed in the Arbelos of Archimedes.
In the same book Pappus defines a type of spiral on the sphere, and he determines
the area bounded by the spiral and an arc of circle, the method being taken
from Archimedes. He shows how a Neusis-construction, used somewhere by
Archimedes, can be reduced to the intersection of two conics.

[1] See A. Rome, *Annales Soc. scient.* Bruxelles, 47 AI, p. 46.

In Book 5, the work of Zenodorus on isoperimetric figures is reproduced, supplemented by some propositions of Pappus himself.

In Book 6, Pappus determines the center of an ellipse, given as the perspective of a circle.

Historically, Book 7 is extremely important, because it gives a survey of a large number of works on geometric analysis and on geometric loci, nearly all of which are lost. As a discovery of his own, Pappus formulates a proposition on the volume of a solid of revolution, which is essentially the same as what we know as the "theorem of Pappus-Guldin".

Book 8 is largely devoted to mechanics, but it also contains a construction of a conic through 5 given points. The motivation for this is the problem of determining the diameter of a cylindrical column from an arbitrary fragment of the column. The book contains moreover a construction of the principal axes of an ellipse when two conjugate diameters are given.

To give the reader some notion of the high level of this powerful work, we shall now discuss thirteen lemmas on the Porisms of Euclid. They contain ideas which were destined to become later on of great importance for projective geometry.

A porism of Euclid.

In the introduction to Book 7 of the Collectio, Pappus says that a number of porisms of Euclid can be summarized into a single proposition, stated as follows:

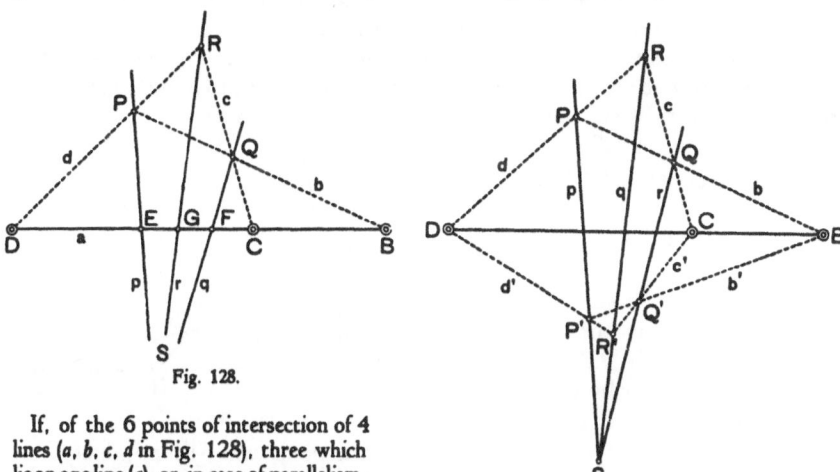

Fig. 128.

Fig. 129.

If, of the 6 points of intersection of 4 lines (a, b, c, d in Fig. 128), three which lie on one line (a), or, in case of parallelism, two points are held fixed, and if two of the remaining three (P and Q) lie on fixed lines (p and q), then the sixth point of intersection (R) will also lie on a fixed line.

Pappus might have added that the three lines p, q, r pass through one point S.

In our diagram variable lines are dotted, fixed lines are drawn in full. If one imagines two different positions of the dotted lines (*b, c, d* and *b', c', d'*), one obtains
the diagram of the well-known *Theorem of Desargues*:

If two triangles *PQR* and *P'Q'R'* are so placed that the points of intersection
of corresponding sides (*b* and *b', c* and *c', d* and *d'*) lie on a line (*a*), then the lines
joining corresponding vertices (*P* and *P', Q* and *Q', R* and *R'*) ·pass through one
point (*S*), and conversely (fig. 129).

We shall not discuss the generalization to more than 4 lines, added by Pappus,
but we shall raise the question: how may Euclid have proved the proposition for
4 lines?

It is an obvious conjecture that the proof can be given by means of the lemmas
which Pappus then develops. This is indeed the case, as we shall see presently.

The theorem on the complete quadrangle.

The diagram of the proposition on the 4 lines can also be interpreted as the
figure of a complete quadrangle *PQRS*, whose sides (*PQ, QR*, etc.) are cut by a
fixed line *a*. If one can prove that the sixth point of intersection *G* is uniquely
determined by the remaining five points *B, C, D, E* and *F*, then it follows immediately that, when *P* moves on the line *ES* and *Q* on the line *FS*, the point *R*
must move on the line *GS*; and this proves the proposition of the four lines.

This proposition, that *G* is uniquely determined by *B, C, D, E* and *F* is sometimes called the *Second Theorem of Desargues* or the *Theorem on the complete quadrangle.*
Pappus formulates this proposition in his own manner, viz. in lemmas I and II,
for the case that one of the sides of the quadrangle is parallel to *a*, and in lemma IV
for the case that all six points *B — G* are at finite distance.

Lemma I runs as follows:

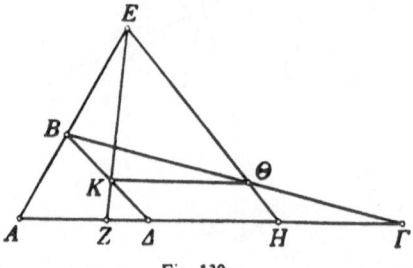

Let figure $AB\Gamma\Delta EZH$ be drawn
and let

(1) $A\Delta : \Delta\Gamma = AZ : ZH$

and let Θ be joined to K; I assert
that ΘK is parallel to $A\Gamma$.

One recognizes in the figure the
complete quadrangle $BE\Theta K$, whose
6 sides are being met by one line. If
5 of the 6 points satisfy the relation

Fig. 130.

(1), then, in our modern terminology, the sixth point will "lie at infinity".

Lemma II is analogous, except that $K\Theta \mathbin{//} A\Gamma$ is part of the hypothesis
Lemma IV says:

Let figure $AB\Gamma\Delta EZH\Theta K\Lambda$ be drawn and let

(2) $AZ \cdot \Delta E : A\Delta \cdot EZ = AZ \cdot B\Gamma : AB \cdot \Gamma Z.$

I assert that the points Θ, H, Z lie on one straight line.

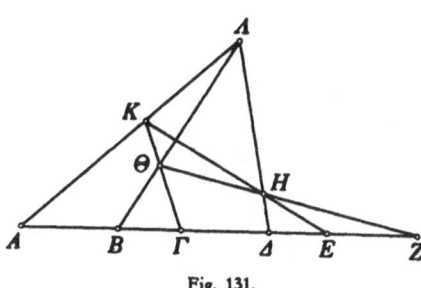

Fig. 131.

One recognizes again the complete quadrilateral ($H\Theta K\Lambda$) whose 6 sides are brought to intersection with one line. In modern terminology, the relation (2) expresses that the cross-ratio ($AZE\Delta$) is equal to the cross-ratio ($ZAB\Gamma$). Nowadays we say that the six points A, B, Γ, Δ, E, Z are in *involution*.

LemmaV refers to the special case in which of the six points two sets, of two points each, coincide. Then (2) is replaced by the proportion

(3) $$AB : B\Gamma = A\Delta : \Delta\Gamma.$$

In this case we speak of *four harmonic points* (fig. 132).

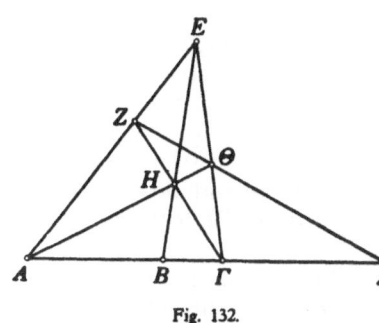

Fig. 132.

Pappus proves his lemmas every time by drawing parallel lines and by cleverly juggling with proportions. In his proofs he might have used the fact that the cross-ratio of four points remains unchanged under projection on another line, for this is exactly what he formulated as Lemma III. It is however true that he only considers the special case in which one of the four points is the point of intersection of the two lines.

Lemma III. Draw the lines ΘE and $\Theta\Delta$ so as to intersect the three lines AB, $A\Gamma$, $A\Delta$. I assert that $\Theta B \cdot \Gamma\Delta : \Theta\Delta \cdot B\Gamma = \Theta E \cdot HZ : \Theta H \cdot ZE$.

Lemma X is the converse of lemma III and lemma XI is concerned with the special case in which one of the transversals is parallel to one of the four lines.

By means of these lemmas Pappus proves the famous

Theorem of Pappus,

which runs as follows:

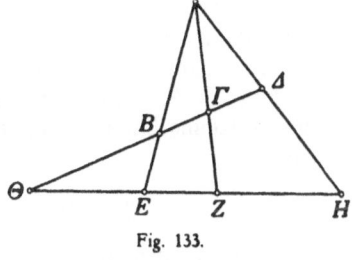

Fig. 133.

If the points E and Z are taken cn the lines AB and $\Gamma\Delta$, and if $A\Delta$, AZ, $B\Gamma$, BZ, $E\Delta$ and $E\Gamma$ are drawn, then the points of intersection H, K, M will lie on one line.

Pappus formulates and proves the proposition first for the case that AB and $\Gamma\Delta$ are parallel (lemma XII) and then for the case that they intersect (lemma XIII).

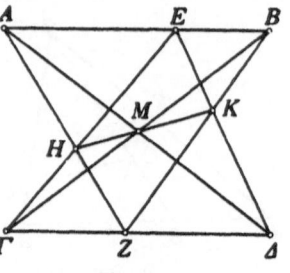

Nowadays the theorem is considered as a special case of the *Theorem of Pascal*, which says that the points of intersection of opposite sides of a hexagon inscribed in a conic lie on a line.

Theon of Alexandria (380 A.D.)

Fig. 134.

was a rather mediocre mathematician. He wrote a commentary on the Almagest [1] and he prepared a new edition of Euclid's Elements with rather unimportant additions and emendations. He acquired his greatest fame not as a mathematician, but as the father of

Hypatia

a very learned woman, heroine of romantic atrocity tales.

She was handsome, she was eloquent, she was charming, she wrote learned commentaries on Diophantus and on Apollonius, and she became professor of Platonic philosophy in Alexandria. The later bishop Synesius calls her his "mother, sister and reverend teacher". She had connections in the highest circles; not only in literary matters, but also in practical affairs her counsel was appreciated. A woman, she moved among men, with freedom and yet with decency. There is a tale that she brought a desperately amorous lover to his senses in a drastic manner (see "Suidas" under Hypatia).

Although on terms of friendship with many Christians, she had remained faithful to the traditions of pagan Hellenism. This became the cause of her downfall. One of her friends was the Roman prefect Orestes, mortal enemy of bishop Cyrillus. In the Christian community she was accused of having incited Orestes against Cyrillus. One day, on her way home, she was hauled from her carriage by the mob, dragged to the church, stripped and assassinated with shards. According to the church history of Socrates, 5th century, her corpse was torn to pieces and burned. This happened in 418.

After Hypatia, Alexandrian mathematics came to an end. Indeed, already in 392, the only remaining library, in the temple of Serapis, had been destroyed by the Alexandria mob.

The Athens school. Proclus Diadochus.

The Academia of Plato, which had access to its own ample financial means, maintained itself for a longer time. The industrious neo-Platonist Proclus (450)

[1] See A. Rome, *Commentaire de Pappus et de Théon d'Alexandrie sur l'Almageste* II (1936).

wrote i.a. a commentary on Euclid, Book I [1] and a very nice opusculum on astronomical hypothesis, [2] After him, Isidore of Alexandria, called the Great, and Damascius of Damascus, became heads of the school. During the time of Damascius, there came to Athens also Simplicius, the excellent commentator of Aristotle, through whom have been preserved for posterity, besides highly important fragments of old philosophers, also the fragment of Eudemus on the lunules of Hippocrates. But in 529, on the order of the Emperor Justinian, the school of Athens, the last rampart of the pagan world, was closed. Simplicius and Damascius emigrated to Persia.

The last center of Greek culture was Constantinople. Here lived

Isidore of Milete and Anthemius of Tralles,

the architects of the noble Hagia Sophia (Plate 28). They were experts not only in architecture, but also in mathematics. Anthemius has written on burning-mirrors.

It is probable that the so-called 15th Book of the Elements, at any rate its last part, comes from the school of Isidore, for the writer speaks of the "famous Isidore" as his teacher. The book contains propositions on regular polyhedra.

After these last flutterings, the history of Greek mathematics dies like a snuffed candle.

[1] French translation by P. Ver Eecke, Bruges 1948.
[2] Proclus, *Hypotiposis*, edited and translated by Manitius, Leipzig, 1909.

INDEX